Mathematics for Engineers

Mathematics for Engineers

Volume 2

L. J. Nicolescu
Doctor in Mathematics
Professor at the Institute of Architecture Bucharest

M. I. Stoka
Doctor in Mathematics

ABACUS PRESS
Tunbridge Wells, Kent

Mathematics for Engineers

Volume 2

© English Edition 1974
ABACUS PRESS
Abacus House, Speldhurst Road, Tunbridge Wells, Kent, England
ISBN 0 85626 004 5

Authorized translation from the Romanian language edition of MATEMATICI PENTRU INGINERI first published by Editura tehnică, Bucharest, 1969

Translated by Dr. L. J. Nicolescu and Dr. A. Georgescu
Translation edited by Dr. John Hammel PhD

Printed in Romania

Foreword

Writing a book of mathematics for engineers, postgraduate or undergraduate, might seem to be a rather ambitious enterprise, but the accelerating development of science and technology necessitates a continual review of the material required by the specialist. This is true of any discipline but is particularly true of mathematics. The advancement of science and technology sets new problems for the mathematician the solution of which enriches the areas of research.

Naturally a book of this kind must be selective if it is not to become merely an unwieldy encyclopedia of the subject. We have tried to cover not only those areas which will be immediately useful to the engineer, but also those which will aid his comprehension of the technical literature.

The first edition of this book, which appears in two volumes, was published in Bucharest by the 'Editura Tehnica' in Romanian, the first volume in 1969 and the second in 1972, the present edition has been revised and there are also some additions such as a section of exercises at the end of each chapter. Some printing errors which occurred in the Romanian edition have now been eliminated.

26 April 1972

L. J. NICOLESCU
M. I. STOKA

CONTENTS OF VOLUME 1

CONTENTS

1

Special Functions

1.1 EULERIAN FUNCTIONS

1.1.1 The function $\Gamma(z)$ defined as an integral

Let us put

$$\Gamma(z) = \int_0^\infty e^{-t} t^{z-1} \, dt, \qquad z = x + iy \tag{1.1}$$

we shall prove the following proposition.

Proposition. *The integral* (1.1) *is summable for* $x > 0$; *the summability is uniform for* $0 < \alpha \leqslant x < A$.

Indeed, for $t \to \infty$ the integral is summable because of the factor e^{-t}.

For $t \to 0$, $|e^{-t}t^{z-1}| \sim t^{x-1}$ hence the summability for $x - 1 > -1$ that is $x > 0$. Let us prove now that the summability is uniform for $0 < \alpha \leqslant x \leqslant A < \infty$.

Indeed in this case we can make the following majoration:

$$\left| \int_0^\infty e^{-t} t^{z-1} \, dt \right| \leqslant \int_0^1 | e^{-t} t^{z-1} | \, dt + \int_1^\infty | e^{-t} t^{z-1} | \, dt \leqslant \int_0^1 t^{\alpha-1} \, dt + $$

$$+ \int_1^\infty t^{A-1} e^{-t} \, dt. \tag{1.2}$$

The last two integrals from relationship (1.2) are finite, hence the above statement is true and thus the proposition is completely proved.

Corollary. $\Gamma(z)$ *is a continuous function of* z *for* $x > 0$.

Indeed, $e^{-t}t^{z-1}$ is a continuous function in t and z for $t > 0$, $x > 0$. Consequently, the corollary follows immediately from Lebesgue's theorem [Vol. 1, Section 5.2.2].

Proposition. $\Gamma(z)$ *is an indefinitely differentiable function of complex variable* z *for* $x > 0$ (*that is holomorphic*).

Indeed we shall formally take the derivative of equation (1.1). We shall have

$$\Gamma'(z) = \int_0^\infty e^{-t} t^{z-1} \ln t \, dt. \tag{1.3}$$

If the above integral is uniformly summable for $0 < \alpha \leqslant x \leqslant A < \infty$ then equation (1.3) is true. But the uniform summability of the integral is proved in the same way as in equation (1.2) taking into account that

$$\int_0^1 t^{\alpha-1} |\ln t| \, dt \quad \text{and} \quad \int_1^\infty t^{A-1} \ln t \; e^{-t} \, dt$$

are finite. The second order derivative is calculated in the same way

$$\Gamma''(z) = \int_0^\infty e^{-t} t^{z-1} (\ln t)^2 \, dt \tag{1.4}$$

and so on.

1.1.2 Properties of the function $\Gamma(z)$

Proposition 1. *We have:*

$$\Gamma(z+1) = z\Gamma(z). \tag{1.5}$$

Indeed, by means of an integration by parts we obtain

$$\Gamma(z+1) = \int_0^\infty e^{-t} t^z \, dt = -t^z e^{-t} \Big|_{t=0}^{t\to\infty} + z \int_0^\infty e^{-t} t^{z-1} \, dt = z\Gamma(z).$$

Corollary 1. *We have the following relation of recurrence:*

$$\Gamma(z+n) = (z+n-1)(z+n-2)\ldots z\Gamma(z), \text{ for } x > 0. \tag{1.6}$$

Corollary 2. *We have*

$$\Gamma(n+1) = n! \tag{1.7}$$

In equation (1.6) we make $z = 1$ and we obtain equation (1.7) by taking into account that $\Gamma(1) = 1$.

Proposition 2. *$\Gamma(z)$ is a meromorphic function in the complex plane; at the points $z = 0, -1, -2, \ldots$, $\Gamma(z)$ has simple poles.*

Indeed, let us consider the quotient

$$\frac{\Gamma(z+n)}{(z+n-1)(z+n-2)\ldots(z+1)z}$$

and notice that it is defined for $z \neq 0, -1, -2, \ldots$ The value of this quotient is independent of n. Indeed, this is immediately seen if we substitute $n + p$ for n

$$\frac{\Gamma(z+n+p)}{(z+n+p-1)\ldots(z+1)z} = \frac{\Gamma(z+n)(z+n+p-1)\ldots(z+n)}{(z+n+p-1)\ldots(z+n)(z+n-1)z} =$$
$$= \frac{\Gamma(z+n)}{(z+n-1)\ldots z}.$$

Consequently, taking into account equation (1.6) one can see that we can define the function $\Gamma(z)$ for every $z \neq 0, -1, -2, \ldots$ be means of

$$\Gamma(z) = \frac{\Gamma(z+n)}{(z+n-1)\ldots(z+1)z} \tag{1.8}$$

where n is any integer with the property $x + n > 0$. Consequently $\Gamma(z)$ is a holomorphic function in the complex plane from which the points $z = 0, -1, -2, \ldots$ have been taken out. We must still show that at these points $\Gamma(z)$ has simple poles. For this let us put $z = -n + u$ in (1.8). Then

$$\Gamma(z) = \frac{\Gamma(u+1)}{u(u-1)(u-2)\ldots(u-n)} \underset{u\to 0}{\widetilde{\quad}} \frac{(-1)^n}{n!\,u} \tag{1.9}$$

from which the above statement follows.

Remark 1. At the pole $z = -n$ the residue of the function $\Gamma(z)$ is $\dfrac{(-1)^n}{n!}$.

Remark 2. From equation (1.9) it follows that:

$$\Gamma(z) \underset{z\to 0}{\widetilde{\quad}} \frac{1}{z}. \tag{1.10}$$

We shall further on see that $\Gamma'(1) = -\gamma$ where γ is Euler's constant. Then for $z \to 0$, $\Gamma(1+z) = 1 - \gamma z + \ldots$ and

$$\Gamma(z) = \frac{\Gamma(z+1)}{z} = \frac{1}{z} - \gamma + \ldots \tag{1.11}$$

Remark 3. If we replace t with t^2 in equation (1.1) we obtain Gauss' integral

$$\Gamma(z) = 2\int_0^\infty e^{-t^2}\, t^{2z-1}\, \mathrm{d}t \qquad \text{for } x > 0. \tag{1.12}$$

In particular, if in (1.12) we make $z = \frac{1}{2}$, we obtain

$$\int_0^\infty e^{-t^2}\, dt = \frac{1}{2}\Gamma\left(\frac{1}{2}\right). \tag{1.13}$$

1.1.3 The function $B(p, q)$

From formula (1.12) it follows that

$$\Gamma(p) = 2\int_0^\infty e^{-u^2} u^{2p-1}\, du,$$

$$\Gamma(q) = 2\int_0^\infty e^{-v^2} v^{2q-1}\, dv,$$

hence we deduce that

$$\Gamma(p)\,\Gamma(q) = 4 \iint\limits_{\substack{u \geqslant 0 \\ v \geqslant 0}} e^{-(u^2+v^2)} u^{2p-1} v^{2p-1}\, du\, dv.$$

By conversion to polar coordinates $u = r\cos\theta$, $v = r\sin\theta$ we obtain

$$\Gamma(p)\,\Gamma(q) = 4 \iint\limits_{\substack{r \geqslant 0 \\ 0 \leqslant \theta \leqslant \frac{\pi}{2}}} e^{-r^2} r^{2(p+q)-1} \cos^{2p-1}\theta \sin^{2q-1}\theta\, dr\, d\theta =$$

$$= 2\int_0^\infty e^{-r^2} r^{2(p+q)-1}\, dr \int_0^{\frac{\pi}{2}} \cos^{2p-1}\theta \sin^{2q-1}\theta\, d\theta =$$

$$= 2\,\Gamma(p+q) \int_0^{\frac{\pi}{2}} \cos^{2p-1}\theta \sin^{2q-1}\theta\, d\theta.$$

Definition. We shall denote by B(p, q) the function

$$B(p,\ q) = \frac{\Gamma(p)\ \Gamma(q)}{\Gamma(p+q)} = 2\int_0^{\frac{\pi}{2}} \cos^{2p-1}\theta\ \sin^{2q-1}\theta\ d\theta. \tag{1.14}$$

Consequence 1. Taking in formula (1.14)

$$p = \frac{r+1}{2}, \qquad q = \frac{1}{2}$$

we obtain the formula

$$\frac{1}{2} B\left(\frac{r+1}{2},\ \frac{1}{2}\right) = \int_0^{\frac{\pi}{2}} \cos^r\theta\ d\theta = \int_0^{\frac{\pi}{2}} \sin^r\theta\ d\theta = W(r) \tag{1.15}$$

which is called the *Wallis integral.*

Consequence 2. If in formula (1.14) we take $p = q = \dfrac{1}{2}$ we obtain

$$B\left(\frac{1}{2},\ \frac{1}{2}\right) = \left(\Gamma\left(\frac{1}{2}\right)\right)^2 = 2\int_0^{\frac{\pi}{2}} d\theta = \pi$$

hence it follows that

$$\Gamma\left(\frac{1}{2}\right) = \sqrt{\pi} \tag{1.16}$$

as

$$\Gamma\left(\frac{1}{2}\right) = \int_0^{\infty} e^{-t}\frac{dt}{\sqrt{t}} > 0.$$

Consequence 3. From formulae (1.13) and (1.16) it follows that

$$\int_0^{\infty} e^{-t^2}\, dt = \frac{1}{2}\Gamma\left(\frac{1}{2}\right) = \frac{1}{2}\sqrt{\pi}\,. \tag{1.17}$$

Consequence 4. From formulae (1.6) and (1.16) it follows that

$$\Gamma\left(n+\frac{1}{2}\right)=\left(n-\frac{1}{2}\right)\left(n-\frac{3}{2}\right)\ \dots\ \frac{1}{2}\Gamma\left(\frac{1}{2}\right)=$$

$$=\frac{1\cdot 3\cdot 5\dots(2n-1)}{2^n}\sqrt{\pi}=\frac{(2n)!}{2^{2n}n!}\sqrt{\pi} \tag{1.18}$$

for integral positive values of n.

Consequence 5. If in formula (1.14) we take $\cos^2\theta = t$ we get

$$B(p,\ q)=\frac{\Gamma(p)\,\Gamma(q)}{\Gamma(p+q)}=\int_0^1 t^{p-1}\,(1-t)^{q-1}\,\mathrm{d}t, \tag{1.19}$$

$$\operatorname{Re} p > 0, \qquad \operatorname{Re} q > 0.$$

Consequence 6. If in formula (1.18) we take $p=\frac{1}{2}$, $q=\frac{1}{2}$ we get

$$B\left(\frac{1}{2},\ \frac{1}{2}\right)=\left(\Gamma\left(\frac{1}{2}\right)\right)^2=\int_0^1\frac{\mathrm{d}t}{\sqrt{t(1-t)}}=\pi. \tag{1.20}$$

Consequence 7. From formula (1.18) we obtain

$$\int_\alpha^\beta (x-\alpha)^{p-1}\,(x-\beta)^{q-1}\,\mathrm{d}x=(\beta-\alpha)^{p+q-1}\frac{\Gamma(p)\,\Gamma(q)}{\Gamma(p+q)}. \tag{1.21}$$

Consequence 8. If in formula (1.18) we take $t=\dfrac{r^2}{1+r^2}$ we obtain

$$\int_0^\infty \frac{2r^{2p-1}}{(1+r)^{p+q}}\,\mathrm{d}r=\frac{\Gamma(p)\,\Gamma(q)}{\Gamma(p+q)}$$

and if we write $2p-1=\alpha$, $p+q=\beta$ the above relationship becomes

$$\int_0^\infty\frac{r^\alpha}{(1+r^2)^\beta}\,\mathrm{d}r=\frac{1}{2}\,\frac{\Gamma\left(\frac{\alpha+1}{2}\right)\Gamma\left(\beta-\frac{\alpha+1}{2}\right)}{\Gamma(\beta)}. \tag{1.22}$$

Consequence 9. If we denote by S_n the area of the unit sphere we shall show that

$$S_n = \frac{2\left(\Gamma\left(\frac{1}{2}\right)\right)^n}{\Gamma\left(\frac{n}{2}\right)} = \frac{2\pi^{\frac{n}{2}}}{\Gamma\left(\frac{n}{2}\right)}.$$

Indeed, we have

$$\Gamma(p_i) = 2\int_0^\infty e^{-u_i^2} u_i^{2p_i-1}\,du_i$$

hence

$$\Gamma(p_1)\,\Gamma(p_2)\,\ldots\,\Gamma(p_n) =$$
$$= 2^n \int\limits_{\substack{u_1 \geqslant 0 \\ \cdots\cdots \\ u_n \geqslant 0}} \ldots \int e^{-(u_1^2+u_2^2+\ldots+u_n^2)}\, u_1^{2p_1-1} \ldots u_n^{2p_n-1}\, du_1 \ldots du_n.$$

If we take $x_i = \xi_i r$ and calculate first the above integral on the sphere of radius r and then with respect to r we obtain

$$2\int_0^\infty e^{-r^2} r^{2(p_1+\ldots+p_n)-1}\,dr \cdot 2^{n-1} \int\limits_{\substack{r=1 \\ \xi_i \geqslant 0}} \ldots \int \xi_1^{2p_1-1} \ldots \xi_n^{2p_n-1}\, dS,$$

$$B(p_1, \ldots, p_n) = \frac{\Gamma(p_1)\ldots\Gamma(p_n)}{\Gamma(p_1+p_2+\ldots+p_n)} = 2^{n-1}\int\limits_{\substack{r=1 \\ \xi_i \geqslant 0}} \ldots \int \xi_1^{2p_1-1} \ldots \xi_n^{2p_n-1}\, dS.$$

If in this formula we take $p_i = \frac{1}{2}$, $(i = 1, 2, \ldots, n)$, we obtain

$$B\left(\frac{1}{2}, \ldots, \frac{1}{2}\right) = \frac{\left(\Gamma\left(\frac{1}{2}\right)\right)^n}{\Gamma\left(\frac{n}{2}\right)} = 2^{n-1}\frac{S_n}{2^n} = \frac{S_n}{2}.$$

1.1.4 The formula of complementaries

Proposition. *We have*

$$\Gamma(z)\,\Gamma(1-z) = \frac{\pi}{\sin \pi z}. \tag{1.23}$$

To prove relationship (1.23) we start from the formula

$$B(z,\ 1-z) = \Gamma(z)\,\Gamma(1-z) = \int_0^1 t^{z-1}(1-t)^{-z}\,\mathrm{d}t$$

in which we take $t = \dfrac{u}{1+u}$ and we obtain

$$B(z,\ 1-z) = \Gamma(z)\,\Gamma(1-z) = \int_0^\infty \frac{u^{z-1}}{1+u}\,\mathrm{d}u. \tag{1.24}$$

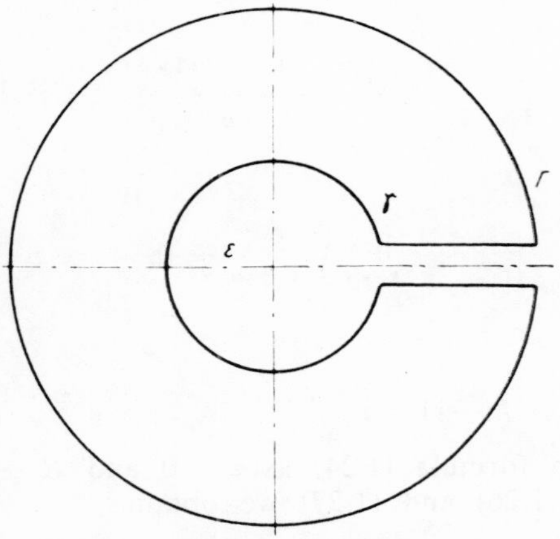

Figure 1.1.

Let us calculate the integral in equation (1.24) by the method of residues [Vol. 1, Section 5.7]. We notice that the function $f(u) = \dfrac{u^{z-1}}{1+u}$ where $0 < x = \operatorname{Re} z < 1$ is multi-valued. The point $u = 0$ is a critical point. We consider the contour in Figure 1.1 and take that branch of our function for which we have $u^{z-1} = e^{(z-1)\ln u}$, with $\ln u$ real on the upper part of the cut.

Applying the theorem of residues for this contour and taking into account that $u = -1$ is the only pole of our function, we have:

$$\int_{\varepsilon}^{R} \frac{u^{z-1}}{1+u}\, du + \int_{\Gamma} \frac{u^{z-1}}{1+u}\, du + e^{2\pi i z} \int_{R}^{\varepsilon} \frac{u^{z-1}}{1+u}\, du + \int_{\gamma} \frac{u^{z-1}}{1+u}\, du =$$

$$= 2\pi i \operatorname{Res} f(-1). \tag{1.25}$$

We notice that on the circle Γ

$$\left| \frac{u^{z-1}}{1+u} \right| \leqslant \frac{e^{(x-1)\ln R + 2\pi |y|}}{|1+u|} \leqslant M \frac{R^{x-1}}{R}$$

where M is a constant and $x - 1 < 0$. Hence

$$\lim_{R \to \infty} \int_{\Gamma} \frac{u^{z-1}}{1+u}\, du = 0. \tag{1.26}$$

Also on the circle γ

$$\left| \frac{u^{z-1}}{1+u} \right| < \frac{e^{(x-1)\ln \varepsilon + 2\pi |y|}}{|1+u|} \leqslant C \varepsilon^{x-1}$$

hence we have

$$\lim_{\varepsilon \to 0} \int_{\gamma} \frac{u^{z-1}}{1+u} = 0. \tag{1.27}$$

Since

$$\operatorname{Res} f(-1) = e^{(z-1)\ln u} \big|_{u=-1} = e^{i\pi(z-1)}$$

taking the limit in formula (1.24) as $\varepsilon \to 0$ and $R \to \infty$ and taking into account formulae (1.26) and (1.27) we obtain

$$\int_{0}^{\infty} \frac{u^{z-1}}{1+u}\, du - e^{2i\pi z} \int_{0}^{\infty} \frac{u^{z-1}}{1+u}\, du = 2\pi i e^{i\pi(z-1)} = -2i\pi e^{i\pi z}$$

that is

$$\int_{0}^{\infty} \frac{u^{z-1}}{1+u}\, du = \frac{-2i\pi e^{i\pi z}}{1 - e^{2\pi i z}} = \pi \frac{2i}{e^{i\pi z} - e^{-i\pi z}} = \frac{\pi}{\sin \pi z}. \tag{1.28}$$

Relationship (1.23) follows from formulae (1.28) and (1.24).
Relationship (1.23) is called the *formula of complementaries.*

Corollary. If in formula (1.23) *we take* $z = \frac{1}{2}$ *we again come across the relationship*

$$\left(\Gamma\left(\frac{1}{2}\right)\right)^2 = \pi$$

that is

$$\Gamma\left(\frac{1}{2}\right) = \sqrt{\pi}\,.$$

Remark. The formula of complementaries has been proved only for $0 < x < 1$. If in formula (1.23) we replace z with $z + 1$ we obtain

$$\Gamma(z+1)\,\Gamma(-z) = z\Gamma(z)\,\Gamma(-z) = -\Gamma(z)\,\Gamma(1-z) = -\frac{\pi}{\sin \pi z}\,.$$

Consequently the formula of complementaries is true for every z so that $x = \operatorname{Re} z$ should not be an integer. From the continuity of the two sides of formula (1.23) its valability results, for every z different from an integer.

1.1.5 The notion of summable series

The notion of summability of a series is an extension of the notion of absolute convergence for the case when the terms of a series depend on an arbitrary set of indices I.

The results we shall give further are valid in the case when the terms of the series are real or complex numbers. For the sake of simplicity we shall suppose they are real.

Definition. Let $(u_i)_{i\in I}$ *be a family of real numbers depending on the set of indices I. We shall say that the series*

$$\sum_{i\in I} u_i$$

is summable and has S as sum, if for every $\varepsilon > 0$, *there is a finite subset of indices* $\mathfrak{J} \subset I$ *so that for every finite subset of indices* $K \supset \mathfrak{J}$ *we have*

$$|S - S_K| \leqslant \varepsilon$$

where

$$S_K = \sum_{i \in K} u_i.$$

In this case we write

$$S = \sum_{i \in I} u_i.$$

Proposition 1. *If* $S = \sum_{i \in I} u_i$ *and* $S' = \sum_{i \in I} u_i$ *then* $S = S'$.

Indeed, for every $\varepsilon > 0$, there is $\mathcal{J}_1 \subset I$ so that $K \supset \mathcal{J}_1$ should imply $| S - S_K | \leqslant \varepsilon$. Similarly for every $\varepsilon > 0$ there is $\mathcal{J}_2 \subset I$ so that $K \supset \mathcal{J}_2$ should imply $| S' - S_K | \leqslant \varepsilon$.

Let us take $K = \mathcal{J}_1 \cup \mathcal{J}_2$; then $| S - S_K | \leqslant \varepsilon$ and $| S' - S_K | \leqslant \varepsilon$ hence $| S - S' | \leqslant 2\varepsilon$. Since ε is arbitrary it follows that $S = S'$.

Remark. Proposition 1 expresses the uniqueness of the sum.

Proposition 2. *If* $\sum_{i \in I} u_i$ *and* $\sum_{i \in I} v_i$ *are two summable series, and if* $S = \sum_{i \in I} u_i$, $T = \sum_{i \in I} v_i$, *then the series* $\sum_{i \in I} (\lambda u_i + \mu v_i)$, *where* λ *and* μ *are constants, is summable and its sum is* $\lambda S + \mu T$.

Indeed, for $\varepsilon > 0$ there is a finite set of indices $\mathcal{J}_1 \subset I$ so that for every finite set of indices $K_1 \supset \mathcal{J}_1$ we shall have

$$| S - S_{K_1} | \leqslant \frac{\varepsilon}{|\lambda| + |\mu|} \text{ where } S_{K_1} = \sum_{i \in K_1} u_i.$$

Also for $\varepsilon > 0$ there is a finite set $\mathcal{J}_2 \subset I$ so that for every finite set of indices $K_2 \supset \mathcal{J}_2$ we have

$$| T - T_{K_2} | \leqslant \frac{\varepsilon}{|\lambda| + |\mu|} \text{ where } T_{K_2} = \sum_{i \in K_2} v_i.$$

Hence for every finite set $K \supset I_1 \cup I_2$ we have

$$| \lambda S - \lambda S_K | \leqslant \frac{|\lambda| \varepsilon}{|\lambda| + |\mu|} \quad \text{and} \quad | \mu T - \mu T_K | \leqslant \frac{|\mu| \varepsilon}{|\lambda| + |\mu|}$$

hence

$$| (\lambda S + \mu T) - (\lambda S_K + \mu T_K) | \leqslant \varepsilon.$$

Proposition 3. *If* $u_i \geqslant 0$ *for every* $i \in I$ *then the necessary and sufficient condition for the series* $\sum_{i \in I} u_i$ *to be summable is that all the partial series* S_K *which correspond to the finite subsets* $K \subset I$ *be bounded by a fixed number* M. *In this case*

$$\sum_{i \in I} u_i = \sup_{K \subset I} (S_K).$$

Indeed, let us suppose that the series is summable. Then, a finite subset $\mathfrak{J} \subset I$ corresponds to $\varepsilon = 1$, so that, for every finite set $K \supset \mathfrak{J}$ we have:

$$S_K \leqslant S + 1.$$

If K is an arbitrary set included in I, and if we take $K_1 = \mathfrak{J} \cup K$, then

$$S_K \leqslant S_{K_1} \leqslant S + 1.$$

It follows that the sums S_K are equibounded. Conversely let us suppose that all the sums S_K are bounded and let B be their upper bound. In that case, for every $\varepsilon > 0$, there is a finite subset $\mathfrak{J} \subset I$ so that

$$B - \varepsilon \leqslant S_{\mathfrak{J}} \leqslant B.$$

It results that for every finite set $K \supset \mathfrak{J}$ we have

$$B - \varepsilon \leqslant S_K \leqslant B$$

hence we deduce that $\sum_{i \in I} u_i$ is summable and its sum is B.

Theorem. *(Cauchy's criterion.) The necessary and sufficient condition or the series* $\sum_{i \in I} u_i$ *to be summable is that to every* $\varepsilon > 0$ *there should correspond a finite* $\mathfrak{J} \subset I$, *so that, for every finite subset of indices* $K \subset I - \mathfrak{J}$, *the following inequality should be verified*

$$| S_K | \leqslant \varepsilon.$$

Let us suppose that $\sum_{i \in I} u_i$ is summable. Then, for every $\varepsilon > 0$, there is a finite set $\mathfrak{J} \subset I$ so that if $K \subset I - \mathfrak{J}$ is finite we have

$$| S_{\mathfrak{J}} - S | \leqslant \frac{\varepsilon}{2} \text{ and } | S_{\mathfrak{J} \cup K} - S | \leqslant \frac{\varepsilon}{2} \cdot$$

But then

$$| S_{\mathfrak{J} \cup K} - S_{\mathfrak{J}} | \leqslant \varepsilon$$

and since

$$S_{\mathfrak{J} \cup K} - S_{\mathfrak{J}} = S_K$$

it follows that

$$| S_K | \leqslant \varepsilon.$$

We will notice that if a series satisfies Cauchy's criterion then each of its partial series satisfies Cauchy's criterion; in particular the partial series formed by the terms $u_i \geqslant 0$ and the partial series formed by the terms $u_i \leqslant 0$.

If a series with terms $\geqslant 0$ (or $\leqslant 0$) satisfies Cauchy's criterion, then it is summable. This follows from proposition 3, since in that case the partial sums are bounded. It follows that if a series satisfies Cauchy's criterion, the series of its positive terms and of its negative terms are summable.

If we write

$$u_i^+ = u_i \text{ for } u_i \geqslant 0, \; u_i^+ = 0 \text{ for } u_i \leqslant 0,$$
$$u_i^- = -u_i \text{ for } u_i \leqslant 0, \; u_i^- = 0 \text{ for } u_i \geqslant 0,$$

then $u_i^+ \geqslant 0$ and $u_i^- \leqslant 0$

$$u_i = u_i^+ - u_i^-$$

and hence the series $\sum_{i \in I} u_i^+$ and $\sum_{i \in I} u_i^-$ are both summable. Hence it follows that the difference of these series, $\sum_{i \in I} u_i$ is summable and thus the theorem has been completely proved.

Corollary 1. The necessary and sufficient condition for the series $\sum_{i \in I} u_i$ *to be summable is that the series* $\sum_{i \in I} | u_i |$ *be summable and then*

$$\left| \sum_{i \in I} u_i \right| \leqslant \sum_{i \in I} | u_i |.$$

Since $\sum_{i \in I} | u_i | = \sum_{i \in I} u_i^+ + \sum_{i \in I} u_i^-$ corollary 1 is deduced immediately.

Corollary 2. If $\sum_{i \in I} v_i$ *is a series with positive terms and if*

$$| u_i | \leqslant v_i$$

then the series $\sum_{i\in I} u_i$ *is summable and in addition*

$$\left|\sum_{i\in I} u_i\right| \leqslant \sum_{i\in I} v_i.$$

This corollary is immediately deduced from proposition 3. Indeed it follows that the partial sums of the series $\sum_{i\in I} v_i$ are bounded, hence the partial sums of the series $\sum_{i\in I} |u_i|$ are bounded, that is the series $\sum_{i\in I} |u_i|$ is summable and thus it follows from corollary 1 that the series $\sum_{i\in I} u_i$ is summable and that

$$\left|\sum_{i\in I} u_i\right| \leqslant \sum_{i\in I} v_i.$$

Proposition 4. *If the series* $\sum_{i\in I} u_i$ *is summable, every partial series* $\sum_{i\in \mathfrak{J}} u_i$ *is summable (where* $\mathfrak{J}\subset I$*). If* $\mathfrak{J}$ *is a finite subset of I so that for every finite set* $K\supset\mathfrak{J}$ *we should have*

$$|S - S_K| \leqslant \varepsilon$$

then the inequality remains true if K is an infinite set countaining $\mathfrak{J}$.

If $\mathfrak{J}_1, \mathfrak{J}_2, \ldots, \mathfrak{J}_n \subset I$ *are disjoint sets and if* $\mathfrak{J} = \mathfrak{J}_1 \cup \mathfrak{J}_2 \cup \ldots \cup \mathfrak{J}_n$ *then*

$$S_{\mathfrak{J}} = S_{\mathfrak{J}_1} \cup S_{\mathfrak{J}_2} \cup \ldots \cup S_{\mathfrak{J}_n}$$

as soon as one of the two sides of the equality makes sense.

Indeed if $\sum_{i\in I} u_i$ is summable, then it verifies Cauchy's criterion hence it is summable. If $\mathfrak{J}$ is a set of indices so that for every finite $K_1\supset\mathfrak{J}$, we have

$$|S - S_{K_1}| \leqslant \varepsilon$$

then if K is an infinite set of indices containing $\mathfrak{J}$, there exists for every $\eta > 0$ a finite set of indices K_1, $\mathfrak{J}\subset K_1\subset K$ so that

$$|S - S_{K_1}| \leqslant \eta.$$

We have

$$|S_K - S_{K_1}| \leqslant \eta, \qquad |S - S_{K_1}| \leqslant \varepsilon$$

therefore

$$| S - S_K | \leqslant \varepsilon + \eta$$

whence

$$| S - S_K | \leqslant \varepsilon,$$

η being arbitrary. The last part of the proposition is obvious.

Proposition 5. *If $\mathcal{J}_n$, $(n = 0, 1, 2, \ldots)$, is a sequence of finite or infinite subsets of I, so that, for every finite subset $\mathcal{J} \subset I$, the set $\mathcal{J}_n$ should contain $\mathcal{J}$ for a sufficiently large value of n, then if $\sum_{i \in I} u_i$, is summable and its sum is S, $S_{\mathcal{J}_n}$ converges to S for $n \to \infty$.*

Indeed, for every $\varepsilon > 0$, there is a finite subset $\mathcal{J} \subset I$ so that, for every K (finite or not) $\supset \mathcal{J}$, we have

$$| S - S_K | \leqslant \varepsilon.$$

We suppose, then, that n_0 exists, so that, for $n \geqslant n_0$, $\mathcal{J}_n \supset \mathcal{J}$ which means that for $n \geqslant n_0$, we have

$$| S - S_{\mathcal{J}_n} | \leqslant \varepsilon.$$

Let us show now that the notion of summability is reduced to the notion of absolute convergence if $I = N$, where N is the set of natural numbers.

Proposition 6. *If $I = N$, then the necessary and sufficient condition for $\sum_{i \in I} u_i$ to be summable is that this series should be absolutely convergent and then its sum in the sense of the summable series theory is identical to its sum in the sense of the convergent series theory.*

Indeed, let us suppose $\sum_{i \in I} u_i$ is summable. Then $\sum_{i \in I} | u_i |$ is also summable. Given $\mathcal{J}_n = \{0, 1, \ldots, n\}$, from proposition 5 applied to the series $\sum_{i \in I} | u_i |$ it follows that the partial sums $\sum_{0 \leqslant \nu \leqslant n} | u_\nu |$ have a limit for $n \to \infty$, hence the series is absolutely convergent. From the same proposition applied to the series $\sum u_i$ it follows that

$$\lim_{n \to \infty} S_{\mathcal{J}_n} = S.$$

Consequently the sums in the sense of the two theories are equal. Conversely if we suppose the series absolutely convergent then the partia sums $\sum_{0 \leqslant \nu \leqslant n} |u_\nu|$ are bounded, hence it follows that all the partial sums $S_{\mathfrak{J}}$ ($\mathfrak{J}$ finite, included in I) with respect to the series $\sum_{i \in I} |u_i|$ are bounded. It follows that the series $\sum |u_i|$ is summable, hence $\sum_{i \in I} u_i$ is summable.

Corollary. In an absolutely convergent series the order of the terms can be changed without altering its sum.

Proposition 7. *If $I = \bigcup_{\alpha \in A} I_\alpha$ where $I_\alpha \cap I_\beta = \varnothing$, if $\alpha \neq \beta$ and if the series $\sum_{i \in I} u_i$ is summable then each of the series $\sum_{i \in I_\alpha} u_i$ is summable, of sum σ_α, the series $\sum_{\alpha \in I} \sigma_\alpha$ is summable and*

$$\sum_{i \in I} u_i = \sum_{\alpha \in A} \sigma_\alpha = \sum_{\alpha \in A} \left(\sum_{i \in I_\alpha} u_i \right).$$

From proposition 4 it follows that every series $\sum_{i \in I_\alpha} u_i$ is summable. But since $\sum_{i \in I} u_i$ is summable it follows that for every $\varepsilon > 0$ there is a finite set of indices $\mathfrak{J} \subset I$ so that, if K is a finite or infinite set of indices containing $\mathfrak{J}$, we have

$$|S - S_K| \leqslant \varepsilon.$$

Let B be the finite subset of A formed by all the indices $\alpha \in A$, so that $I_\alpha \cap \mathfrak{J} \neq \varnothing$. Then, for every finite subset $C \subset A$ containing B, $\bigcup_{\alpha \in C} I_\alpha$ is a subset K of I containing $\mathfrak{J}$, so that

$$\left| S - \sum_{\substack{i \in \cup I_\alpha \\ \alpha \in C}} u_i \right| \leqslant \varepsilon.$$

But from proposition 4 it follows that

$$\sum_{\substack{i \in \cup I_\alpha \\ \alpha \in C}} u_i = \sum_{\alpha \in C} \sigma_\alpha$$

where C is finite. Hence

$$\left| S - \sum_{\alpha \in C} \sigma_\alpha \right| \leqslant \varepsilon$$

and thus $\sum_{\alpha \in C} \sigma_\alpha$ is summable and has the sum S.

Proposition 8. *If $u_i \geqslant 0$ for $i \in I$, and if we agree to denote by ∞ the sum of a divergent series, then*

$$\sum_{i \in I} u_i = \sum_{\alpha \in A} \Big(\sum_{i \in I\alpha} u_i \Big)$$

and the value of the both sides of the equality is finite or $+\infty$.

1.1.6 The notion of infinite product

We shall now introduce the notion of indefinitely multipliable products which can be reduced to the familiar notion of convergence in case $I = N$ in the same way as for the previous case concerning the series.

Definition. We shall say that a product $\prod_{i \in I} u_i$ is multipliable if all the factors u_i are different from zero and if there is a number $P \neq 0$ with the following property: for every $\varepsilon > 0$, there is a finite set of indices $\mathfrak{J} \subset I$ so that for every finite set $K \supset \mathfrak{J}$

$$\left| \frac{\prod_{i \in K} u_i}{P} - 1 \right| \leqslant \varepsilon.$$

The number P is the value of the infinite product.

We shall see below that the properties of the multipliable products are analogous to the properties of the summable series with one difference namely that 0 is replaced by 1.

Proposition 1. *If $\prod_{i \in I} u_i$ is to be multipliable it is necessary that u_i should tend to 1, or in other words, for every $\varepsilon > 0$ there should correspond a finite set $\mathfrak{J} \subset I$ so that for $i \notin \mathfrak{J}$*

$$| u_i - 1 | \leqslant \varepsilon.$$

Proposition 2. *(Cauchy's criterion.) The necessary and sufficient condition for $\prod_{i \in I} u_i$ to be multipliable is that $u_i \neq 0$ for every $i \in I$, and that to*

$\varepsilon > 0$ there should correspond a finite set $\mathcal{J} \subset I$ so that for every finite set K, disjoint from $\mathcal{J}$ we have

$$\left| \prod_{i \in K} u_i - 1 \right| \leqslant \varepsilon.$$

Let A be a set of indices and $(I_\alpha)_{\alpha \in A}$ a partition of the set I in finite or infinite subsets mutually disjoint. Then

Proposition 3. *If $\prod_{i \in I} u_i$ is multipliable, $\prod_{i \in I_\alpha} u_i$ is multipliable for every $\alpha \in A$. Let us denote by $P_\alpha = \prod_{i \in I_\alpha} u_i$, then $\prod_{\alpha \in A} P_\alpha$ is multipliable and*

$$\prod_{i \in I} u_i = \prod_{i \in A} P_\alpha = \prod_{\alpha \in A} \left(\prod_{i \in I_\alpha} u_i \right). \tag{1.29}$$

Remark. Relationship (1.29) is true even if the product $\prod_{i \in I} u_i$ is not multipliable, provided that $u_i \geqslant 1$ for every $i \in I$, if we agree to denote by $+\infty$ the value of this divergent product. If $0 < u_i \leqslant 1$ for every $i \in I$, then we agree to denote by 0 the value of the infinite product.

We leave the proof of the three previous propositions to the reader.

Remark. We can pass from infinite products to series by taking the logarithms. But ln u_i is not uniquely determined if u_i is a complex number or a negative number. From proposition 1 we know that if the infinite product is multipliable then $u_i \to 0$ hence $|u_i - 1| < 1$ except for a finite number of factors which we can leave aside.

But ln z is a continuous function for $|z - 1| < 1$. Let us take the determination of the function ln z which is zero for $z = 1$. Then

Theorem. *The product $\prod_{i \in I} u_i$ is multipliable if, and only if, the series $\sum_{i \in I} \ln u_i$ is summable.*

Let us suppose that the series $\sum_{i \in K} \ln u_i$ is summable. From the theorem in Section 1.1.5 it follows that there is a finite set $\mathcal{J} \subset I$ so that for every set K finite and disjoint from $\mathcal{J}$ we have

$$\left| \sum_{i \in K} \ln u_i \right| < \eta_1.$$

But using the continuity of the exponential e^{ξ} it follows that for $\varepsilon > 0$ there exists η so that $|e^{\xi} - 1| \leqslant \varepsilon$ as soon as $|\xi| < \eta$.

Taking $\eta_1 = \eta$

$$\left|\sum_{i\in K} \ln u_i\right| < \eta \text{ implies that } \left|\prod_{i\in K} u_i - 1\right| = \left|e^{\sum\limits_{i\in K} u_i} - 1\right| \leqslant \varepsilon,$$

hence the condition is sufficient.

Let us suppose now that the product $\prod\limits_{i\in I} u_i$ is multipliable. From proposition 2 it follows that for every $\eta > 0$ there is a finite set $\mathcal{J} \subset I$ so that for every finite set K disjoint from $\mathcal{J}$, $\left|\prod\limits_{i\in K} u_i - 1\right| < \eta_1$. But using the continuity of the function $\ln \xi$ for $|\xi - 1| < 1$ it follows that for every $\varepsilon > 0$ there exists $\eta > 0$ so that $|\xi - 1| < \eta$ implies $|\ln \xi| < \varepsilon$. Then taking $\eta_1 = \eta$ it follows that $\left|\prod\limits_{i\in K} u_i - 1\right| < \eta$ implies that $\left|\sum\limits_{i\in K} u_i\right| < \varepsilon$ hence the necessity of the condition.

Corollary. The infinite product $\prod\limits_{i\in I} (1 + v_i)$ *is multipliable if and only if the series* $\sum\limits_{i\in I} v_i$ *is summable and the elements* $1 + v_i$ *are different from zero.*

We notice first that if the product $\prod\limits_{i\in I} (1 + v_i)$ is multipliable then $1 + v_i \to 1$ hence $v_i \to 0$. But if this condition is satisfied then it follows from the previous theorem that the product $\prod\limits_{i\in I} (1 + v_i)$ is multipliable if and only if the series $\sum\limits_{i\in I} \ln (1 + v_i)$ is summable. But $\ln (1 + v) \sim v$ for $v \to 0$ and since, in the case of the summability, we are allowed to replace the general term by an equivalent term, the corollary follows.

1.1.7 Stirling's formula

Let there be the function

$$\Gamma(x + 1) = \int_0^{\infty} e^{-t}\, t^x \, \mathrm{d}t. \tag{1.30}$$

Let us consider the function $e^{-t}\, t^x$ (for a fixed x) and take the logarithmic derivative of this function with respect to t. We obtain $-1 + \dfrac{x}{t}$. This function is < 0 for $0 \leqslant t \leqslant x$, zero for $t = x$ and positive for $t > x$. Consequently the function $e^{-t}\, t^x$ has a maximum for $t = x$. This maximum is $x^x\, e^{-x}$. We make the change of variable $t = x + u$ in the integral of equality (1.30). We obtain

$$\Gamma(x+1) = x^x\, e^{-x} \int_{-x}^{\infty} e^{-u + x \ln\left(1 + \frac{u}{x}\right)}\, du. \tag{1.31}$$

If u is small compared with x we can write

$$-u + x \ln\left(1 + \frac{u}{x}\right) = -u + x\left[\frac{u}{x} - \frac{u^2}{2x^2} + O\left(\frac{u^3}{x^3}\right)\right] = -\frac{u^2}{2x^2} + O\left(\frac{|u|^3}{x^2}\right).$$

Let us fix x and consider the function $e^{-u + x \ln\left(1 + \frac{u}{x}\right)}$ as a function of u. For $u = 0$ this function has a maximum equal to 1. Let us take $u = \sqrt{x}\, \nu$. We have

$$\Gamma(x+1) = x^x\, e^{-x} \sqrt{x} \int_{-\sqrt{x}}^{\infty} e^{-\nu\sqrt{x} + x \ln\left(1 + \frac{\nu}{\sqrt{x}}\right)}\, d\nu. \tag{1.32}$$

Let us take

$$h(x, \nu) = e^{-\nu\sqrt{x} + x \ln\left(1 + \frac{\nu}{\sqrt{x}}\right)} \tag{1.33}$$

then we can prove the following lemma.

Lemma. *For $\nu \leqslant 0$, $h(x, \nu)$ is an increasing function in x and for $\nu > 0$, $h(x, \nu)$, is a decreasing function in x.*

Let us first notice that for $\nu \leqslant -\sqrt{x_1}$, $x_1 \leqslant x_2$ implies $h(x_1, \nu) \leqslant h(x_2, \nu)$ since $h(x_1, \nu) = 0$ and $h(x_2, \nu) \geqslant 0$. We have to prove that $\ln h(x_1, \nu)$ is an increasing function of $\sqrt{x}$, for $0 \geqslant \nu \geqslant -\sqrt{x}$ and a decreasing function of $\sqrt{x}$ for $\nu \geqslant 0$. In other words

$$\frac{\partial}{\partial\sqrt{x}} \ln h(x, \nu) \geqslant 0 \text{ for } -\sqrt{x} < \nu \leqslant 0,$$

$$\frac{\partial}{\partial\sqrt{x}} \ln h(x, \nu) \leqslant 0 \text{ for } \nu \geqslant 0.$$

But $h(x, 0)$ is identically equal to 1 hence $\frac{\partial}{\partial\sqrt{x}}\ln h(x, \nu) = 0$ for $\nu = 0$. Consequently it will be sufficient to show that $\frac{\partial}{\partial\sqrt{x}}\ln h(x, \nu)$ is a decreasing function of ν for $\nu > -\sqrt{x}$ that is

$$\frac{\partial}{\partial\nu}\left(\frac{\partial}{\partial\sqrt{x}}\ln h(x, \nu)\right) \leqslant 0 \text{ for } \nu > -\sqrt{x}.$$

If we reverse the integration order we get

$$\frac{\partial}{\partial\sqrt{x}}\cdot\frac{\partial}{\partial\nu}\ln h(x, \nu) = -\frac{\nu^2}{(\nu+\sqrt{x})^2} \leqslant 0.$$

Corollary 1. $h(x,\nu)$ tends to a summable function $h(\nu)$ for $x \to \infty$.

Indeed, for $\nu \leqslant 0$ the function $h(x, \nu)$ is dominated by its limit $h(\nu)$. For $\nu \geqslant 0$, $h(x, \nu)$ is dominated for $x \geqslant 1$ by $h(1, \nu)$. The corollary results by applying Lebesgue's theorem * separately for $\nu \leqslant 0$ and $\nu \geqslant 0$.

Corollary 2. $\lim_{x\to\infty} h(x, \nu) = h(\nu) = e^{-\nu^2/2}$.

Indeed, for ν fixed and $x \to \infty$ (thus $\nu > -x$),

$$-\nu\sqrt{x} + x\ln\left(1+\frac{\nu}{\sqrt{x}}\right) = -\nu\sqrt{x} + x\left[\frac{\nu}{\sqrt{x}} - \frac{\nu^2}{2x} + O\left(\frac{|\nu|^3}{x^{3/2}}\right)\right] =$$

$$= -\frac{\nu^2}{2} + O\left(\frac{|\nu^3|}{x^{3/2}}\right).$$

From the previous lemma and the two corollaries we have the following:

Theorem. *(Stirling's formula.) We have*

$$\Gamma(x+1) = \int_0^\infty e^{-t}t^x\,dt \sim x^x e^{-x}\sqrt{2\pi x} \tag{1.34}$$

for $x \to \infty$.

---

* *Lebesgue's theorem.* Let $F(x) = \int_a^b f(x, t)\,dt$. If f is separately continuous in x at the point $x = x_0$ for almost all the values of t, for every integration interval, and if $f(x, t)$ is dominated in its module by a summable function $g(t) \geqslant 0$, then F is continuous at the point x_0.

Indeed, taking into account formulae (1.32) and (1.33) and denoting by

$$I(x) = \int_{-\sqrt{x}}^{\infty} h(x, \nu)\, d\nu \tag{1.35}$$

we have to show that integral $I(x)$ converges to $\sqrt{2\pi}$ for x real $\to \infty$. But from corollaries 1 and 2 it follows that $h(x, \nu)$ tends to the summable function $h(\nu) = e^{-\nu^2/2}$. Since

$$\int_{-\infty}^{\infty} e^{-\frac{\nu^2}{2}}\, d\nu = \sqrt{2\pi} \tag{1.36}$$

the theorem is proved completely.

1.1.8 The expansion of the function $1/\Gamma(x)$ into an infinite product

Theorem. *The function* $\dfrac{1}{\Gamma(x)}$ *can be expanded into the infinite product*

$$\frac{1}{\Gamma(x)} = e^{\gamma x} x \prod_{n=1}^{\infty} \left[\left(1 + \frac{x}{n}\right) e^{-\frac{x}{n}}\right] \tag{1.37}$$

where γ *is Euler's constant.*

Indeed, it follows from Stirling's formula that

$$\frac{\Gamma(x+t)}{\Gamma(t)} \sim \frac{(x+t)^{x+t} e^{-x-1} \sqrt{2\pi(x+t)}}{t^t e^{-t} \sqrt{2\pi t}} \sim t^x \left(1 + \frac{x}{t}\right)^{x+t} e^{-x}.$$

But

$$\lim_{t\to\infty} \left(1 + \frac{x}{t}\right)^{x+t} = \lim_{t\to\infty} \left(1 + \frac{x}{t}\right)^{x} \left(1 + \frac{x}{t}\right)^{t} = 1 \cdot e^x$$

that is for $t \to \infty$

$$\frac{\Gamma(x+t)}{\Gamma(t)} = t^x. \tag{1.38}$$

In particular, for $t = n$ formula (1.38) becomes

$$\frac{\Gamma(x+n)}{\Gamma(n)} = \frac{(x+n-1)\,(x+n-2)\dots(x+1)\,\Gamma(x)}{(n-1)\,(n-2)\dots 1} =$$

$$= \left(1+\frac{x}{n-1}\right)\left(1+\frac{x}{n-2}\right)\dots(1+x)x\,\Gamma(x). \tag{1.39}$$

From formulae (1.38) and (1.39) we obtain

$$\frac{1}{\Gamma(x)} = \lim_{n\to\infty}\left[x\left(1+\frac{x}{1}\right)\left(1+\frac{x}{2}\right)\dots\left(1+\frac{x}{n-1}\right)n^{-x}\right]. \tag{1.40}$$

But

$$n^{-x} = e^{-x}\ln n = e^{-x}\left(1+\frac{1}{2}+\dots+\frac{1}{n-1}\right)+\gamma_n x, \tag{1.41}$$

where γ_n tends for $n\to\infty$ to Euler's constant γ. From formulae (1.40) and (1.41) we deduce that

$$\frac{1}{\Gamma(x)} = e^{\gamma x}x\lim\left[\left(1+\frac{x}{1}\right)e^{-\frac{x}{1}}\left(1+\frac{x}{2}\right)e^{-\frac{x}{2}}\dots\left(1+\frac{x}{n-1}\right)e^{-\frac{x}{n-1}}\right]$$

therefore we obtained the expansion of formulae (1.37).

Our problem now is to determine the domain in which the product

$$e^{\gamma z}z\prod_{n=1}^{\infty}\left(1+\frac{z}{n}\right)e^{-\frac{z}{n}} \tag{1.42}$$

is multipliable.

We have the following proposition.

Proposition. *The product (1.42) is multipliable for every value of z, for which no factor is zero that is for $z \neq 0, -1, -2, \dots, -n, \dots$, uniformly on any bounded set of the complex plane, which does not contain these points* $\left(\textit{that is the points at which } \frac{1}{\Gamma(z)} = 0\right)$.

Indeed, we have

$$\ln\left[\left(1+\frac{z}{n}\right)e^{-\frac{z}{n}}\right] = \ln\left(1+\frac{z}{n}\right)-\frac{z}{n} = \frac{z}{n}+O\left(\frac{|z|^2}{n^2}\right)-\frac{z}{n} = O\left(\frac{|z|^2}{n^2}\right)$$

and the proposition follows from theorem of Section 1.16.

Corollary. Product (1.42) *represents a function of a complex variable holomorphic in every bounded domain which does not contain the points* $z = 0, -1, -2, \ldots, -n, \ldots$ *coinciding with* $\frac{1}{\Gamma(z)}$ *for* $z = x > 0$ *which is meromorphic, hence product* (1.42) *coincides with* $\frac{1}{\Gamma(z)}$ *for every value of* z, *that is for every complex* z:

$$\frac{1}{\Gamma(z)} = e^{\gamma z} z \prod_{n=1}^{\infty} \left(1 + \frac{z}{n}\right) e^{-\frac{z}{n}}. \tag{1.43}$$

The expansion into an infinite product of the function $\sin z$. We shall write formula (1.23) in a different way:

$$\frac{\sin \pi z}{\pi} = \frac{1}{\Gamma(z)} \cdot \frac{1}{\Gamma(1-z)} = -\frac{1}{z} \cdot \frac{1}{\Gamma(z)\,\Gamma(-z)}\,. \tag{1.44}$$

Then if in equation (1.44) we replace $\Gamma(z)$ and $\Gamma(-z)$ by the infinite products (1.42) we obtain

$$\sin \pi z = \pi z \prod_{n=1}^{\infty} \left[\left(1 + \frac{z}{n}\right) e^{-\frac{z}{n}} \left(1 - \frac{z}{n}\right) e^{\frac{z}{n}}\right] = \pi z \prod_{n=1}^{\infty} \left(1 - \frac{z^2}{n^2}\right). \tag{1.45}$$

Proposition. *The infinite product from formula* (1.45) *is multipliable for every* $z \neq 0, 1, 2, \ldots, -1, -2, \ldots$ *uniformly for* z *bounded.*

Indeed, this results at once from the fact that $\sum_{n=1}^{\infty} \frac{1}{n^2} < \infty$.

Consequence 1. If we agree to give the value 0 to the product for z integral it follows that for any z

$$\sin \pi z = \pi z \prod_{n=1}^{\infty} \left(1 - \frac{z^2}{n^2}\right). \tag{1.46}$$

Consequence 2. From equation (1.46) we deduce by an obvious change of variable

$$\sin z = z \prod_{n=1}^{\infty} \left(1 - \frac{z^2}{\pi^2 n^2}\right). \tag{1.47}$$

1.1.9 The decomposition into partial fractions of the function cot z

Let us take the logarithmic derivative of expression (1.47). Then

$$\cot z = \frac{1}{z} + \sum_{n=1}^{\infty} \left(\frac{1}{z-n} + \frac{1}{z+n} \right) \cdot \tag{1.48}$$

The series from the right-hand side of equality (1.48) is obviously summable. This expression is a decomposition into partial fractions of the function cot z.

Remark. From equation (1.48) it is immediately clear that $z = n\pi$ are simple poles for cot z with residues equal to 1.

In order to study the function $\psi(z) = \frac{\Gamma'(z)}{\Gamma(z)}$, let us take the logarithmic derivative of the two sides of formula (1.43). We have

$$-\frac{\Gamma'(z)}{\Gamma(z)} = \frac{1}{z} + \sum_{k=1}^{\infty} \left(\frac{1}{z+k} - \frac{1}{k} \right) + \gamma. \tag{1.49}$$

Proposition. *The series from the right-hand side of formula* (1.49) *is uniformly convergent for* z *bounded and different from* 0, —1, —2, . . .

Indeed, the modulus of the general term of the series, is

$$\left| \frac{1}{z+k} - \frac{1}{k} \right| = \left| \frac{-z}{k(z+k)} \right| = \left| \frac{z}{k^2 \left(1 + \frac{z}{k}\right)} \right|$$

and is dominated by $\frac{2 \sup |z|}{k^2}$ for $|k| > 2 \sup |z|$ hence the proposition.

Remark 1. Formula (1.49) represents in fact a decomposition into partial fractions of the function $-\frac{\Gamma'}{\Gamma}$ whose poles are $z = 0, -1, -2, \ldots$ with the residue equal to 1.

Let us take

$$\psi(z) = \frac{\Gamma'(z)}{\Gamma(z)} \cdot \tag{1.50}$$

Remark 2. The following formula can be proved

$$\psi(z) = -\gamma + \int_0^1 \frac{1-(1-t)^{z-1}}{t}\,dt, \text{ for } \operatorname{Re} z > 0.$$

The consequences of formula (1.49) *are as follows:*

(i) If in formula (1.49) we take $z = 1$, we obtain

$$-\Gamma'(1) = 1 + \left[\left(\frac{1}{2} - 1\right) + \left(\frac{1}{3} - \frac{1}{2}\right) + \ldots\right] + \gamma$$

that is

$$\Gamma'(1) = -\gamma. \tag{1.51}$$

(ii) If in formula (1.3) we take $z = 1$ and take formula (1.51) into account, we obtain

$$\Gamma'(1) = \int_0^\infty e^{-t} \ln t\,dt = -\gamma. \tag{1.52}$$

(iii) If in formula (1.49) we take $z = 2$, we obtain

$$-\frac{\Gamma'(2)}{\Gamma(2)} = \frac{1}{2} + \left(\frac{1}{3} - 1\right) + \left(\frac{1}{4} - 1\right) + \ldots + \gamma$$

that is

$$-\frac{\Gamma'(2)}{\Gamma(2)} = -1 + \gamma. \tag{1.52'}$$

(iv) If we take into account equation (1.3), we obtain

$$\Gamma'(2) = \int_0^\infty e^{-t}\, t \ln t\,dt = 1 - \gamma. \tag{1.52''}$$

(v) In formula (1.49) we take $z = p$, where p is an integer > 2. Then

$$\Gamma'(p) = (p-1)!\left(1 + \frac{1}{2} + \frac{1}{3} + \ldots + \frac{1}{p-1} - \gamma\right). \tag{1.53}$$

(vi) If in formula (1.3) we take $z = p$ and consider equation (1.53) we obtain

$$\Gamma'(p) = (p-1)\,!\left(1 + \frac{1}{2} + \frac{1}{3} + \ldots + \frac{1}{p-1} - \gamma\right) =$$
$$= \int_0^\infty e^{-t}\, t^{p-1} \ln t \, dt. \tag{1.54}$$

(vii) Let us take the derivative in formula (1.49). We have

$$\frac{\Gamma\Gamma'' - \Gamma'^2}{\Gamma^2} = \frac{1}{z^2} + \sum_{k=1}^{\infty}\left(\frac{1}{z+k}\right)^2 = \sum_{k=0}^{\infty}\left(\frac{1}{z+k}\right)^2. \tag{1.55}$$

Obviously the series (1.55) converges uniformly for z bounded and different from 0, —1, —2,. . .

(viii) If in equation (1.55) we take $z = x$ real and different from 0, —1, —2, . . . we obtain

$$\frac{\Gamma\Gamma'' - \Gamma'^2}{\Gamma^2} = \sum_{k=0}^{\infty} \frac{1}{(x+k)^2}. \tag{1.56}$$

From equation (1.56) we have

$$\Gamma(x)\,\Gamma''(x) - \Gamma'^2(x) > 0. \tag{1.57}$$

1.1.10 The graphical representation of the function $y = \Gamma(x)$

Let us notice first that for $x > 0$, it follows from formula (1.4) that $\Gamma''(x) > 0$. Consequently Γ is a convex function. In this case it can be monotonous or it can have a minimum. But as $\Gamma(1) = \Gamma(2) = 1$, it follows that it has a minimum at a point, x_0, $1 < x_0 < 2$, and thus for $x < x_0$ it is decreasing and for $x > x_0$ increasing. Let us take the logarithmic derivative of the relationship $\Gamma(x+1) = x\Gamma(x)$. We have

$$\frac{\Gamma'(x+1)}{\Gamma(x+1)} = \frac{\Gamma'(x)}{\Gamma(x)} + \frac{1}{x}.$$

If we now take the usual derivative of the preceding equation we have

$$\frac{\Gamma''(x+1)\,\Gamma(x+1) - \Gamma'^2(x+1)}{\Gamma^2(x+1)} = \frac{\Gamma''(x)\,\Gamma(x) - \Gamma'^2(x)}{\Gamma^2(x)} - \frac{1}{x^2}. \tag{1.58}$$

From formula (1.58) it follows that if we increase x by one unit, $\Gamma''\Gamma - \Gamma'^2$ decreases and consequently it increases if we reduce x by one unit. But for $x > 0$, taking into account formulae (1.1), (1.3) and (1.4) we have

$$\Gamma''\Gamma - \Gamma'^2 = \int_0^\infty e^{-t}\, t^{x-1} (\ln t)^2\, dt \int_0^\infty e^{-t}\, t^{x-1}\, dt - \left(\int_0^\infty e^{-t}\, t^{x-1} \ln t\, dt\right)^2 \cdot \tag{1.59}$$

From equation (1.59) and Schwartz's inequality (Vol. 1, Section 5.3.1) we have $\Gamma''\Gamma - \Gamma'^2 > 0$. As $\Gamma''\Gamma - \Gamma'^2$ increases when x decreases by one unit if follows that $\Gamma''\Gamma - \Gamma'^2 > 0$, when $x < 0$. But then, for every real value of x, $\Gamma''\Gamma > \Gamma'^2 > 0$ consequently

$$\Gamma''(x)\, \Gamma(x) > 0.$$

Hence $\Gamma''(x)$ and $\Gamma(x)$ have the same sign. Consequently at the points x at which $\Gamma(x) > 0$, since $\Gamma''(x) > 0$, Γ is convex. At the points where $\Gamma(x) < 0$, it follows that $\Gamma''(x) < 0$ and Γ is concave. Consequently $y = \Gamma(x)$ will have the form shown graphically in Figure 1.2.

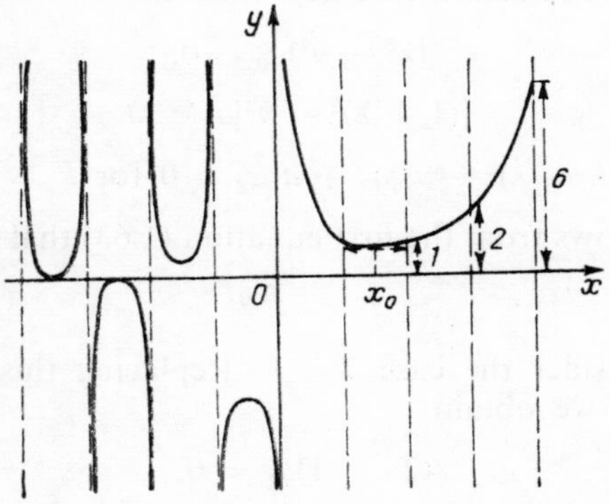

Figure 1.2.

1.2 BESSEL'S FUNCTIONS

1.2.1 Bessel's equation. Bessel's functions

Definition. The differential equation of second order

$$y'' + \frac{1}{x}\, y' + \left(1 - \frac{\nu^2}{x^2}\right) y = 0. \tag{1.60}$$

where x is real or complex and ν is a given complex number, is called Bessel's equation.

We shall suppose that x is real and positive. Almost all the results expressed below are true when x is any complex number. We shall also suppose that Re $\nu \geqslant 0$.

Let us find the solution of equation (1.60) in the form

$$y = x^{\lambda} \sum_{k=0}^{\infty} a_k x^k \tag{1.61}$$

with $a_0 \neq 0$. By differentiating twice series (1.61) term by term and introducing the derivatives y', y'' into equation (1.60), we obtain

$$\sum_{k=0}^{\infty} [(k+\lambda)^2 - \nu^2] a_k x^{k+\lambda-2} + \sum_{k=0}^{\infty} a_k x^{k+\lambda} = 0. \tag{1.62}$$

In order to satisfy this equality it is necessary that all the coefficients of $x^{k+\lambda}$, ($k = 0, 1, 2, \ldots$) should be null, that is

$$(\lambda^2 - \nu^2) a_0 = 0, \tag{1.63}$$

$$[(1+\lambda)^2 - \nu^2] a_1 = 0$$

$$[(k+\lambda)^2 - \nu^2] a_k + a_{k-2} = 0 \text{ for } k \geqslant 2.$$

Since $a_0 \neq 0$, it follows from the first equation (1.63) that

$$\lambda = \pm \nu.$$

Let us first consider the case $\lambda = \nu$. Replacing this value in the other relationships (1.63) we obtain

$$(2\nu + 1) a_1 = 0, \tag{1.63}$$

$$k(k + 2\nu) a_k + a_{k-2} = 0, \quad k \geqslant 2. \tag{1.65}$$

From relationships (1.64) and (1.65) it follows that the coefficients with odd indices are null. Taking $k = 2n$, $n \geqslant 1$ in equation (1.65) we obtain

$$a_{2n} = \frac{(-1)^n a_0}{2^{2n} n! (\nu + 1)(\nu + 2) \ldots (\nu + n)}, \quad n \geqslant 1.$$

Since $a_0 \neq 0$ we can write

$$a_0 = \frac{1}{2^{\nu} \Gamma(\nu + 1)} \cdot$$

It follows that

$$a_{2n} = \frac{1}{2^{\nu}} \cdot \frac{(-1)^n}{2^{2n}\, n!\ \Gamma(\nu + n + 1)}$$

and replacing this value in equation (1.61) we get the solution of the Bessel equation in the case where $\lambda = \nu$.

$$J_{\nu}(x) = \left(\frac{x}{2}\right)^{\nu} \sum_{n=0}^{\infty} \frac{(-1)^n \left(\frac{x}{2}\right)^{2n}}{n!\ \Gamma(n + \nu + 1)} \cdot \tag{1.66}$$

Proposition. *The convergence radius of the series* (1.66) *is infinite.*

Corollary. The series of the right-hand side of relationship (1.66) *is uniformly convergent in every bounded interval.*

Definition. The function J_{ν} (x) given by the relationship (1.66) *is called the Bessel function of index* ν.

Let us now proceed to the case $\lambda = -\nu$. Replacing this value in equation (1.63) we obtain

$$\begin{aligned} &(-2\nu + 1)\, a_1 = 0, \\ &k\,(k - 2\nu)\, a_k + a_{k-2} = 0, \qquad k \geqslant 2. \end{aligned} \tag{1.67}$$

Let us now consider the case where ν is different from an integer and put

$$J_{-\nu}(x) = \left(\frac{x}{2}\right)^{-\nu} \sum_{n=0}^{\infty} \frac{(-1)^n \left(\frac{x}{2}\right)^{2n}}{n!\ \Gamma(n - \nu + 1)} \cdot \tag{1.68}$$

It is immediately clear that the coefficients of $x^{2n-\nu}$ verify relationships (1.67), that is $J_{-\nu}(x)$ is a solution of equation (1.60).

Proposition. *If ν is not an integer then the functions $J_{\nu}(x)$ and $J_{-\nu}(x)$ are linearly independent.*

Indeed, since Re $\nu > 0$, it follows that $J_{\nu}(x)$ converges to 0 as $x \to 0$ and $J_{-\nu}(x)$ tends to the infinity.

Corollary. If ν is not an integer then the general solution of equation (1.60) *is*

$$y = AJ_{\nu}(x) + BJ_{-\nu}(x).$$

Let us now study the case where ν is an integer $p \geqslant 1$.

Since the poles of the function Γ are $n = 0, 1, \ldots, p-1$, formula (1.68) becomes

$$J_{-p}(x) = \left(\frac{x}{2}\right)^{-p} \sum_{n=p}^{\infty} \frac{(-1)^n \left(\frac{x}{2}\right)^{2n}}{n!\,\Gamma(n-p+1)}. \tag{1.69}$$

Replacing the summing index with $k = n - p$ we get

$$J_{-p}(x) = (-1)^p \left(\frac{x}{2}\right)^{p} \sum_{k=0}^{\infty} \frac{(-1)^k \left(\frac{x}{2}\right)^{2k}}{k!\,\Gamma(k+p+1)} = (-1)^p J_p(x). \tag{1.70}$$

Proposition. *The functions $J_p(x)$ and $J_{-p}(x)$ are not linearly independent.*

This follows from relationship (1.70). Consequently we shall have to find a solution linearly independent from $J_p(x)$ in a different way.

1.2.2 Neumann's functions

Definition. We shall call the function

$$N_\nu(x) = \frac{\cos \nu\pi J_\nu(x) - J_{-\nu}(x)}{\sin \nu\pi}, \qquad \nu \neq \textit{integer} \tag{1.71}$$

a Neumann function of index ν.

If $\nu = p$ (integer)

$$N_p(x) = \lim_{\nu \to p} N_\nu(x). \tag{1.72}$$

Remark. For every real x, the mapping $\nu \to J_\nu(x)$ is an analytic function of ν, the numerator and the denominator of the expression (1.71) are hence analytic functions of ν.

Then in equation (1.72) we can apply Hôspital's rule and we obtain

$$N_p(x) = \lim_{\nu \to p} \frac{\frac{\partial}{\partial \nu}[J_\nu(x) \cos \nu\pi - J_{-\nu}(x)]}{\pi \cos \nu\pi},$$

that is

$$\pi N_p(x) = \lim_{\nu \to p} \left[\frac{\partial}{\partial \nu} J_\nu(x) - (-1)^p \frac{\partial}{\partial \nu} J_{-\nu}(x) \right]. \tag{1.73}$$

Theorem. *For p a positive integer, the function $N_p(x)$ is a solution of equation* (1.60) *linearly independent from $J_p(x)$.*

Indeed, let us differentiate the following relationships with respect to ν.

$$\frac{d^2}{dx^2} J_\nu(x) + \frac{1}{x} \frac{d}{dx} J_\nu(x) + \left(1 - \frac{\nu^2}{x^2}\right) J_\nu(x) = 0,$$

$$\frac{d^2}{dx^2} J_{-\nu}(x) + \frac{1}{x} \frac{d}{dx} J_{-\nu}(x) + \left(1 - \frac{\nu^2}{x^2}\right) J_{-\nu}(x) = 0.$$

We obtain

$$\frac{d^2}{dx^2} \frac{\partial}{\partial \nu} J_\nu(x) + \frac{1}{x} \frac{d}{dx} \frac{\partial}{\partial \nu} J_\nu(x) + \left(1 - \frac{\nu^2}{x^2}\right) \frac{\partial}{\partial \nu} J_\nu(x) = \frac{2\nu}{x^2} J_\nu(x),$$

$$\frac{d^2}{dx^2} \frac{\partial}{\partial \nu} J_{-\nu}(x) + \frac{1}{x} \frac{d}{dx} \frac{\partial}{\partial \nu} J_{-\nu}(x) + \left(1 - \frac{\nu^2}{x^2}\right) \frac{\partial}{\partial \nu} J_{-\nu}(x) = \frac{2\nu}{x^2} J_{-\nu}(x).$$

If we multiply the second relationship above by $(-1)^p$ and subtract the result from the first, we obtain

$$\frac{d^2}{dx^2} F_\nu(x) + \frac{1}{x} \frac{d}{dx} F_\nu(x) + \left(1 - \frac{\nu^2}{x^2}\right) F_\nu(x) =$$
$$= \frac{2\nu}{\pi x^2} [J_\nu(x) - (-1)^p J_{-\nu}(x)], \tag{1.74}$$

where we have

$$F_\nu(x) = \frac{1}{\pi} \left[\frac{\partial}{\partial \nu} J_\nu(x) - (-1)^p \frac{\partial}{\partial \nu} J_{-\nu}(x) \right].$$

If $\nu \to p$ in equation (1.74) and take into account equations (1.70) and (1.73), we obtain

$$\frac{d^2}{dx^2} N_p(x) + \frac{1}{x} \frac{d}{dx} N_p(x) + \left(1 - \frac{p^2}{x^2}\right) N_p(x) = 0.$$

We have to show that the functions $J_p(x)$ and $N_p(x)$ are linearly independent. One can notice that if $x \to 0$, then $J_0(x) \to 1$ and $J_p(x) \to 0$. We shall have to calculate the limits of the functions $N_0(x)$ and $N_p(x)$ as

$x \to 0$. To this purpose we calculate first $\frac{\partial}{\partial \nu} J_\nu(x)$ and $\frac{\partial}{\partial \nu} J_{-\nu}(x)$.

We have

$$\frac{\partial}{\partial \nu} J_\nu(x) = \left(\ln \frac{x}{2}\right) J_\nu(x) + \left(\frac{x}{2}\right)^\nu \sum_{k=0}^{\infty} (-1)^k \left(\frac{x}{2}\right)^{2k} \frac{1}{k!} \frac{\partial}{\partial \nu} \frac{1}{\Gamma(\nu + k + 1)}, \tag{1.75}$$

$$\frac{\partial}{\partial \nu} J_{-\nu}(x) = \left(-\ln \frac{x}{2}\right) J_{-\nu}(x) + \left(\frac{x}{2}\right)^{-\nu} \sum_{k=0}^{\infty} (-1)^k \left(\frac{x}{2}\right)^{2k} \frac{1}{k!} \frac{\partial}{\partial \nu} \frac{1}{\Gamma(-\nu + k + 1)}.$$

But

$$\frac{\partial}{\partial \nu} \frac{1}{\Gamma(\nu + k + 1)} = -\frac{\Gamma'(\nu + k + 1)}{\Gamma^2(\nu + k + 1)}.$$

If we consider formula (1.49), we have

$$-\frac{\Gamma'(\nu + k + 1)}{\Gamma(\nu + k + 1)} = \frac{1}{\nu + k + 1} + \sum_{n=1}^{\infty} \left[\frac{1}{\nu + k + 1 + n} - \frac{1}{n}\right] + \gamma. \tag{1.76}$$

Let ν tend to p. We distinguish two cases:

(i) If $p = 0$, then the right-hand side of equation (1.76) tends to γ if $k = 0$ and to $-\sum_{n=1}^{k} \frac{1}{n} + \gamma$ if $k \neq 0$. Consequently, in this case $\frac{\partial}{\partial \nu} J_\nu(x)$ will tend to

$$\left(\ln \frac{x}{2} + \gamma\right) J_0(x) - \sum_{k=0}^{\infty} \left\{(-1)^k \left(\frac{x}{2}\right)^{2k} \frac{1}{(k!)^2} \sum_{n=1}^{k} \frac{1}{n}\right\} \tag{1.77}$$

and $\frac{\partial}{\partial \nu} J_\nu(x)$ will tend to

$$-\ln \frac{x}{2} J_0(x) - \gamma J_0(x) + \sum_{k=1}^{\infty} \left[(-1)^k \left(\frac{x}{2}\right)^{2k} \frac{1}{k!} \sum_{n=1}^{k} \frac{1}{n}\right]. \tag{1.78}$$

Considering (1.73), (1.77) and (1.78) we have

$$N_0(x) = \frac{2}{\pi} \left\{\left(\ln \frac{x}{2} + \gamma\right) J_0(x) - \left[\sum_{k=1}^{\infty} (-1)^k \left(\frac{x}{2}\right)^{2k} \frac{1}{(k!)^2} \sum_{n=1}^{k} \frac{1}{n}\right]\right\}. \tag{1.79}$$

(ii) If $p > 0$, then the right-hand side of relationship (1.76) tends to $-\sum_{n=1}^{n+k} \frac{1}{n} + \gamma$ for every k and hence $\frac{\partial}{\partial \nu} J_\nu(x)$ will tend, as $\nu \to p$, to

$$\left(\ln \frac{x}{2} + \gamma\right) J_p(x) + \left(\frac{x}{2}\right)^p \sum_{k=0}^{\infty} \left\{(-1)^k \left(\frac{x}{2}\right)^{2k} \frac{1}{k!} \frac{(-1)^k}{\Gamma(p+k+1)} \sum_{n=1}^{p+k} \frac{1}{n}\right\}. \tag{1.80}$$

We shall now consider the terms which correspond to $k \geqslant p$ in the series of the right-hand side of the second relationship from (1.75).

As $\nu \to p$ the expression

$$\left(\frac{x}{2}\right)^{-\nu} \sum_{k=p}^{\infty} (-1)^k \left(\frac{x}{2}\right)^{2k} \frac{1}{k!} \frac{1}{\Gamma(-\nu+k+1)} \frac{\Gamma'(-\nu+k+1)}{\Gamma(-\nu+k+1)} \tag{1.81}$$

tends to

$$(-1)^p \left[-\gamma J_p(x) + \left(\frac{x}{2}\right)^p \sum_{k=1}^{\infty} (-1)^k \left(\frac{x}{2}\right)^{2k} \frac{1}{k!} \frac{1}{\Gamma(k+p+1)} \sum_{n=1}^{k} \frac{1}{n}\right]. \tag{1.82}$$

From the series mentioned above we have

$$\left(\frac{x}{2}\right)^{-\nu} \sum_{k=0}^{p-1} (-1)^k \left(\frac{x}{2}\right)^{2k} \frac{1}{k!} \frac{\Gamma'(-\nu+k+1)}{\Gamma^2(-\nu+k+1)}. \tag{1.83}$$

From formulae (1.9) and (1.10) it follows that for $0 \leqslant k \leqslant p-1$ the expression $\frac{\Gamma'(-\nu+k+1)}{\Gamma^2(-\nu+k+1)}$ will tend to $(-1)^{p-k}(p-k-1)!$ as $\nu \to p$ and consequently the expression (1.83) will tend to

$$(-1)^p \left(\frac{x}{2}\right)^{-p} \sum_{k=0}^{p-1} \left(\frac{x}{2}\right)^{2k} \frac{1}{k!} (p-k-1)! \tag{1.84}$$

Consequently if $\nu \to p > 0$, taking into account equations (1.73), (1.75), (1.80), (1.83) and (1.84), we obtain

$$\pi N_p(x) = 2\left(\ln \frac{x}{2} + \gamma\right) J_p(x) - \left(\frac{x}{2}\right)^{-p} \sum_{k=1}^{p-1} \left(\frac{x}{2}\right)^{2k} \frac{1}{k!} (p-k-1)! -$$

$$-\left(\frac{x}{2}\right)^p \frac{1}{p!} \left(\sum_{n=1}^{p} \frac{1}{n}\right) - \left(\frac{x}{2}\right)^p \sum_{k=1}^{\infty} \frac{(-1)^k \left(\frac{x}{2}\right)^{2k}}{k!\,(p+k)!} \left(\sum_{n=1}^{k} \frac{2}{n} + \sum_{n=k+1}^{p+k} \frac{1}{n}\right). \tag{1.85}$$

Taking into account equations (1.79) and (1.85), it follows that $N_0(x) \to \to -\infty$ like $\ln \frac{x}{2}$ as $x \to 0$ and for $p \geqslant 1$, $N_p(x) \to -\infty$ like $-(x)^{-p}$. Hence the functions $J_p(x)$ and $N_p(x)$ are linearly independent for every p integral and $\geqslant 0$ and the general solution of equation (1.60) is

$$y = AJ_p(x) + BN_p(x). \tag{1.86}$$

We shall prove the following proposition.

Proposition. *Between the functions $N_p(x)$ and $N_{-p}(x)$ there is the relationship*

$$N_{-p}(x) = (-1)^p\, N_p(x). \tag{1.87}$$

Indeed, it results from equation (1.73)

$$\pi N_{-p}(x) = \lim_{\nu \to -p} \left[\frac{\partial}{\partial \nu} J_\nu(x) - (-1)^p \frac{\partial}{\partial \nu} J_{-\nu}(x)\right],$$

whence taking $\nu = -\lambda$ we get

$$\pi N_{-p}(x) = \lim_{\lambda \to p} (-1)^p \left[\frac{\partial}{\partial \lambda} J_\lambda(x) - (-1)^p \frac{\partial}{\partial \lambda} J_{-\lambda}(x)\right] = (-1)^p \pi\, N_p(x)$$

that is, relationship (1.87).

1.2.3 Hankel's functions

Let ν be a complex number; we shall give the following definition.

Definition. The functions

$$\begin{aligned} H_\nu^{(1)} &= J_\nu + iN_\nu, \\ H_\nu^{(2)} &= J_\nu - iN_\nu \end{aligned} \tag{1.88}$$

are called Hankel's functions of index ν.

Proposition 1. *Hankel's functions are the solutions of equation* (1.60).

Proposition 2. *Between Hankel's functions there are the following relationships*

$$H_{-\nu}^{(1)} = e^{i\nu\pi} H_{\nu}^{(1)},$$
$$H_{-\nu}^{(2)} = e^{-i\nu\pi} H_{\nu}^{(2)}, \tag{1.89}$$

valid for every ν. *If* $\nu = p$ *integral the relationships* (1.89) *are written*

$$H_{-p}^{(k)} = (-1)^p H_p^{(k)}, \quad k = 1, 2, \ldots$$

We leave the proof of the relationships (1.89) to the reader.

Remark. If we take $H_\nu^{(1)} = e^{ix}$ and $H_\nu^{(2)} = e^{-ix}$ we get $J_\nu = \cos x$, $N_\nu = \sin x$.

1.2.4 The integral representation of Bessel's functions

Theorem. *The function* $J_\nu(x)$ *can be expressed by formula*

$$J_\nu(x) = \frac{\left(\frac{x}{2}\right)^\nu}{\Gamma\left(\nu + \frac{1}{2}\right)\sqrt{\pi}} \int_{-1}^{1} (1 - t^2)^{\nu - \frac{1}{2}} e^{-itx}\, dt \tag{1.90}$$

valid for $\text{Re}\left(\nu + \frac{1}{2}\right) > 0$.

To establish formula (1.90) we shall first consider the complex function $f_\nu(t)$ of real variable t defined in the following way:

$$f_\nu(t) = \begin{cases} (1 - t^2)^{\nu - \frac{1}{2}} & \text{for } |t| < 1, \\ 0 & \text{for } |t| \geqslant 1. \end{cases}$$

The function $f_\nu(t)$ is summable for $\text{Re}\left(\nu + \frac{1}{2}\right) > 0$. The Fourier transform of this function (see Vol. 1, Section 6.2.2) is

$$Tf_\nu = \int_{-1}^{1} (1 - t^2)^{\nu - \frac{1}{2}} e^{-itx}\, dx. \tag{1.91}$$

Let us first show that the function

$$Z_\nu(x) = x^\nu T f_\nu$$

is a solution of equation (1.60). We notice that for every integer $m > 0$, if $\text{Re}\left(\nu + \frac{1}{2}\right) > 0$ then $t^m f_\nu(t)$ is summable. It results that Tf_ν is indefinitely differentiable. In particular, we have the following relationships which are obvious

$$\begin{aligned} (Tf_\nu)' &= T(-it f_\nu(t)), \\ (Tf_\nu)'' &= T(-t^2 f_\nu(t)). \end{aligned} \tag{1.92}$$

Replacing $Z_\nu(x)$ into equation (1.60) we obtain

$$Z_\nu'' + \frac{1}{x} Z_\nu' + \left(1 - \frac{\nu^2}{x^2}\right) Z_\nu = x^\nu (Tf_\nu + (Tf_\nu)'') + (2\nu + 1)\, x^{\nu-1} (Tf_\nu)'. \tag{1.92'}$$

A simple calculation will give us, if we take into account the second relationship (1.92)

$$Tf_\nu + (Tf_\nu)'' = Tf_{\nu+1}. \tag{1.93}$$

On the other hand, from the relationship

$$f_{\nu+1}' = -(2\nu + 1)\; tf_\nu(t) \tag{1.94}$$

it follows that $f_{\nu+1}'$ is summable and

$$xTf_{\nu+1} = -i\; Tf_{\nu+1}' \tag{1.95}$$

By considering equations (1.93) and (1.95) it follows that the right-hand side of equation (1.92′) can be written as follows

$$\begin{aligned} x^\nu Tf_{\nu+1} + (2\nu + 1)\; x^{\nu-1} (Tf_\nu)' &= x^{\nu-1} [-i\; Tf_{\nu+1}' (2\nu + 1)\; T(-itf_\nu)] = \\ &= x^{\nu-1} (-iT(-2\nu + 1)\; tf_\nu) + (2\nu + 1)\, T(-itf_\nu) = \\ &= -ix^{\nu-1}\, T[-(2\nu + 1)\, tf_\nu + (2\nu + 1)\; tf_\nu] = 0, \end{aligned}$$

which proves that Z_ν is a solution of equation (1.60). But as $x \to 0$, $Z_\nu(x)$ behaves like x, therefore the function $Z_\nu(x)$ is proportional to $J_\nu(x)$. If we denote the proportionality coefficient by a_ν, we can write

$$Tf_\nu = a_\nu \frac{1}{2^\nu} \sum_{k=0}^{\infty} \frac{(-1)^k \left(\frac{x}{2}\right)^{2k}}{k!\, \Gamma(\nu + k + 1)}\,. \tag{1.96}$$

If in equation (1.96) we take $x = 0$, we obtain

$$\int_{-1}^{1} (1 - t^2)^{\nu - \frac{1}{2}} dt = a_\nu \frac{1}{2^\nu \Gamma(\nu + 1)} \cdot \tag{1.97}$$

If in formula (1.18) we take $p = \left(\frac{1}{2}\right)$, $q = \nu + \frac{1}{2}$ then the left-hand side of formula (1.97) is just $B\left(\frac{1}{2}, \nu + \frac{1}{2}\right)$. Taking into account formulae (1.14) and (1.17) we obtain

$$a_\nu = 2^\nu \sqrt{\pi}\, \Gamma\left(\nu + \frac{1}{2}\right) \cdot \tag{1.98}$$

But from equation (1.96) we have

$$x^\nu T f_\nu = a_\nu \sum_{k=0}^{\infty} \frac{(-1)^k \left(\frac{x}{2}\right)^{\nu + 2k}}{k!\, \Gamma(\nu + k + 1)} = a_\nu J_\nu(x). \tag{1.99}$$

If in equation (1.99) we replace a by its expression (1.95) we get

$$J_\nu(x) = \frac{\left(\frac{x}{2}\right)^\nu}{\sqrt{\pi}\, \Gamma\left(\nu + \frac{1}{2}\right)} T f_\nu = \frac{\left(\frac{x}{2}\right)^\nu}{\sqrt{\pi}\, \Gamma\left(\nu + \frac{1}{2}\right)} \int_{-1}^{1} (1 - t^2)^{\nu - \frac{1}{2}} e^{-itx}\, dt \tag{1.99'}$$

and thus the theorem is proved completely.

Corollary 1. The function $J_\nu(x)$ has the following integral representation

$$J_\nu(x) = \left(\frac{x}{2}\right)^\nu \frac{1}{\sqrt{\pi}\, \Gamma\left(\nu + \frac{1}{2}\right)} \int_{-1}^{1} (1 - t^2)^{\nu - \frac{1}{2}} \cos tx\, dt. \tag{1.100}$$

Indeed,

$$\int_{-1}^{1} (1 - t^2)^{\nu - \frac{1}{2}} \sin tx\, dt = 0$$

since the function to be integrated is odd as a product of an even with an odd function, hence

$$J_\nu(x) = \left(\frac{x}{2}\right)^\tau \frac{1}{\sqrt{\pi}\,\Gamma\left(\nu + \frac{1}{2}\right)} \int_{-1}^{1} (1-t^2)^{\nu - \frac{1}{2}} e^{itx} \mathrm{d}t =$$

$$= \left(\frac{x}{2}\right)^\nu \frac{1}{\sqrt{\pi}\,\Gamma\left(\nu + \frac{1}{2}\right)} \int_{-1}^{1} (1-t^2)^{\nu - \frac{1}{2}} \cos tx \; \mathrm{d}t.$$

Corollary 2. The function $J_\nu(x)$ has the following integral representation:

$$J_\nu(x) = \frac{\left(\frac{x}{2}\right)^\nu}{\sqrt{\pi}\,\Gamma\left(\nu + \frac{1}{2}\right)} \int_0^\pi \sin^{2\nu}\theta \; e^{-ix\cos\theta} \, \mathrm{d}\theta. \tag{1.101}$$

Indeed, if in equation (1.99′) we make the change of variable $t = \cos\theta$ we get formula (1.101) valid for $\mathrm{Re}\left(\nu + \frac{1}{2}\right) > 0$.

1.2.5 Recurrence formulae

Proposition. *Between the functions J_ν and $J_{\nu+1}$ there is the relationship*

$$-\frac{1}{x}\frac{\mathrm{d}}{\mathrm{d}x}\left(x^{-\nu} J_\nu(x)\right) = x^{-(\nu+1)} J_{\nu+1}(x). \tag{1.102}$$

Indeed, in order to prove this relationship it will be sufficient to establish it for Re ν sufficiently large since both sides in equation (1.102) are analytic functions of ν. By analytic continuation it will follow that formula (1.102) is true for every ν. But from equation (1.98) it follows by considering $\mathrm{Re}\left(\nu + \frac{1}{2}\right) > 0$ that

$$a_{\nu+1} = 2^{\nu+1}\sqrt{\pi}\,\Gamma\left(\nu + 1 + \frac{1}{2}\right) = \left(\nu + \frac{1}{2}\right) 2^{\nu+1}\sqrt{\pi}\,\Gamma\left(\nu + \frac{1}{2}\right) = 2\left(\nu + \frac{1}{2}\right) a_\nu$$

that is

$$\frac{1}{a_\nu} = \frac{2\nu + 1}{a_{\nu+1}}. \tag{1.103}$$

Therefore taking into account equations (1.92), (1.94) and (1.103) we obtain

$$\frac{-iTf'_\nu + 1}{a_{\nu+1}} = \frac{iT[-(2\nu+1)tf_\nu]}{a_{\nu+1}} = \frac{2\nu+1}{a_{\nu+1}} T(-itf_\nu) =$$

$$= \frac{2\nu+1}{a_{\nu+1}} (Tf_\nu)' = \frac{(Tf_\nu)'}{a_\nu}.$$

From the above relationship, taking into account equation (1.95) one can deduce that

$$-x \frac{Tf_{\nu+1}}{a_{\nu+1}} = \frac{\mathrm{d}}{\mathrm{d}x}\left(\frac{Tf_\nu}{a_\nu}\right) \tag{1.104}$$

hence, if we take into account equation (1.96)

$$-x(x^{-(\nu+1)} J_{\nu+1}(x)) = \frac{\mathrm{d}}{\mathrm{d}x}(x^{-\nu} J_\nu(x))$$

which is exactly the same as equation (1.102).

Corollary 1. We have the following relationship

$$J'_\nu(x) = \frac{\nu}{x} J_\nu(x) - J_{\nu+1}(x). \tag{1.105}$$

The above relationship is obtained by taking the derivative of equation (1.103).

Corollary 2.

$$J'_0(x) = -J_1(x). \tag{1.106}$$

The above relationship is obtained if we take $\nu = 0$ in equation (1.102).

Proposition. *Between the functions J_ν and $J_{\nu-1}$ we have the relation*

$$\frac{1}{x}\frac{\mathrm{d}}{\mathrm{d}x}(x^\nu J_\nu(x)) = x^{\nu-1} J_{\nu-1}(x). \tag{1.107}$$

As in the case of the previous proposition, it is sufficient to prove formula (1.107) for ν sufficiently large. If we suppose $\operatorname{Re}\left(\nu - 1 + \frac{1}{2}\right) > 0$ as in the previous proposition we obtain

$$\frac{2\nu - 1}{a_\nu} = \frac{1}{a_{\nu-1}} \cdot$$

On the other hand we have

$$(2\nu - 1) f_\nu - t f'_\nu = (2\nu - 1) f_{\nu-1}$$

that is

$$\frac{T[(2\nu - 1) f_\nu - t f'_\nu]}{a_\nu} = \frac{T f_{\nu-1}}{a_{\nu-1}} \cdot \tag{1.108}$$

The left-hand side of equation (1.108) is

$$x \frac{\mathrm{d}}{\mathrm{d}x}\left(\frac{T f_\nu}{a_\nu}\right) + 2\nu \frac{T f_\nu}{a_\nu}$$

hence

$$x (x^{-\nu} J_\nu)' = -2\nu x^{-\nu} J_\nu + x^{-\nu+1} J_{\nu-1}$$

that is

$$J'_\nu(x) = -\frac{\nu}{x} J_\nu(x) + J_{\nu-1}(x), \tag{1.109}$$

which does not differ from relationship (1.107) where we take the derivative.

Example. (1) Show that

$$J_{\frac{1}{2}}(x) = \sqrt{\frac{2}{\pi x}} \sin x. \tag{1.110}$$

In formula (1.90) we take $\nu = \frac{1}{2}$ and we obtain

$$J_{\frac{1}{2}}(x) = \sqrt{\frac{x}{2\pi}} \int_{-1}^{1} e^{-itx} \, \mathrm{d}t = \sqrt{\frac{x}{2\pi}} \frac{1}{x} \frac{e^{ix} - e^{-ix}}{i} = \sqrt{\frac{2}{\pi x}} \sin x.$$

(2) Show that

$$J_{-\frac{1}{2}}(x) = \sqrt{\frac{2}{\pi x}} \cos x. \tag{1.111}$$

We have from equation (1.107)

$$x^{-\frac{1}{2}} J_{-\frac{1}{2}}(x) = \frac{1}{x} \cdot \frac{d}{dx}\left(\sqrt{\frac{2}{\pi}} \sin x\right) = \frac{1}{x}\sqrt{\frac{2}{\pi}} \cos x$$

hence equation (1.111) follows.

(3) The following relationships are self evident

$$N_{-\frac{1}{2}}(x) = J_{\frac{1}{2}}(x) = \sqrt{\frac{2}{\pi x}} \sin x,$$

$$N_{\frac{1}{2}}(x) = J_{-\frac{1}{2}}(x) = \sqrt{\frac{2}{\pi x}} \cos x.$$

1.2.6 Bessel's functions of integral order

Let us consider the function $e^{ix \sin \theta}$, where x is a real positive parameter and

$$e^{ix \sin \theta} = \sum_{n=-\infty}^{\infty} a_n(x)\, e^{in\theta} \tag{1.112}$$

with

$$a_n(x) = \frac{1}{2\pi} \int_0^{2\pi} e^{ix \sin \theta}\, e^{-in\theta}\, d\theta \tag{1.113}$$

its expansion in terms of Fourier series.

Let us consider the expansion of the function $e^{ix \sin \theta}$ in a power series

$$e^{ix \sin \theta} = \sum_{k=0}^{\infty} \frac{i^k x^k \sin^k \theta}{k!}\, . \tag{1.114}$$

If we fix x, then for $0 < \theta < 2\pi$ the series in equation (1.114) is normally convergent. Consequently, if in equation (1.114) we multiply both sides by $e^{-in\theta}$ and integrate term by term we obtain

$$a_n(x) = \sum_{k=0}^{\infty} \left(\frac{x}{2}\right)^k \frac{1}{k!} \frac{1}{2\pi} \int_0^{2\pi} (e^{i\theta} - e^{-i\theta})^k\, e^{-in\theta}\, d\theta \tag{1.115}$$

Let us suppose $n \geqslant 0$; then if one expands $(e^{i\theta} - e^{-i\theta})^k$ according to the binomial formula one can immediately see that the integral of the right-hand side of expression (1.115) is null except for the case when $k = n + 2\,m$, with $m \geqslant 0$. In this case it is equal to

$$(-1)^m\, C_m^{n+2m} \frac{1}{2\pi} \int_0^{2\pi} d\theta = (-1)^m \frac{(n+2m)!}{m!(n+m)!}\,. \tag{1.116}$$

From (1.116) and (1.115) we obtain

$$a_n(x) = \left(\frac{x}{2}\right)^n \sum_{m=0}^{\infty} (-1)^m \left(\frac{x}{2}\right)^m \frac{1}{m!(n+m)!} = J_n(x). \tag{1.117}$$

If $n < 0$ then by analogy we find

$$a_n(x) = J_n(x). \tag{1.118}$$

From the last two formulae, taking equation (1.113) into account, we have

$$J_n(x) = \frac{1}{2\pi} \int_0^{2\pi} e^{ix \sin\theta}\, e^{-in\theta}\, d\theta. \tag{1.119}$$

Consequence. By taking into account equation (1.119) formula (1.112) can be written

$$e^{ix \sin\theta} = \sum_{n=-\infty}^{\infty} J_n(x)\, e^{in\theta}$$

or

$$e^{ix \sin\theta} = J_0(x) + 2 \sum_{n=1}^{\infty} J_{2n}(x) \cos 2n\theta + 2i \sum_{n=0}^{\infty} J_{2n+1}(x) \sin (2n+1)\theta,$$

hence

$$\cos (x \sin\theta) = J_0(x) + 2 \sum_{n=1}^{\infty} J_{2n}(x) \cos 2n\theta, \tag{1.120}$$

$$\sin (x \sin\theta) = 2 \sum_{n=0}^{\infty} J_{2n+1}(x) \sin (2n+1)\,\theta. \tag{1.121}$$

1.2.7 Asymptotic expansion of Bessel's functions

We shall examine the behaviour of the function $J_\nu(x)$, with ν real, as $x \to \infty$.
Let us set

$$u_\nu(x) = \sqrt{x}\, J_\nu(x).$$

It is immediately obvious that the function $u_\nu(x)$ is the solution of the differential equation

$$u'' + \left(1 - \frac{\nu^2 - \frac{1}{4}}{x^2}\right) u = 0.$$

Consequently, if $x \to \infty$ we can suppose that $u_\nu(x)$ will tend to a trigonometric function which is a solution of the differential equation

$$u'' + u = 0.$$

One can prove rigorously that for $x \to \infty$

$$J_\nu(x) = \sqrt{\frac{2}{\pi x}} \cos\left(x - (2\nu + 1)\frac{\pi}{4}\right) + O\left(\frac{1}{x^{3/2}}\right). \tag{1.122}$$

Consequently the function $J_\nu(x)$ behaves as $x \to \infty$ like the function

$$\sqrt{\frac{2}{\pi x}} \cos\left(x - (2\nu + 1)\frac{\pi}{4}\right).$$

If $\nu = +\frac{1}{2}$ it follows from relationships (1.110) and (1.111) that $J_\nu(x)$ is equal to $\sqrt{\frac{2}{\pi x}} \sin x$ and to $\sqrt{\frac{2}{\pi x}} \cos x$ if $\nu = -\frac{1}{2}$, for every value of x. From relationships (1.122) and (1.71) it follows that Neumann's function $N_\nu(x)$ can be written for $x \to \infty$.

$$N_\nu(x) = \sqrt{\frac{2}{\pi x}} \sin\left(x - (2\nu + 1)\frac{\pi}{4}\right) + O\left(\frac{1}{x^{3/2}}\right). \tag{1.123}$$

From formulae (1.123), (1.122) and (1.87) we have for $x \to \infty$

$$H_\nu^1(x) = \sqrt{\frac{2}{\pi x}}\, e^{i\left(x - (2\nu+1)\frac{\pi}{4}\right)} + O\left(\frac{1}{x^{3/2}}\right)$$

and

$$H_\nu^2(x) = \sqrt{\frac{2}{\pi x}}\, e^{-i\left(x-(2\nu+1)\frac{\pi}{4}\right)} + O\left(\frac{1}{x^{3/2}}\right).$$

1.2.8 The zeroes of Bessel's functions

Theorem. *For $x \to \infty$, the zeroes of the function $J_\nu(x)$ are approximately the zeroes of the function* $\cos\left(x - (2\nu+1)\frac{\pi}{4}\right)$.

Indeed, from equation (1.122) it follows that the zeroes of the function $J_\nu(x)$ can be found only in the neighbourhood of the zeroes of the function $\cos\left(x - (2\nu+1)\frac{\pi}{4}\right)$. One has to show that in some neighbourhood of a zero of the function $\cos\left(x - (2\nu+1)\frac{\pi}{4}\right)$ there is only one zero of the function $J_\nu(x)$. But $\cos\left(x - (2\nu+1)\frac{\pi}{4}\right) = 0$ for $x = (2\nu+1)\frac{\pi}{4} + (2n+1)\frac{\pi}{2}$, n integral. Consequently for given $\varepsilon > 0$ one can determine $\eta > 0$ independent of n so that

$$\left| x - (2\nu+1)\frac{\pi}{4} - (2n+1)\frac{\pi}{2} \right| < \eta$$

should imply

$$\left| \cos\left(x - (2\nu+1)\frac{\pi}{4}\right) - \cos\,(2n+1)\frac{\pi}{2} \right| < \varepsilon.$$

Let us consider the closed interval

$$I_n = \left[(2\nu+1)\frac{\pi}{4} + (2n+1)\frac{\pi}{2} - \eta,\ (2\nu+1)\frac{\pi}{4} + (2n+1)\frac{\pi}{2} + \eta\right].$$

When x ranges over the closed interval I_n, the function $\cos\left(x - (2\nu+1)\frac{\pi}{4}\right)$ varies from the value $(-1)^n\varepsilon$ to the value $(-1)^{n+1}\varepsilon$, consequently the function changes its sign and thus has an odd number of zeroes. To prove that is has only one zero it is sufficient to show that in the interval I_n, $J_n'(x) \neq 0$.

But it follows from formula (1.122) that:

$$J'_\nu(x) = -\sqrt{\frac{2}{\pi x}}\sin\left(x - (2\nu+1)\frac{\pi}{4}\right) + O\left(\frac{1}{x^{3/2}}\right)$$

for $x \to \infty$, that is the zeroes of the derivative $J'_\nu(x)$ belong to some neighbourhood of the zeroes of the function $\sin\left(x - (2\nu+1)\frac{\pi}{4}\right)$. But since in a sufficiently small neighbourhood of a zero of the function $\sin\left(x - (2\nu+1)\frac{\pi}{4}\right)$ there is no zero of the function $\cos\left(x - (2\nu+1)\frac{\pi}{4}\right)$ the theorem is proved completely.

Remark. The zeroes of the function $J_{\frac{1}{2}}(x)$ are given by formula (1.110) since $J_{\frac{1}{2}}(x) = 0$ implies $x = n\pi$, with n a positive integer.

Formula (1.111) gives the zeroes of the function $J_{-\frac{1}{2}}(x)$ namely

$$J_{-\frac{1}{2}}(x) = 0 \text{ implies } x = (n+1)\frac{\pi}{2}, \ n \text{ integral} \geqslant 0.$$

1.2.9 Kelvin's function

If we write

$$I_\nu(x) = e^{-i\nu\frac{\pi}{2}} J_\nu(ix) = \left(\frac{x}{2}\right)^\nu \sum_{k=0}^{\infty} \left(\frac{x}{2}\right)^{2k} \frac{1}{2k!\,\Gamma(\nu+k+1)}$$

we can give the following definition.

Definition. The function

$$K_\nu(x) = \frac{\pi}{2\sin\nu\pi}(I_{-\nu}(x) - I_{+\nu}(x))$$

is called Kelvin's function of index ν.

Proposition 1. *For every* ν

$$K_{-\nu}(x) = K_\nu(x).$$

Proposition 2. *For n integral*

$$I_{-n} = I_n.$$

1.3 EXERCISES

1. Calculate the integral

$$\int_0^\infty \frac{x^m}{(a + bx^n)^p}\, dx, \quad a > 0,\ b > 0,\ xp > m + 1 > 0.$$

2. Prove Euler's formula

$$\int_0^\infty t^{z-1}\, e^{-\lambda t \cos \alpha} \cos (\lambda t \sin \alpha)\, dt = \frac{\Gamma(z)}{\lambda^z} \cos z\alpha$$

$$\left(\lambda > 0,\ t > 0, -\frac{\pi}{2} < \alpha < \frac{\pi}{2}\right).$$

3. Show that

$$\frac{\Gamma'(1)}{\Gamma(1)} - \frac{\Gamma'\left(\frac{1}{2}\right)}{\Gamma\left(\frac{1}{2}\right)} = 2 \ln 2.$$

4. Solve the differential equation

$$x^2y'' + (1 - 2\nu)\, xy' + \nu^2 (x^{2\nu} + 1 - \nu^2)\, y = 0.$$

5. Solve the differential equation

$$y'' - xy' = 0.$$

2

Elements of Distribution Theory

2.1 DEFINITION OF DISTRIBUTIONS

2.1.1 Vector space $\mathfrak{D}$

Let φ be a function defined in R^n with complex values.

Definition 1. The smallest closed set K on whose complement φ is null is called the support of the function φ.

In other words the support of the function φ is the closure of the set of points x for which $\varphi(x) \neq 0$.

Definition 2. We say that the support of the function φ is bounded if there is a bounded set $K \subset R^n$ on whose complement φ is identically zero.

Remark. If the support of the function φ is bounded, since it is a closed set by definition, it follows that it is compact.

Notation. Let us denote by $\mathfrak{D}$ the set of complex functions defined on R^n, indefinitely differentiable, with compact support.

Proposition 1. *The set $\mathfrak{D}$ is a vector space.*

Indeed, if $\varphi_1, \varphi_2 \in \mathfrak{D}$ then $\varphi_1 + \varphi_2 \in \mathfrak{D}$. If λ is a complex number and $\varphi \in \mathfrak{D}$ then $\lambda\varphi \in \mathfrak{D}$.

Remark 1. The set $\mathfrak{D}$ is a ring (or an algebra) for the usual multiplication since if $\varphi_1, \varphi_2 \in \mathfrak{D}$ then $\varphi_1 \varphi_2 \in \mathfrak{D}$.

Proposition 2. *If $\varphi \in \mathfrak{D}$ and ψ is an indefinitely differentiable function with arbitrary support, then $\varphi\psi \in \mathfrak{D}$ and the support of the function $\varphi\psi$ is contained in the intersection of the supports of φ and ψ.*

Definition. *If d is a strictly positive number, the set K_d of the points which are situated at a distance $\leqslant d$ from the set K, is called the neighbourhood of order d of a set K.*

Remark. It can be easily noticed that K_d is a closed bounded set which contains K.

We can now give the following theorem.

Theorem of approximation. *Given $\varepsilon > 0$, for every continuous function f with the bounded support K, there is $\varphi \in \mathfrak{D}$ so that $\| f - \varphi \| < \varepsilon$. In addition, the function φ can be chosen so that its support be contained in an arbitrary neighbourhood of K.*

In order to prove this theorem we have to show that given the function f with the support K and given two real positive numbers ε and d, there is $\varphi \in \mathfrak{D}$ with its support contained in K_d so that

$$\| f - \varphi \| = \sup_{x \in R^n} | f(x) - \varphi(x) | \leqslant \varepsilon.$$

Let us consider the function

$$\theta_1(x) = \begin{cases} 0, & \text{if } r \geqslant 1 \\ \exp\left(-\dfrac{1}{1-r^2}\right), & \text{if } r < 1 \end{cases}$$

where $r = | x | = \sqrt{\sum_{i=1}^{n} x_i^2}$. It is easy to see that $\theta_1 \in \mathfrak{D}$. Let us write

$$\theta = \frac{1}{c} \theta_1 \left(\frac{x}{a}\right). \tag{2.1}$$

It is immediately seen that the support of the function θ is the closed ball $| x | \leqslant a$. If the constant c has the value

$$c = \int_{R^n} \cdots \int \theta_1 \left(\frac{x}{a}\right) \, dx = a^n \int_{R^n} \cdots \int \theta_1 (y) \, dy > 0,$$

then

$$\int_{R^n} \cdots \int \theta (x) \, dx = 1. \tag{2.2}$$

Let us take

$$\varphi(x) = \int_{R^n} \cdots \int f(x-\xi)\,\theta(\xi)\,\mathrm{d}\xi = \int_{R^n} \cdots \int f(\xi)\,\theta(x-\xi)\,\mathrm{d}\xi \tag{2.3}$$

and let us show that if the constant a is sufficiently small then φ fulfils the conditions of the theorem.

Let us consider a point $x \notin K_a$. Then $|x-\xi| > a$ for every $\xi \in K$. But since the second integral from formula (2.3) is calculated on the support K of the function f and since the support of the function θ is a closed ball with the radius a, it follows that $\theta(x-\xi) = 0$ for every $\xi \in K$ hence $\varphi(x) = 0$.

Consequently if $x \in \complement K_a$ then $\varphi(x) = 0$, whence it follows that the support of φ is contained in K_d if $a \leqslant d$.

Let us show that the function φ is indefinitely differentiable. We notice that the product $f(\xi)\,\theta(x-\xi)$ has derivatives with respect to x of every order, and its derivatives are continuous with respect to x' and ξ. We then notice that the set K, on which the last integral from formula (2.3) is calculated, is bounded, therefore we can take the derivative with respect to x as many times as we wish and hence φ is indefinitely differentiable.

We now consider formula (2.2). Then

$$f(x) - \varphi(x) = \int_{R^n} \cdots \int (f(x) - f(x-\xi))\,\mathrm{d}\xi. \tag{2.4}$$

The function f is continuous and has a bounded support, hence it is uniformly continuous. Consequently, to an arbitrary $\varepsilon > 0$ there corresponds $\eta > 0$ so that $|f(x) - f(x-\xi)| \leqslant \varepsilon$ as soon as $|\xi| < \eta$. But in formula (2.4) the integral is calculated on the support of the function θ hence on the sphere $|\xi| \leqslant a$. Hence $a \leqslant \eta$ implies $|f(x) - f(x-\xi)| \leqslant \varepsilon$. Therefore taking into account formula (2.2) we have

$$|f(x) - \varphi(x)| \leqslant \varepsilon.$$

Thus for $a \leqslant d$ and $a \leqslant \eta$, φ fulfils the conditions of the theorem.

2.1.2 The notion of convergence in the space $\mathfrak{D}$

Definition. We say that a sequence $\{\varphi_j\}$ of functions from $\mathfrak{D}$ converges to a function $\varphi \in \mathfrak{D}$ if the following conditions are fulfilled:

(a) The supports of the functions φ_j, $(j = 1, 2, \ldots)$ are contained in the same bounded set.

(b) The derivatives of every order of the functions φ_j converge uniformly for $j \to \infty$ to the corresponding derivatives of φ.

Remark. We underline the fact that in the above definition it is not the uniform convergence with respect to all differentiation orders which is required, but the uniform convergence with respect to each order of differentiation separately.

2.1.3 Distributions — definition, examples

Definition. A continuous linear functional in the vector space $\mathfrak{D}$ is called a distribution and is denoted by T.

We recall [Vol. 1, Section 3.1.6] that a mapping T is called a linear functional if, for every $\varphi \in \mathfrak{D}$, we can find a complex number $T(\varphi)$ which we shall also denote by $\langle T, \varphi \rangle$ with the properties

$$T(\varphi_1 + \varphi_2) = T(\varphi_1) + T(\varphi_2),$$

$$T(\alpha\varphi) = \alpha T(\varphi), \ \alpha \text{ complex constant.}$$

The linear functional T is continuous if the convergence in $\mathfrak{D}$ of a sequence $\{\varphi_j\}$ to φ implies the convergence of the sequence of complex numbers $\{T(\varphi_j)\}$ to the complex number $T(\varphi)$.

Notation. We denote by $\mathfrak{D}'$ the set of distributions defined in $\mathfrak{D}$.

Proposition. *$\mathfrak{D}'$ is a vector space.*

Indeed this follows, immediately, if the sum $T_1 + T_2$ of two distributions and the product αT of a complex number with a distribution are defined as

$$\langle T_1 + T_2, \ \varphi \rangle = \langle T_1, \ \varphi \rangle + \langle T_2, \ \varphi \rangle,$$

$$\langle \alpha T, \ \varphi \rangle = \alpha \langle T, \ \varphi \rangle.$$

Remark. $\mathfrak{D}'$ is the topological dual of the space $\mathfrak{D}$ (Vol. 2, Section 3.3.7).

Example of a distribution. Let f be a locally summable function, that is summable on every bounded set and let $\varphi \in \mathfrak{D}$. Let us take

$$\langle T_f, \ \varphi \rangle = \int \underset{R^n}{\cdots} \int f(x)\, \varphi(x) \ \mathrm{d}x. \tag{2.5}$$

Proposition. *The function f defines a distribution by formula* (2.5).

Indeed, let us first show that the integral in formula (2.5) makes sense. The integral in formula (2.5) is in fact taken on the support of φ, which is a bounded set. Since f is locally summable it follows that f is summable on the support of φ and as the function φ is continuous we deduce that $f\varphi$ is a summable function. On the other hand the value of the integral is a linear functional of φ. Let us show that it is also continuous on $\mathfrak{D}$. Therefore let us suppose that φ_j converges in $\mathfrak{D}$ to φ and denote by K the bounded set which contains the supports of all the functions φ_j. We have

$$|\langle T_f, \varphi_j \rangle - \langle T_f, \varphi \rangle| \leqslant \left(\int \underset{K}{\cdots} \int | f(x) | \, \mathrm{d}x \right) \max | \varphi - \varphi_j |.$$

Since $\max | \varphi - \varphi_j | \to 0$ as $j \to \infty$ the difference in the left-hand side of the above relationship tends to 0 and thus the proposition is completely proved.

Let us consider two functions f and g locally summable. Then

Proposition. $T_f = T_g$ *if and only if* $f = g$ *almost everywhere.*

Indeed, if $f = g$ almost everywhere then $\langle T_f, \varphi \rangle = \langle T_g, \varphi \rangle$ for every $\varphi \in D$. Conversely, if we write $h = f - g$, we have to show that

$$\int \underset{R^n}{\cdots} \int h(x) \, \varphi(x) \, \mathrm{d}x = 0 \text{ for every } \varphi \in \mathfrak{D}$$

implies $h(x) = 0$ almost everywhere.

The proof will be made in three steps. First we shall show that

$$\int \underset{R^n}{\cdots} \int h(x) \, \psi(x) \, \mathrm{d}x = 0 \tag{2.6}$$

for every continuous function ψ with bounded support.

From the approximation theorem (Section 2.1.1) it follows that for every $\varepsilon > 0$ and $d > 0$ there is a function $\varphi \in \mathfrak{D}$ so that $| \varphi - \psi | \leqslant \varepsilon$ and denoting by K the support of ψ, φ has the support in K_d. We have

$$\left| \int \underset{R^n}{\cdots} \int (\varphi(x) - \psi(x)) \, h(x) \, \mathrm{d}x \right| \leqslant$$

$$\leqslant \int \underset{K_d}{\cdots} \int | \varphi(x) - \psi(x) | \, | h(x) | \, \mathrm{d}x \leqslant \varepsilon \int \underset{K_d}{\cdots} \int | h(x) | \, \mathrm{d}x. \tag{2.7}$$

But since from the hypothesis $\int_{R^n} \ldots \int h(x)\, \varphi(x)\, dx = 0$, relationship (2.7) becomes

$$\left| \int_{R^n} \ldots \int \psi(x)\, h(x)\, dx \right| \leqslant \varepsilon \int_{K_d} \ldots \int |\, h(x) \,|\, dx. \tag{2.8}$$

As $\varepsilon \to 0$, for d fixed, we obtain formula (2.6). Let us now consider a function $\chi(x)$ measurable and bounded with bounded support. We shall show that

$$\int_{R^n} \ldots \int h(x)\, \chi(x)\, dx = 0. \tag{2.9}$$

Since χ is measurable on R^n it follows that there is a sequence of continuous functions χ_j which for $j \to \infty$, converges to χ almost everywhere. Since χ has a bounded support we can suppose that the functions χ_j have the support in the same bounded set independent of j; otherwise the functions χ_j can be replaced by $\alpha\chi_j$, where α is a continuous function equal to 1 on the support of χ and with bounded support.

As the function χ is bounded we suppose the functions χ_j bounded by $M = \sup |\chi|$. Otherwise taking $\chi_j(x) = \rho_j(x)\, e^{i\omega_j(x)}$ the functions χ_j can be replaced by $\sigma_j\, e^{i\omega_j(x)}$ where $\sigma_j(x) = \inf\,(M, \rho_j(x))$.

Since these two conditions are fulfilled and since the functions χ_j are continuous functions with bounded support, we have from formula (2.6)

$$\int_{R^n} \ldots \int h(x)\, \chi_j(x)\, dx = 0. \tag{2.10}$$

The function h is locally summable hence $M\,|\, h(x) \,|$ is locally summable and since it has its support contained in a bounded set it follows that it is summable. But the sequence of functions $\{\chi_j\}$ converges almost everywhere to $\chi(x)$ and $|\, h(x)\, \chi_j(x) \,|$ is bounded by $M\,|\, h(x) \,|$ for arbitrary j, hence from Lebesgue's theorem [Vol. 1, Section 5.2.2] we have

$$\int_{R^n} \ldots \int h(x)\, \chi_j(x)\, dx \to \int_{R^n} \ldots \int h(x)\, \chi(x)\, dx \tag{2.11}$$

as $j \to \infty$. From formulae (2.11) and (2.10) we obtain formula (2.9).

Now we pass to the last step of the proof. Let us denote by $\chi(x)$ the function defined as

$$\chi(x) = \begin{cases} 0 \text{ for } |\, x \,| > a \text{ or } h(x) = 0, \\ e^{-i\omega(x)} \text{ for } |\, x \,| \leqslant a \text{ and } h(x) \neq 0, \text{ where } h(x) = r(x)\, e^{i\omega(x)}. \end{cases}$$

It is easily seen that the function χ is measurable, has the module 1 or 0 and its support is contained in the closed ball $|x| \leqslant a$. Hence from formula (2.9).

$$\int_{R^n} \dots \int h(x)\, \chi(x)\, dx = \int_{|x| \leqslant a} \dots \int r(x)\, dx = \int_{|x| \leqslant a} \dots \int |h(x)|\, dx = 0. \quad (2.12)$$

But $|h(x)| \geqslant 0$ and taking formula (2.12) into account we have $h(x) = 0$ almost everywhere in the closed ball $|x| \leqslant a$. As a is arbitrary it follows that h is zero almost everywhere and thus the proposition is proved completely.

Important remark. If we deal only with classes of locally summable functions and if two locally summable, almost everywhere equal, functions, are considered as identical then the distributions are a generalisation of the notion of locally summable functions.

Consequently we shall identify a locally summable function f defined almost everywhere with the functional T_f which it defines and we shall write

$$\langle f, \varphi \rangle = \langle T_f, \varphi \rangle = \int_{R^n} \dots \int f(x)\, \varphi(x)\, dx.$$

In particular the functional for which to each function φ there corresponds its integral $\int_{R^n} \dots \int \varphi(x)\, dx$, defines a distribution which we identify with $f = 1$.

Let us now give some examples of distributions.

Example 1. Dirac's distribution is

$$\langle \delta_0, \varphi \rangle = \varphi(0).$$

Dirac's distribution δ_a at the point $a \in R^n$ is

$$\langle \delta_a, \varphi \rangle = \varphi(a).$$

If we consider the distributions $T \in \mathfrak{D}'$ as distributions of electric or magnetic charge or of material mass, then Dirac's distribution δ_a represents the mass $+1$ at the point $a \in R^n$.

Example 2. If D is an arbitrary partial derivative, the following distributions can be defined

$$\langle T, \varphi \rangle = D\varphi(a).$$

Example 3. Let f be a locally summable function, D a partial differentiation symbol of an arbitrary order with respect to $x_1, \ldots, x_n$, then the functional

$$\langle T, \varphi \rangle = \int_{R^n} \ldots \int f(x)\, \mathrm{D}\varphi(x)\, \mathrm{d}x = \langle f, \mathrm{D}\varphi \rangle$$

defines a distribution.

Remark. The distribution associated to a locally summable function f can be considered as a distribution of charges defined by the density f. In physics the expressions $\langle T, \varphi \rangle$ are calculated, where φ does not belong to the space $\mathfrak{D}$. But each distribution T can be extended as a functional on a set larger than $\mathfrak{D}$, which depends on T. Thus δ_0 can be extended to all the functions continuous at the origin.

Example 4. If we consider the distribution associated to a locally summable function f as a distribution of charge defined by the density f, in order to calculate the total charge we calculate $\langle T, 1 \rangle$ and we obtain

$$\langle T, 1 \rangle = \begin{cases} \int_{R^n} \ldots \int f(x)\, \mathrm{d}x, & \text{for } T = f, \\ +1 & \text{for } T = \delta_a. \end{cases}$$

Example 5. In order to calculate the moment of inertia with respect to the origin, we calculate $\langle T, r^2 \rangle$ where $r^2 = \sum_{i=1}^{n} x_i^2$ and we obtain

$$\langle T, r^2 \rangle = \begin{cases} \int_{R^n} \ldots \int f(x)\, r^2\, \mathrm{d}x & \text{for } T = f, \\ |a|^2 & \text{for } T = \delta_a. \end{cases}$$

Example 6. In order to calculate in R^n the newtonian potential at a point $b \in R^n$ we calculate $\left\langle T, \dfrac{1}{|x-b|} \right\rangle$ and we obtain

$$\left\langle T, \frac{1}{|x-b|} \right\rangle = \begin{cases} \int_{R^n} \ldots \int \dfrac{f(x)\, \mathrm{d}x}{|x-b|} & \text{for } T = f, \\ \dfrac{1}{|a-b|} & \text{for } T = \delta_a. \end{cases}$$

Example 7. (1) Determine the mathematical distribution which corresponds to a distribution of electric charges defined by a dipole of electric moment $+1$ located at $0 \in R$.

It is known that the dipole is a limit as $\varepsilon \to 0$ of a system T_ε of two masses $\frac{1}{\varepsilon}$, $-\frac{1}{\varepsilon}$ placed at the points 0, ε. To this system we can attach the distribution

$$\langle T_\varepsilon, \varphi \rangle = \frac{1}{\varepsilon} \varphi(\varepsilon) - \frac{1}{\varepsilon} \varphi(0) = \frac{\varphi(\varepsilon) - \varphi(0)}{\varepsilon}$$

hence

$$\lim_{\varepsilon \to 0} \langle T_\varepsilon, \varphi \rangle = \varphi'(0).$$

Thus we can define the dipole by

$$\langle T, \varphi \rangle = \varphi'(0)$$

(2) Let us consider in R^n a dipole of electric moment $\mathfrak{M}$ directed in a given direction, located at a point $a \in R^n$. Then the distribution corresponding to this dipole will be

$$\langle T, \varphi \rangle = \mathfrak{M} \mathrm{D}\, \varphi(a).$$

Example 8. In quantum physics instead of the distribution δ_0 we use Dirac's function defined thus:

$$\delta(x) = \begin{cases} 0 \text{ for } x \neq 0, \\ \infty \text{ for } x = 0 \end{cases}$$

which verifies the relationship

$$\int \cdots \int_{R^n} \delta(x)\, \mathrm{d}x = 1.$$

Clearly the above relation is incorrect because $\delta(x) = 0$ almost everywhere hence its Lebesgue integral is zero. However physicists use $\delta(x)$ as a 'function'.

The distribution δ_a is used in physics in the form $\delta(x - a)$ and the formula

$$\langle \delta_a, \varphi \rangle = \varphi(a)$$

is written as

$$\int \cdots \int_{R^n} \delta(x - a)\, \varphi(x)\, \mathrm{d}x = \varphi(a).$$

2.2 DIFFERENTIATION OF DISTRIBUTIONS

2.2.1 The support of a distribution

Definition 1. We say that a distribution T is zero in an open set Ω from R^n, if for every function $\varphi \in \mathfrak{D}$, having the support in Ω, $\langle T, \varphi \rangle = 0$.

Definition 2. *We say that a distribution T is defined in an open set Ω, if it is a linear functional continuous in the subspace of $\mathfrak{D}$ consisting of functions φ with the support contained in Ω.*

Let $\{\Omega_i\}$ be a family of open sets depending on a set I of indices. Let $\{T_i\}$ be a family of distributions depending on the same set of indices with the following properties.

(1) The distribution T_i is defined on the set Ω_i.

(2) If $\Omega_i \cap \Omega_j \neq \varnothing$, then T_i and T_j coincide on this intersection.

We shall admit that there exists a distribution T and only one defined on Ω and coinciding with T_i on each open set.

In this case we see that if $T_i = T = 0$ for every $i \in I$, in other words if the distribution T is zero on every open set Ω_i then T is zero on the union Ω of the sets of the family $\{\Omega_j\}$. Consequently the union of all open sets where a distribution T is zero is an open set, namely it is the greatest open set with this property.

We can now give the definition of the support of a distribution.

Definition. *We call support of a distribution T, the complement of the largest open set where T is zero; in other words, the support of the distribution T is the smallest closed set on whose complement T is zero.*

Example. The support of the distribution δ_a is reduced to the point a.

From the above definition we immediately deduce the following:

Proposition. *The necessary and sufficient condition that a point should belong to the support of T is that T be different from zero in every neighbourhood of this point.*

Remark. If T is a continuous function, its support as a distribution coincides with its support as a function.

2.2.2 Derivative of a distribution

We shall define the derivative $\frac{\partial T}{\partial x_i}$ of a distribution T on R^n with respect to x_i so that if T is reduced to a function f, its derivative be reduced to the usual derivative of the function f with respect to x_i.

Proposition. *Let us consider a function f continuously differentiable and write*

$$\left\langle \frac{\partial f}{\partial x_1}, \varphi \right\rangle = \int \cdots_{R^n} \int \frac{\partial f}{\partial x_1} \varphi \, dx. \tag{2.13}$$

Then

$$\left\langle \frac{\partial f}{\partial x_1}, \varphi \right\rangle = -\left\langle f, \frac{\partial \varphi}{\partial x_1} \right\rangle. \tag{2.14}$$

Indeed, we can apply Fubini's theorem (Vol. 1, Section 5.2.2) to the integral in formula (2.13) since the integrands are continuous functions and the integral is over a bounded set. Consequently

$$\int \cdots_{R^n} \int \frac{\partial f}{\partial x_1} \varphi \, dx = \int \cdots_{x_2, \ldots, x_n} \int dx_2 \ldots dx_n \int_{-\infty}^{\infty} \frac{\partial f}{\partial x_1} \varphi \, dx_1. \tag{2.15}$$

But an integration by parts yields

$$\int_{-\infty}^{\infty} \frac{\partial f}{\partial x_1} \varphi \, dx_1 = f \varphi \Big|_{-\infty}^{\infty} - \int_{-\infty}^{\infty} f \frac{\partial \varphi}{\partial x_1} dx_1.$$

The first term in the right-hand side of the above relationship is zero since φ is with bounded support and relationship (2.15) becomes

$$\int \cdots_{x_2, \ldots, x_n} \int dx_2 \ldots dx_n \int_{-\infty}^{\infty} \frac{\partial f}{\partial x_1} \varphi \, dx_1 =$$

$$= -\int \cdots_{R^n} \int f \frac{\partial \varphi}{\partial x_1} dx_1 = -\left\langle f, \frac{\partial \varphi}{\partial x_1} \right\rangle$$

whence taking into account formulae (2.15) and (2.13) we obtain formula (2.14) and thus the proposition is proved.

We can now give the following definition.

Definition. *We say that* $\frac{\partial T}{\partial x_1}$ *is the derivative of the distribution* T *on* R^n *with respect to* x_1 *if*

$$\left\langle \frac{\partial T}{\partial x_1}, \varphi \right\rangle = -\left\langle T, \frac{\partial \varphi}{\partial x_1} \right\rangle. \tag{2.16}$$

Proposition. *The derivative* $\frac{\partial T}{\partial x_1}$ *of the distribution* T *is a distribution.*

To prove the proposition we have to show that $\frac{\partial T}{\partial x_1}$ is a linear continuous functional in the space $\mathfrak{D}$. It is obvious that $\frac{\partial T}{\partial x_1}$ is a linear functional of φ.

We have to prove that it is also continuous. We suppose that φ_j converges to φ in $\mathfrak{D}$, as $j \to \infty$; then according to the definition of the convergence in $\mathfrak{D}$, it follows that $\frac{\partial \varphi_j}{\partial x_1}$ converges to $\frac{\partial \varphi}{\partial x_1}$ in $\mathfrak{D}$, and since T is a distribution it follows that $\left\langle T, \frac{\partial \varphi_j}{\partial x_1} \right\rangle$ converges to $\left\langle T, \frac{\partial \varphi}{\partial x_1} \right\rangle$. By taking formula (2.16) into account it follows that $\left\langle \frac{\partial T}{\partial x_1}, \varphi_j \right\rangle$ converges to $\left\langle \frac{\partial T}{\partial x_1}, \varphi \right\rangle$ and hence $\frac{\partial T}{\partial x_1}$ is a distribution and the proposition is proved completely.

The derivatives $\frac{\partial T}{\partial x_i}$, $i = 2, 3, \ldots, n$, are defined in the same way. Let us now try to define a derivative of second-order $\frac{\partial^2 T}{\partial x_i \, \partial x_j}$. Applying formula (2.16) twice we have

$$\left\langle \frac{\partial^2 T}{\partial x_i \, \partial x_j}, \varphi \right\rangle = -\left\langle \frac{\partial T}{\partial x_j}, \frac{\partial \varphi}{\partial x_i} \right\rangle = \left\langle T, \frac{\partial^2 \varphi}{\partial x_j \partial x_i} \right\rangle, \tag{2.17}$$

$$\left\langle \frac{\partial^2 T}{\partial x_j \partial x_i}, \varphi \right\rangle = -\left\langle \frac{\partial T}{\partial x_i}, \frac{\partial \varphi}{\partial x_j} \right\rangle = \left\langle T, \frac{\partial^2 \varphi}{\partial x_i \partial x_j} \right\rangle. \tag{2.18}$$

But the function φ has continuous second-order derivatives, hence

$$\frac{\partial^2 \varphi}{\partial x_j \partial x_i} = \frac{\partial^2 \varphi}{\partial x_i \partial x_j}$$

therefore

$$\frac{\partial^2 T}{\partial x_i \partial x_j} = \frac{\partial^2 T}{\partial x_j \partial x_i}. \tag{2.19}$$

Let $p = (p_1, p_2, \ldots, p_n)$ be a system of n integers $\geqslant 0$. Let us denote by

$$\mathrm{D}^p = \left(\frac{\partial}{\partial x_1}\right)^{p_1} \left(\frac{\partial}{\partial x_2}\right)^{p_2} \cdots \left(\frac{\partial}{\partial x_n}\right)^{p_n}.$$

If $|p| = p_1 + p_2 + \ldots + p_n$ then

$$\langle \mathrm{D}^p T, \varphi \rangle = (-1)^{|p|} \langle T, \mathrm{D}^p \varphi \rangle. \tag{2.20}$$

We have the following proposition.

Proposition. *Every distribution T has derivatives of arbitrary order and the order of differentiation can be reversed.*

Corollary. Every continuous function or locally summable function has derivatives of every order.

Indeed the successive derivatives are defined as derivatives of distributions. Generally they are not functions.

In particular, if f is continuously differentiable its derivative as distribution coincides with its usual derivative, according to formula (2.14).

Let $D = \sum_p A_p D^p$ be a differential operator with constant coefficients.

Definition. We call adjoint differential operator the operator ${}^tD$ *defined by*

$$ {}^tD = \sum_p (-1)^{|p|} A_p D^p. \tag{2.21} $$

Remark. Formula (2.21) yields

$$ \langle DT, \varphi \rangle = \langle T, {}^tD\,\varphi \rangle. \tag{2.22} $$

It is obvious that

$$ {}^t{}^tD = D $$

and hence

$$ \langle {}^tD\,T, \varphi \rangle = \langle T, D\varphi \rangle. \tag{2.23} $$

2.2.3 Applications

Application 1. Given $D = \Delta = \sum \frac{\partial^2}{\partial x_i^2}$ then from equation (2.23) we have

$$ \langle \Delta T, \varphi \rangle = \langle T, \Delta\varphi \rangle. $$

Application 2. Let us consider Heaviside's function or the unit function $u(x)$ (Vol. 1, example 1, Section 7.2.2). From formula (2.16) we have

$$ \langle u, \varphi \rangle = -\langle u, \varphi' \rangle = -\int_{-\infty}^{\infty} u(x)\,\varphi'(x)\,dx = -\int_0^{\infty} \varphi'(x)\,dx = $$

$$ = -\varphi(x)\Big|_{x=0}^{x\to\infty} = \varphi(0) = \langle \delta_0, \varphi \rangle $$

hence

$$u' = \delta_0. \tag{2.24}$$

We notice that the discontinuity of u appears in the form of punctual mass in its derivative. Let us take the successive derivatives of δ_0. We have

$$\langle \delta_0', \varphi \rangle = -\langle \delta_0, \varphi' \rangle = -\varphi'(0),$$

δ_0 is therefore a dipole of moment equal to -1, located at the origin

$$\langle \delta_0^{(m)}, \varphi \rangle = (-1)^{(m)} \varphi^{(m)}(0). \tag{2.25}$$

From the previous considerations we see that the 'incorrect' calculations used by the physicists [Vol. 1, Section 7.4.2] are justified in the language of the theory of distributions.

Let us consider a function f indefinitely differentiable for $x < 0$ and $x > 0$, so that each of its derivatives should have a limit to the left and a limit to the right at the point $x = 0$. Let us denote by σ_m the difference between the limit to the left and the limit to the right of the mth order derivative at the point $x = 0$.

Notation. Let $f', f'', \ldots$ be the derivatives of f in the sense of distributions and let $\{f'\}, \{f''\}, \ldots$ be the distributions represented by functions equal to the usual derivatives for $x < 0$ and $x > 0$ and not defined for $x = 0$.

Example. If $f = u$, then $f' = \delta_0$ and $\{f'\} = 0$.

Proposition. *The following formulae are true:*

$$f' = \{f'\} + \sigma_0\,\delta_0,$$
$$f'' = \{f''\} + \sigma_0\,\delta_0 + \sigma_1\,\delta_0, \tag{2.26}$$

....

$$f^{(m)} = \{f^{(m)}\} + \sigma_1\delta_0^{(m-1)} + \sigma_2\delta_0^{(m-2)} + \ldots + \sigma_{m-1}\,\delta_0.$$

Indeed we have

$$\langle f', \varphi \rangle = -\langle f, \varphi' \rangle = -\int_{-\infty}^{\infty} f(x)\,\varphi'(x)\,dx = -\int_{-\infty}^{0} f(x)\,\varphi'(x)\,dx -$$

$$-\int_{0}^{\infty} f(x)\,\varphi'(x)\,dx. \tag{2.27}$$

But

$$-\int_0^\infty f(x)\,\varphi'(x)\,dx = -\Big[f(x)\,\varphi(x)\Big]_{x=0}^\infty + \int_0^\infty f'(x)\,\varphi(x)\,dx =$$

$$= f(+0)\,\varphi(0) + \int_0^\infty f'(x)\,\varphi(x)\,dx \tag{2.28}$$

and

$$-\int_{-\infty}^0 f(x)\,\varphi'(x)\,dx = -f(-0)\,\varphi(0) + \int_{-\infty}^0 f'(x)\,\varphi(x)\,dx. \tag{2.29}$$

Taking the sum of formulae (2.28) and (2.29) and taking formula (2.27) into account we obtain

$$-\langle f,\ \varphi'\rangle = \sigma_0\,\varphi(0) + \int_{-\infty}^\infty f'(x)\,\varphi(x)\,dx \tag{2.30}$$

or

$$f' = \{f'\} + \sigma_0\ \delta_0. \tag{2.31}$$

From formula (2.31) we see that the discontinuity of f appears in its derivative in the form of a punctual mass. The other formulae (2.26) are obtained by differentiating equation (2.30) a.s.o.

Application 1. We have

$$[u(x)\ \cos\ x]' = -\,u(x)\ \sin x + \delta_0,$$

$$[u(x)\ \sin\ x]' = u(x)\ \cos\ x.$$

We have to recall some definitions and results from analysis in order to understand application 2.

Let f be a function defined in $[\,a,\ b\,]$, summable in $(a,\ c-\varepsilon)$ and $(c+\varepsilon,\ b)$ for every $\varepsilon > 0$, not necessarily summable in $(a,\ b)$.

Definition. *We say that the integral of f is convergent in the main value of Cauchy if* $\lim_{\varepsilon\to 0}\int_a^{c-\varepsilon} f(x)\,dx$ and $\lim_{\varepsilon\to 0}\int_{c+\varepsilon}^{b} f(x)\,dx$ exist and is denoted by

$$vp\int_a^b f(x)\,dx = \lim_{\varepsilon\to 0}\left(\int_a^{c-\varepsilon} f(x)\,dx + \int_{c+\varepsilon}^{b} f(x)\,dx\right).$$

We state, without proof, the following proposition.

Proposition. *For* $vp\int_a^b f(x)\,dx$ *to exist it is necessary and sufficient that f should be in the neighbourhood of* $x = c$ *the sum of an antisymmetric function* $(f_1(c+u) = -f_1(c-u))$ *and of a symmetric function* $(f_2(c+u) = f_2(c-u))$ *so that* $\lim_{\varepsilon\to 0}\int_{c+\varepsilon}^{b} f(x)\,dx$ *should exist.*

Remark. The antisymmetric and symmetric functions become odd and even respectively for $c = 0$ and by the change of variable $x = c + u$ we can always transform these functions into odd and even functions respectively.

Remark. Every function f can be expressed uniquely in the neighbourhood of $x = 0$, as the sum of an even function and of an odd function, namely

$$f(x) = \frac{f(x) + f(-x)}{2} + \frac{f(x) - f(-x)}{2} = f_1(x) + f_2(x).$$

But f_1 is odd, hence its integral in a symmetric interval is zero, that is

$$a \leqslant -\alpha < 0 < \alpha \leqslant b \text{ implies } \int_{-\alpha}^{\varepsilon} f_1(x)\,dx + \int_{\varepsilon}^{\alpha} f_1(x)\,dx = 0.$$

As the function f_2 is even we have

$$\int_{-\alpha}^{\varepsilon} f_2(x)\,dx + \int_{\varepsilon}^{\alpha} f_2(x)\,dx = 2\int_{\varepsilon}^{\alpha} f_2(x)\,dx.$$

We obtain thus the following proposition.

Proposition. *The integral of f exists in main value if and only if* $\mathrm{i\,m}_{\to 0} \int_{\varepsilon}^{\alpha} f_2(x)\, \mathrm{d}x$ *exists.*

Example. If $f(x) = \dfrac{\varphi(x)}{x - c}$ where $\varphi(x)$ is continuous in the neighbourhood of $x = c$, and differentiable at the point $x = a$, then $vp \int_a^b f(x)\, \mathrm{d}x$ exists since $f(x)$ can be written

$$f(x) = \frac{\varphi(c)}{x - c} + \frac{\varphi(x) - \varphi(c)}{x - c},$$

where the first function in the right-hand side is antisymmetric and the second function is bounded in the neighbourhood of $x = c$.

Definition. *We say that* $vp \int_{-\infty}^{\infty} f(x)\, \mathrm{d}x$ *exists if f is summable in every finite interval and we write*

$$vp \int_{-\infty}^{\infty} f(x)\, \mathrm{d}x = \lim_{B \to \infty} \int_{-B}^{B} f(x)\, \mathrm{d}x.$$

Proposition. *The necessary and sufficient condition that* $vp \int_{-\infty}^{\infty} f(x)\, \mathrm{d}x$ *should exist is that* $f = f_1 + f_2$ *with* f_1 *even and* f_2 *odd and that* $\int_0^{\infty} f_2(x)\, \mathrm{d}x$ *should exist.*

We can now give application 2.

Application 2. Let us consider the distribution $vp \dfrac{1}{x}\cdot$ The function $\dfrac{1}{x}$ does not define a distribution as it is not summable in the neighbourhood of $x = 0$.

But from the above considerations it follows that the integral

$$vp \int_{-\infty}^{\infty} \frac{\varphi(x)}{x} dx = \lim_{\varepsilon \to 0} \int_{|x|>\varepsilon} \frac{\varphi(x)}{x} dx \tag{2.32}$$

makes sense for $\varphi \in \mathfrak{D}$. In fact the above integral makes sense even if φ has a bounded support and is once differentiable in the origin. By formula (2.32) we defined a linear functional of φ. Let us show that it is also continuous. Therefore we consider a sequence $\{\varphi_j\}$ convergent to φ in $\mathfrak{D}$. We can always suppose $\varphi = 0$, since otherwise we consider the sequence $\{\varphi_j - \varphi\}$ which converges to 0 in $\mathfrak{D}$. Let $(-A, A)$ be an interval chosen so as to contain all the supports of the functions φ_j. We have

$$vp \int_{-\infty}^{\infty} \frac{\varphi_j(x)}{x} dx = \varphi_j(0)\, vp \int_{-A}^{A} \frac{dx}{x} + vp \int_{-A}^{A} \frac{\varphi_j(x) - \varphi_j(0)}{x} dx.$$

Since $\frac{1}{x}$ is an odd function, the first term of the right-hand side is zero and if we apply the mean value formula to the second term we obtain

$$\left| vp \int_{-A}^{A} \frac{\varphi_j(x) - \varphi_j(0)}{x} dx \right| \leqslant 2\, A \max |\varphi_j'|.$$

Thus the second term tends to 0 as $j \to \infty$ because by hypothesis φ_j tends to 0 in $\mathfrak{D}$. Consequently the functional defined by formula (2.32) is linear and continuous, in other words we defined above a distribution which shall be denoted by $vp \frac{1}{x}$. We have

$$\left\langle vp \frac{1}{x}, \varphi \right\rangle = vp \int_{-\infty}^{\infty} \frac{\varphi(x)}{x} dx.$$

Example. In quantum mechanics the following distributions are currently used:

$$\delta_0^+ = \frac{\delta_0}{2} + \frac{1}{2\pi i}\, vp \frac{1}{x},$$

$$\delta_0^- = \frac{\delta_0}{2} - \frac{1}{2\pi i}\, vp \frac{1}{x}.$$

We remark that $\delta_0 = \delta_0^+ + \delta_0^-$.

Application 3. The function $\ln|x|$ is locally summable. Consequently it defines a distribution and we have

$$\langle(\ln|x|)', \varphi\rangle = -\langle \ln|x|, \varphi'\rangle = -\int_{-\infty}^{\infty} \ln|x| \, \varphi'(x)\, dx =$$

$$= -\int_{-\infty}^{0} \ln|x| \, \varphi'(x)\, dx - \int_{0}^{\infty} \ln|x| \, \varphi'(x)\, dx.$$

We have

$$-\int_{0}^{\infty} \ln x \, \varphi'(x)\, dx = -\lim_{\varepsilon\to 0}\int_{\varepsilon}^{\infty} \ln x \, \varphi'(x)\, dx =$$

$$= \lim_{\varepsilon\to 0}\Big[-\ln x \, \varphi(x)\Big]_{\varepsilon}^{\infty} + \int_{\varepsilon}^{\infty} \frac{\varphi(x)}{x}\, dx = \lim_{\varepsilon\to 0}\left(\ln \varepsilon\varphi(\varepsilon) + \int_{0}^{\infty} \frac{\varphi(x)}{x}\, dx\right). \quad (2.33)$$

But as

$$|\varphi(\varepsilon) - \varphi(0)| \leqslant \varepsilon \max|\varphi'|$$

and since $\varepsilon \ln \varepsilon \to 0$ as $\varepsilon \to 0$ it follows that $\ln \varepsilon[\varphi(\varepsilon) - \varphi(0)] \to 0$ as $\varepsilon \to 0$. Hence formula (2.33) becomes

$$-\int_{0}^{\infty} \ln x\varphi'(x)\, dx = \lim_{\varepsilon\to 0}\Bigg[\ln \varepsilon\varphi(0) + \ln \varepsilon(\varphi(\varepsilon) - \varphi(0)) +$$

$$+ \int_{\varepsilon}^{\infty} \frac{\varphi(x)}{x}\, dx\Bigg] = \lim_{\varepsilon\to 0}\left[\ln \varepsilon\varphi(0) + \int_{\varepsilon}^{\infty} \frac{\varphi(x)}{x}\, dx\right]. \quad (2.34)$$

In the same way we obtain

$$-\int_{-\infty}^{\infty} \ln|x| \, \varphi'(x)\, dx = \lim_{\varepsilon\to 0}\left[-\ln \varepsilon\varphi(0) + \int_{-\infty}^{-\varepsilon} \frac{\varphi(x)}{x}\, dx\right]. \quad (2.35)$$

Taking the sum of formulae (2.34) and (2.35) we obtain

$$-\int_{-\infty}^{\infty} \ln |x| \varphi'(x) \, dx = \lim_{\varepsilon \to 0} \int_{|x| \geq \varepsilon} \frac{\varphi(x)}{x} \, dx = vp \int_{-\infty}^{\infty} \frac{\varphi(x)}{x} \, dx$$

hence

$$(\ln |x|)' = vp \frac{1}{x} \cdot \tag{2.36}$$

Application 4. Let S be a regular surface in R^n on which a crossing sense is established. Let f be an indefinitely differentiable function in the complement of S. If we suppose that each partial derivative of f has a limit at each point $x_0 \in S$ as $x \to x_0$ on one or the other side of the surface S then the difference between these limits represents the jump of the corresponding partial derivative. If the crossing sense of the surface changes then this difference changes its sign.

Let us denote by $D^p f$ (application 2) a derivative of f in the sense of the distributions and by $\{D^p f\}$ the distribution represented by the usual function, defined for $x \notin S$.

Let us consider the point $\bar{x}$ of intersection of the surface S with the parallel to the axis x_1, with the coordinates $x_2, \ldots, x_n$ and denote by σ_0 the discontinuity of f at the crossing of the surface S at this point, in the sense of the axis x_1.

Let us calculate

$$\int_{-\infty}^{\infty} f(x) \frac{\partial \varphi}{\partial x_1} \, dx.$$

We have

$$\left\langle \frac{\partial f}{\partial x_1}, \varphi \right\rangle = -\left\langle f, \frac{\partial \varphi}{\partial x_1} \right\rangle = -\int \ldots \int_{R^n} f(x) \frac{\partial \varphi}{\partial x_1} \, dx =$$

$$= -\int_{x_2, \ldots, x_n} \int dx_2 \ldots dx_n \int_{-\infty}^{\infty} f(x) \frac{\partial \varphi}{\partial x_1} \, dx_1 =$$

$$= \int_{x_2, \ldots, x_n} \ldots \int dx_2 \ldots dx_n \left[\sigma_0 \varphi + \int_{-\infty}^{\infty} \frac{\partial f}{\partial x_1} \varphi \, dx_1 \right] \tag{2.37}$$

where the value of φ in the product $\sigma_0\varphi$ is calculated at $\bar{x}$. Formula (2.37) becomes

$$\left\langle \frac{\partial f}{\partial x_1}, \varphi \right\rangle = \int_S \dots \int \sigma_0 \varphi \, dx_2 \dots dx_n + \int_{R^n} \dots \int \frac{\partial f}{\partial x_1} \varphi \, dx. \qquad (2.38)$$

If we denote by θ_1 the angle between the axis x_1 and the normal to the surface S directed in the increasing sense of x_1 then

$$\int_S \dots \int \sigma_0 \, \varphi \, dx_2 \dots dx_n = \int_S \dots \int \sigma_0 \varphi \cos \theta_1 \, dS.$$

The following distribution is underlined

$$\langle T, \varphi \rangle = \int_S \dots \int \sigma_0 \varphi \cos \theta_1 \, dS = (\sigma_0 \cos \theta_1) \, \delta_{(S)} \qquad (2.39)$$

corresponding to masses located on the surface S with the surface density $\sigma_0 \cos \theta_1$. Taking formula (2.39) into account and replacing x_1 by x_i in formula (2.38) we obtain

$$\frac{\partial f}{\partial x_i} = \left\{ \frac{\partial f}{\partial x_i} \right\} + (\sigma_0 \cos \theta_i) \, \delta_{(S)} \, . \qquad (2.40)$$

If in formula (2.40) we differentiate with respect to x_i we obtain

$$\frac{\partial^2 f}{\partial x_i^2} = \left\{ \frac{\partial^2 f}{\partial x_i^2} \right\} + \frac{\partial}{\partial x_i} [(\sigma_0 \cos \theta_i) \, \delta_{(S)}] + \sigma_i \cos \theta_i \, \delta_{(S)} \, , \qquad (2.41)$$

where σ_i is the jump of the derivative $\dfrac{\partial f}{\partial x_i}$ at the crossing of the surface S. From formula (2.41) we can deduce the laplacian $\Delta f = \sum_{i=1}^{n} \dfrac{\partial^2 f}{\partial x_i^2}$.

Remark. We notice that $\sum_{i=1}^{n} \sigma_i \cos \theta_i$ is the jump σ_i of the function $\sum_i \cos \theta_i \dfrac{\partial f}{\partial x_i}$. But $\sum_{i=1}^{n} \cos \theta_i \dfrac{\partial f}{\partial x_i} = \dfrac{df}{d\nu}$, i.e. the normal derivative of the function f. Consequently the jump σ_ν is independent of the sense of the normal. Indeed if we change the sense of the normal, the sign of the jump changes too, but the sign of the normal derivative changes as well and thus the jump of the normal derivative remains the same.

2.2.4 Green's formula

Let us consider the expression

$$\left\langle \sum_i \frac{\partial}{\partial x_i}(\sigma_0 \cos \theta_i \, \delta_{(S)}), \varphi \right\rangle = -\int \dots_S \int \sum_i \cos \theta_i \frac{\partial \varphi}{\partial x_i} \sigma_0 \, dS =$$

$$= -\int \dots_S \int \frac{d\varphi}{d\nu} \sigma_0 \, dS. \tag{2.42}$$

We notice that it does not depend on the sense of the normal. If the latter changes then the sign of σ_0 and the sign of $\frac{d\varphi}{d\nu}$ change as well. Considering the previous remark and formula (2.42) and using the notations from formula (2.39) we obtain

$$\Delta f = \{\Delta f\} + \sigma_\nu \, \delta_{(S)} + \frac{d}{d\nu}(\sigma_0 \, \delta_{(S)}). \tag{2.43}$$

Proposition. *(Green's formula.) Let S be a closed surface and V the volume bounded by this surface, then if f is zero outside V we have*

$$\int \dots_V \int (f\Delta\varphi - \varphi\Delta f) \, dx + \int \dots_S \int \left(f \frac{d\varphi}{d\nu_i} - \varphi \frac{df}{d\nu_i} \right) dS = 0 \tag{2.44}$$

where ν_i is the normal directed towards the interior of the surface S.

Indeed, taking into account formula (2.43) and the above hypothesis we have

$$\langle \Delta f, \varphi \rangle = \langle f, \Delta\varphi \rangle = \int \dots_V \int f \Delta\varphi \, dx = \langle \{\Delta f\}, \varphi \rangle +$$

$$+ \left\langle \frac{df}{d\nu_i} \delta_{(S)}, \varphi \right\rangle + \left\langle \frac{d(f\delta_{(S)})}{d\nu_i}, \varphi \right\rangle = \int \dots_V \int \Delta f \, \varphi \, dx +$$

$$+ \int \dots_S \int \frac{df}{d\nu_i} \varphi \, dS - \int \dots_S \int f \frac{d\varphi}{d\nu_i} \, dS, \tag{2.45}$$

whence Green's formula (2.44) is immediately deduced.

Remark. Formula (2.44) is nothing else than the generalisation of the second formula of Green [Vol. 1, Section 1.3.4].

Example 1. Calculate the laplacian $\Delta \frac{1}{r^{n-2}}$, where $r = \sqrt{\sum_i x_i^2}$, in R^n.

Let us first show that $\frac{1}{r^{n-2}}$ is a harmonic function in the complement of the origin. Indeed, if f depends only on r then

$$\frac{\partial f}{\partial x_i} = \frac{df}{dr} \frac{\partial r}{\partial x_i} = \frac{df}{dr} \frac{x_i}{r},$$

$$\frac{\partial^2 f}{\partial x_i^2} = \frac{d^2 f}{dr^2} \left(\frac{x_i}{r}\right)^2 + \frac{df}{dr}\left(\frac{1}{r} - \frac{x_i^2}{r^3}\right),$$

hence

$$\Delta f = \frac{d^2 f}{dr^2} + \frac{n-1}{r} \frac{df}{dr}.$$

Consequently if $f(r)$ is harmonic then it verifies the Euler type equation

$$\frac{d^2 f}{dr^2} + \frac{n-1}{r} \frac{df}{dr} = 0$$

which has the solutions

$$f = \frac{A}{r^{n-2}} + B, \text{ if } n \neq 2,$$

$$f = A \ln \frac{1}{r} + B, \text{ if } n = 2,$$

where A and B are constants.

We notice that the functions $\frac{1}{r^{n-2}}$ and $\ln \frac{1}{r}$ which have a singularity in the origin, are summable in the neighbourhood of the origin because $n - 2 < n$, consequently they define distributions.

Let us calculate their laplacian. We have

$$\left\langle \Delta \frac{1}{r^{n-2}}, \varphi \right\rangle = \left\langle \frac{1}{r^{n-2}}, \Delta\varphi \right\rangle = \int \underset{R^n}{\cdots} \int \frac{1}{r^{n-2}} \Delta\varphi \, dx =$$

$$= \lim_{\varepsilon \to 0} \int \underset{r \geqslant \varepsilon}{\cdots} \int \frac{1}{r^{n-2}} \Delta\varphi \, dx.$$

In order to calculate the integral $\int_{r\geqslant\varepsilon}\dots\int \frac{1}{r^{n-2}}\,\Delta\varphi\,\mathrm{d}x$ we can apply Green's formula with respect to the volume V bounded by the sphere S_ε with the radius $\varepsilon = r$. But the same integral can be calculated as follows

$$\int_{r\geqslant\varepsilon}\dots\int \frac{1}{r^{n-2}}\,\Delta\varphi\,\mathrm{d}x = \langle \rho_\varepsilon, \Delta\varphi\rangle = \langle \Delta\rho_\varepsilon, \varphi\rangle$$

where

$$\rho_\varepsilon = \begin{cases} 0, & \text{if } r < \varepsilon, \\ \dfrac{1}{r^{n-2}}, & \text{if } r > \varepsilon. \end{cases}$$

If we consider that the sense of the crossing of the surface S_ε is the same as the sense of the normal which we supposed directed towards the interior of the domain then $\dfrac{\mathrm{d}}{\mathrm{d}\nu_i} = \dfrac{\mathrm{d}}{\mathrm{d}r}$.

Let us apply formula (2.45) to $\langle \rho_\varepsilon, \Delta\varphi\rangle$ and take into account that $\left\{\Delta \dfrac{1}{r^{n-2}}\right\} = 0.$ We have

$$\langle \rho_\varepsilon, \Delta\varphi\rangle = \langle \Delta\rho_\varepsilon, \varphi\rangle = \int_{r\geqslant\varepsilon}\dots\int \frac{1}{r^{n-2}}\,\Delta\varphi\,\mathrm{d}x = -\int_{r=\varepsilon}\dots\int \frac{1}{\varepsilon^{n-2}}\frac{\mathrm{d}\varphi}{\mathrm{d}r}\,\mathrm{d}S + $$
$$+\int_{r=\varepsilon}\dots\int \frac{-(n-2)}{\varepsilon^{n-1}}\,\varphi\,\mathrm{d}S. \tag{2.46}$$

The function $\dfrac{\mathrm{d}\varphi}{\mathrm{d}r} = \sum_i \dfrac{x_i}{r}\dfrac{\partial\varphi}{\partial x_i}$ is bounded, and the area of the integration surface is $S_n\,\varepsilon^{n-1}$, where S_n is the area of the sphere of radius 1 in R^n. Consequently the first integral of the right-hand side of formula (2.46) tends to 0 as $\varepsilon \to 0$. Let us consider now the second integral. We have

$$\int_{r=\varepsilon}\dots\int \frac{-(n-2)}{\varepsilon^{n-1}}\,\varphi\,\mathrm{d}S = \int_{r=\varepsilon}\dots\int \frac{-(n-2)}{\varepsilon^{n-1}}\,\varphi(0)\,\mathrm{d}S +$$
$$+\int_{r=\varepsilon}\dots\int \frac{-(n-2)}{\varepsilon^{n-1}}\,(\varphi(x)-\varphi(0))\,\mathrm{d}S. \tag{2.47}$$

The first integral of the right-hand side of the above formula is equal to $-(n-2)\,\varphi(0)\,S_n$.

We make the following majoration in the second integral

$$|\,\varphi(x)-\varphi(0)\,| \leqslant |\,x\,|\sqrt{n}\max_{\substack{|x|\leqslant r\\ i=1,\dots,n}}\left|\frac{\partial\varphi}{\partial x_i}\right| = \varepsilon\sqrt{n}\max\left|\frac{\partial\varphi}{\partial x_i}\right|$$

hence

$$\left| \int_{r=\varepsilon} \cdots \int \frac{-(n-2)}{\varepsilon^{n-1}} (\varphi(x) - \varphi(0))\, dS \right| \leqslant$$

$$\leqslant \varepsilon \sqrt{n} \max \left| \frac{\partial \varphi}{\partial x_i} \right| \frac{n-2}{\varepsilon^{n-1}} \int_{r=\varepsilon} \cdots \int dS = \varepsilon \sqrt{n} \max \left| \frac{\partial \varphi}{\partial x_i} \right| \frac{n-2}{\varepsilon^{n-1}} \varepsilon^{n-1} S_n \to 0$$

for $\varepsilon \to 0$. Consequently formula (2.47) becomes $\left\langle \Delta \frac{1}{r^{n-2}}, \varphi \right\rangle = -(n-2) S_n \varphi(0)$

that is [Section 1.1.3, Consequence 9]

$$\Delta \frac{1}{r^{n-2}} = -(n-2) S_n \delta_0 = -(n-2) \frac{2\pi^{\frac{n}{2}}}{\Gamma\left(\frac{n}{2}\right)} \delta_0. \tag{2.48}$$

Particular cases. (1) For $n = 1$ we have, taking into account that $\Gamma\left(\frac{1}{2}\right) = \sqrt{\pi}$ (See Chapter 1)

$$\Delta |x| = 2\, \delta_0.$$

(2) For $n = 3$,

$$\Delta \frac{1}{r} = -4\pi\delta_0.$$

2.3 THE MULTIPLICATION OF DISTRIBUTIONS

2.3.1 The product of an indefinitely differentiable function and a distribution

Remark. If f is a locally summable function then f defines a distribution, but f^2 may no longer be locally summable and hence f^2 no longer defines a distribution.

Consequently the multiplication ST of two arbitrary distributions S and T is not always possible.

Example. The function $f(x) = \frac{1}{\sqrt{|x|}}$ is locally summable but $f^2 = \frac{1}{|x|}$ is not locally summable.

In the following we shall define the multiplication of a distribution by an indefinitely differentiable function in the usual sense.

Let us first suppose that T is a locally summable function and α an indefinitely differentiable function, then

$$\langle \alpha f, \varphi \rangle = \int_{R^n} \dots \int (\alpha(x) f(x)) \varphi(x) \, dx = \int_{R^n} \dots \int f(x) (\alpha(x) \varphi(x)) \, dx = \langle f, \alpha\varphi \rangle.$$

This yields the following proposition.

Proposition. *If α is an indefinitely differentiable function and T is an arbitrary distribution, then there is a distribution αT defined by*

$$\langle \alpha T, \varphi \rangle = \langle T, \alpha\varphi \rangle. \tag{2.49}$$

Indeed let us notice that from Leibniz's formula with respect to the successive derivatives of a product it follows that $\alpha\varphi$ is an indefinitely differentiable function. The support of the function $\alpha\varphi$ is included in the support of the function φ and consequently is bounded. The right-hand side of formula (2.49) depends linearly on φ. Let us now consider a sequence $\{\varphi_j\}$ convergent to 0 in $\mathfrak{D}$. Then all the functions of the sequence will have the support in a fixed bounded set. Also from Leibniz's formula it follows that $\alpha\varphi_j \to 0$ in $\mathfrak{D}$ and consequently the supports of these functions remain included in a bounded set and the proposition is proved.

Example 1. Let us calculate the product $\alpha\delta_0$. We have

$$\langle \alpha\delta_0, \varphi \rangle = \langle \delta_0, \alpha\varphi \rangle = \alpha(0) \varphi(0) = \langle \alpha(0)\delta_0, \varphi \rangle$$

that is

$$\alpha\delta_0 = \alpha(0) \delta_0.$$

In particular

$$x\delta_0 = 0.$$

Example 2. The product $\alpha\delta_0'$. We have

$$\langle \alpha\delta_0', \varphi \rangle = \langle \delta_0', \alpha \varphi \rangle = -(\alpha\varphi)'_{x=0} = -\alpha(0) \varphi'(0) - \alpha'(0) \varphi(0) =$$
$$= \langle \alpha(0) \delta_0' - \alpha'(0) \delta_0, \varphi \rangle$$

hence we have

$$\alpha\delta_0' = \alpha(0) \delta_0' - \alpha'(0) \delta_0.$$

In particular

$$x\delta_0' = \delta_0.$$

Considering the particular case in Example 1 we have

$$x^2\delta_0' = 0.$$

In general

$$x\delta_0^{(m)} = -m\delta_0^{(m-1)}.$$

Let us now consider a distribution T on R, and prove the following important result.

Proposition 1. *The necessary and sufficient condition that*

$$xT = 0 \tag{2.50}$$

is that

$$T = C\delta_0$$

that is that the distribution T be proportional to δ_0.

We saw that $x\delta_0 = 0$. Let us now suppose that T is a distribution which verifies relationship (2.50). Then

$$\langle xT, \varphi\rangle = \langle T, x\varphi\rangle = 0. \tag{2.51}$$

From relationship (2.51) it follows that T is zero on every function $\psi \in \mathfrak{D}$ of the form $\psi = x\varphi$ with $\varphi \in \mathfrak{D}$. But obviously a function $\psi \in \mathfrak{D}$ has the form $x\varphi$ if and only if it is zero at the origin. Indeed, if it has the form $x\varphi$ it is obviously zero at the origin.

Conversely if $\psi(0) = 0$, the function $\varphi = \dfrac{\psi}{x}$ is indefinitely differentiable everywhere.

Let $\tau \in \mathfrak{D}$ be a function which takes the value 1 at the origin. Then every function $\Phi \in \mathfrak{D}$ can be written

$$\Phi = \lambda\tau + \psi$$

where

$$\lambda = \Phi(0) \text{ and } \psi(0) = 0.$$

But $T(\psi) = 0$ from formula (2.51), hence

$$T(\Phi) = \lambda T(\tau) = C\Phi(0),$$

where $C = T(\tau)$. Hence

$$T = C\,\delta_0$$

and thus the proposition is proved.

Corollary 1. From the above proposition it follows that δ_0 is, with the exception of the constant vector, the only eigenvector of the operator of multiplication by x, corresponding to the eigenvalue $\mu = 0$.

Corollary 2. Proposition 1 remains true if x is replaced by an arbitrary function α indefinitely differentiable which has the origin as unique root of order 1.

Indeed $\frac{x}{\alpha}$ is indefinitely differentiable and from $\alpha T = 0$ we have $xT = \frac{x}{\alpha}\alpha T = 0$ and conversely.

Proposition 2. *The product αT of an indefinitely differentiable function and a distribution T is differentiable according to the formula*

$$\frac{\partial}{\partial x_i}(\alpha T) = \frac{\partial \alpha}{\partial x_i}\,T + \alpha\,\frac{\partial T}{\partial x_i}\,\cdot$$

Indeed

$$\left\langle \frac{\partial}{\partial x_i}(\alpha T),\ \varphi \right\rangle = -\left\langle \alpha T,\ \frac{\partial \varphi}{\partial x_i} \right\rangle = -\left\langle T,\ \alpha\,\frac{\partial \varphi}{\partial x_i} \right\rangle \cdot \tag{2.52}$$

But

$$\left\langle \frac{\partial \alpha}{\partial x_i}\,T,\ \varphi \right\rangle = \left\langle T,\ \frac{\partial \alpha}{\partial x_i}\,\varphi \right\rangle \tag{2.53}$$

and

$$\left\langle \alpha\,\frac{\partial T}{\partial x_i},\ \varphi \right\rangle = \left\langle \frac{\partial T}{\partial x_i},\ \alpha\varphi \right\rangle = -\left\langle T,\ \frac{\partial}{\partial x_i}(\alpha\varphi) \right\rangle \cdot \tag{2.54}$$

Since

$$\frac{\partial}{\partial x_i}(\alpha\varphi) = \alpha\,\frac{\partial \varphi}{\partial x_i} + \varphi\,\frac{\partial \alpha}{\partial x_i}$$

from the linearity of T we obtain

$$\left\langle T, \frac{\partial}{\partial x_i}(\alpha\varphi)\right\rangle = \left\langle T, \alpha\frac{\partial\varphi}{\partial x_i}\right\rangle + \left\langle T, \varphi\frac{\partial\alpha}{\partial x_i}\right\rangle. \tag{2.55}$$

From formulae (2.52), (2.53), (2.54) and (2.55) we have the differentiation formula of the product αT.

2.4 THE CONVERGENCE OF DISTRIBUTIONS

2.4.1 Sequences and series of distributions

Definition 1. *We say that the distributions T_j converge to the distribution T for $j \to \infty$, if, for every $\varphi \in \mathfrak{D}$, the complex numbers $\langle T_j, \varphi\rangle$ converge towards the complex number $\langle T, \varphi\rangle$ for $j \to \infty$.*

Definition 2. *We say that a series $\sum_{i \in I} T_i$ is summable and its sum is T, if for every $\varphi \in \mathfrak{D}$ the numeric series $\sum_{i \in I} \langle T_i, \varphi\rangle$ is summable and its sum is $\langle T, \varphi\rangle$.*

In particular if $I = N$ is the set of the natural numbers, then the series $\sum_{i \in N} T_i$ is convergent and its sum is T, if for every $\varphi \in \mathfrak{D}$ the series $\sum_{i \in N} \langle T_i, \varphi\rangle$ is convergent and its sum is $\langle T, \varphi\rangle$.

Proposition 1. *If $\{T_j\}$ is a sequence of distributions so that for $j \to \infty$ the sequence $\{\langle T_j, \varphi\rangle\}$ has a limit for every $\varphi \in \mathfrak{D}$ then the sequence $\{T_j\}$ has a limit in $\mathfrak{D}'$.*

Indeed let us suppose that $\langle T, \varphi\rangle$ is the limit of the sequence $\langle T_j, \varphi\rangle$. This limit obviously defines T as a linear functional in the space $\mathfrak{D}$. If T is also continuous then T is a distribution and the statement of the proposition follows from definition 1. We shall not give here the proof of the continuity of T. It can be found in Ref. 49.

Proposition 2. *If the functions f_j converge almost everywhere to a function f for $j \to \infty$ and if there is a function $g \geqslant 0$ locally summable so that for every i, $|f_i| \leqslant g$, then the distributions f_j converge to the distribution f.*

The proof of this proposition follows immediately from Lebesgue's theorem [Vol. 1, Section 5.2.2].

Indeed according to this theorem it follows that the sequence of integrals $\int \underset{R^n}{\cdots} \int f_j \varphi \, dx$ converges for $j \to \infty$ to $\int \underset{R^n}{\cdots} \int f \varphi \, dx$.

Corollary. From the above proposition it immediately follows that given a sequence $\{f_j\}$ of locally summable functions converging to f for $j \to \infty$, uniformly on each bounded set, then the distributions $\{f_j\}$ converge to the distribution f.

Proposition. *If the distributions T_j converge for $j \to \infty$ to a distribution T, then the derivatives T_j' converge to T'.*

Indeed let us suppose that the sequence $\{T_j\}$ converges to T and show that the sequence $\{T_j'\}$ converges to T'. We have by definition

$$\langle T_j', \quad \varphi \rangle = -\langle T_j, \quad \varphi' \rangle.$$

But from definition 1 it follows that $\langle T_j, \quad \varphi' \rangle \to \langle T, \quad \varphi' \rangle$. As $\langle T, \varphi' \rangle = -\langle T', \varphi \rangle$ it follows that $\lim_{j \to \infty} T_j' = T'$.

Corollary. From the above proposition it follows that every series of distributions, summable or convergent, is differentiable term by term under the sign Σ.

Remark. Given a sequence $\{f_j\}$ of continuous and differentiable functions, uniformly convergent to a function f, it does not follow that f' exists and if f' exists it does not follow that $\{f_j'\}$ converges to f'.

But from the above proposition it follows that f' always exists in the sense of the distributions and that the sequence $\{f_j'\}$ converges to f' in $\mathfrak{D}'$.

Proposition. *Let $\{f_j\}$ be a sequence of functions with the following properties:*

(i) *For every j, $f_j \geqslant 0$ for $|x| \leqslant k$, $k > 0$ fixed.*

(ii) *The sequence $\{f_j\}$ converges to 0, uniformly in every set*

$$0 < a \leqslant |x| < \frac{1}{a} < \infty$$

(iii) *The sequence $\left\{ \int \underset{|x| \leqslant a}{\cdots} \int f_j(x) \, dx \right\}$ converges to 1 as $j \to \infty$ for every a.*

In this case $\{f_j\}$ *converges to* δ_0 *as* $j \to \infty$.

Indeed, let $\varphi \in \mathfrak{D}$. Then

$$\langle f_j, \varphi \rangle = \int \dots \int_{R^n} f_j(x)\ \varphi(x)\ dx = \int_{|x| \leqslant a} \dots \int \varphi(0)\ f_j(x)\ dx +$$

$$+ \int_{|x| \leqslant a} \dots \int (\varphi(x) - \varphi(0))\ f_j(x)\ dx + \int_{|x| > a} \dots \int \varphi(x)\ f_j(x)\ dx.$$

But

$$|\varphi(x) - \varphi(0)| < |x| \sqrt{n} \max_{i=1,\dots,n} \left| \frac{\partial \varphi}{\partial x_i} \right|$$

and if we denote by M the maximum of the modulus of the first-order derivatives of φ we have

$$\left| \int_{|x| \leqslant a} \dots \int (\varphi(x) - \varphi(0))\ f_j(x)\ dx \right| \leqslant a \sqrt{n} M \int_{|x| \leqslant a} \dots \int |f_j(x)|\ dx.$$

Given $a \leqslant k$, then

$$\int_{|x| \leqslant a} \dots \int f_j(x)\ dx \leqslant \int_{|x| \leqslant k} \dots \int f_j(x)\ dx$$

and since the last integral converges to 1 as $j \to \infty$, it follows that it is dominated by a constant K. Therefore

$$\left| \int_{|x| \leqslant a} \dots \int (\varphi(x) - \varphi(0))\ f_j(x)\ dx \right| \leqslant a \sqrt{n}\ KM < \frac{\varepsilon}{3}$$

f we choose $a \leqslant \dfrac{\varepsilon}{3\sqrt{n}\,KM}$. Let us choose a so that $a \leqslant k$, $a \leqslant \dfrac{\varepsilon}{3\sqrt{n}\,KM}$ and small enough for the support of φ to be contained in $|x| \leqslant \dfrac{1}{a}$. Then from (ii) it follows that $f_j \to 0$ uniformly for $a \leqslant |x| \leqslant \dfrac{1}{a}$ hence $\int_{|x|>a} \dots \int \varphi(x)\ f_j(x)\ dx \to 0$ as $j \to \infty$, that is a value j_1 of j, exists so that for $j \geqslant j_1$

$$\left| \int_{|x| > a} \dots \int \varphi(x)\ f_j(x)\ dx \right| < \frac{\varepsilon}{3}.$$

Finally

$$\int_{|x|\leqslant a}\cdots\int \varphi(0)\, f_j(x)\, dx = \varphi(0)\int_{|x|\leqslant a}\cdots\int f_j(x)\, dx \to \varphi(0)$$

for $j \to \infty$, consequently j_2 exists so that for $j \geqslant j_2$

$$\left|\int_{|x|\leqslant a}\cdots\int \varphi(0)\, f_j(x)\, dx - \varphi(0)\right| \leqslant \frac{\varepsilon}{3}.$$

Consequently if $j_0 = \sup(j_1, j_2)$ it follows that as $j \geqslant j_0$ we have

$$|\langle f_j, \varphi\rangle - \langle \delta_0, \varphi\rangle| \leqslant \varepsilon$$

and thus the proposition is proved.

Example 1. Let us show that

$$f_\varepsilon = \begin{cases} 0 & \text{for } |x| > \varepsilon, \\ \dfrac{n}{\varepsilon^n S_n} & \text{for } |x| < \varepsilon \end{cases}$$

converges to δ_0 as $\varepsilon \to 0$.

It will be sufficient to show that the conditions of the previous proposition are fulfilled. Indeed,

(i) $f_\varepsilon \geqslant 0$ everywhere,

(ii) for $|x| \geqslant a$, f_ε is zero as soon as $\varepsilon < a$,

(iii) $\displaystyle\int_{|x|\leqslant a}\cdots\int f_\varepsilon(x)\, dx = \int_{|x|\leqslant \varepsilon\leqslant a}\cdots\int f_\varepsilon(x)\, dx = \frac{n}{\varepsilon^n S_n} V_\varepsilon,$

where V_ε is the volume of the sphere of radius ε that is exactly $\dfrac{\varepsilon^n S_n}{n}$, hence

$$\int_{|x|\leqslant a}\cdots\int f_\varepsilon(x)\, dx = 1.$$

Example 2. In the same way as in the previous example let us show that $f_\varepsilon = \dfrac{1}{\varepsilon^n(\sqrt{2\pi})^n} e^{-\frac{r^2}{2\varepsilon^2}}$ converges to δ_0 as $\varepsilon \to 0$.

(i) $f_\varepsilon \geqslant 0$;

(ii) if $|x| \geqslant a$ then

$$f_\varepsilon \leqslant \frac{1}{\varepsilon^n(\sqrt{2\pi})^n} e^{-\frac{a^2}{2\varepsilon^2}} \to 0 \text{ as } \varepsilon \to 0;$$

(iii) $$\int_{|x|\leqslant a}\cdots\int \frac{1}{\varepsilon^n(\sqrt{2\pi})^n}\, e^{-\frac{r^2}{2\varepsilon^2}}\, dx = \int_{|x|\leqslant \frac{a}{\varepsilon}}\cdots\int \frac{1}{(\sqrt{2\pi})^n}\, e^{-\frac{r^2}{2}}\, dx \rightarrow$$

$$\rightarrow \int_{R^n}\cdots\int \frac{1}{\sqrt{2\pi}}\, e^{-\frac{r^2}{2}}\, dx = 1 \text{ as } \varepsilon \rightarrow 0.$$

Example 3. $f_\varepsilon = \frac{1}{n}\, e^{-\pi\frac{r^2}{\varepsilon^2}}$ tends to δ_0 as $\varepsilon \rightarrow 0$. Conditions (i) and (ii) are obviously fulfilled. Condition (iii) is fulfilled since

$$\int_{R^n}\cdots\int e^{-\pi r^2}\, dx = 1.$$

2.4.2 Applications to trigonometric series

Proposition. *For a trigonometric series* $\sum_{k=-\infty}^{\infty} a_k e^{2i\pi kx}$ *to be summable in* $\mathfrak{D}'$ *it is sufficient that its coefficients be dominated in the modulus for* $k \rightarrow \infty$ *by* $A\,|\,k\,|^\alpha$ *with* $\alpha \geqslant 0$ *conveniently chosen.*

Indeed, let us consider the series

$$\sum_{k=-\infty}^{\infty} \frac{a_k}{(2i\pi k)^{\beta+2}}\, e^{2i\pi kx} \tag{2.56}$$

with $\beta \geqslant \alpha$. Since the general term of this series is dominated by $\frac{A}{k^2}$ it results that the series is uniformly summable on the real line and hence its sum is a continuous function f. Let us differentiate expression (2.56), $\beta + 2$ times in the sense of the distributions. We have

$$\sum_{k=-\infty}^{\infty} a_k e^{2i\pi kx} = a_0 + f^{(\beta+2)}$$

and the proposition is proved.

Corollary. *From the above proposition it is seen that the sum of the trigonometric series is a distribution of period 1, namely it is the sum of a constant a_0 and of the derivative in the sense of the distributions of a continuous periodic function f.*

Example. It can be shown that the series $\sum_{k=-\infty}^{\infty} e^{2ik\pi x}$ converges to the periodic distribution $\sum_{l=-\infty}^{\infty} \delta_{(l)}$ which consists of masses equal to l at each point whose abscissa is an integer. By differentiation we obtain

$$\sum_{k=-\infty}^{\infty} (2i\pi k)^m e^{2i\pi kx} = \sum_{l=-\infty}^{\infty} \delta_{(l)}^{(m)}$$

We give below, without proof, the following proposition.

Proposition. *Every periodic distribution can be expanded in a trigonometric series in $\mathfrak{D}'$. The necessary and sufficient condition of convergence of the trigonometric series is that its coefficients a_k be dominated by $A \mid k \mid^{\alpha}$, with α conveniently chosen for $k \to \infty$.*

2.5 DISTRIBUTIONS WITH BOUNDED SUPPORT

2.5.1 The space $\mathfrak{E}$

Notation. Let $\mathfrak{E}$ be the space of the complex functions on R^n, indefinitely differentiable with arbitrary support.
Remark. Let T be a distribution with bounded support K and let $\varphi \in \mathfrak{E}$. If we consider a function $\alpha \in \mathfrak{D}$ equal to one in a neighbourhood of the support K of T, then $\alpha\varphi \in \mathfrak{D}$ and hence $\langle T, \alpha\varphi \rangle$ makes sense. We shall prove the following proposition.

Proposition. *The number $\langle T, \alpha\varphi \rangle$ is independent of the function α.*
Indeed, let $\beta \in \mathfrak{D}$ be another function equal to 1 in a neighbourhood of K then $(\alpha - \beta)\, \varphi \in \mathfrak{D}$ and the support of the function $(\alpha - \beta)\, \varphi$ is included in the complement of K. Consequently

$$\langle T, \alpha\varphi \rangle - \langle T, \beta\varphi \rangle = \langle T, (\alpha - \beta)\varphi \rangle = 0$$

and thus the proposition is proved completely.

Definition. *For every $\varphi \in \mathfrak{E}$ we write*

$$\langle T, \varphi \rangle = \langle T, \alpha\varphi \rangle. \tag{2.57}$$

2.5.2 The convergence in $\mathscr{E}$

Definition. We say that a sequence of functions $\{\varphi_j\}$ *from* $\mathscr{E}$ *converges in* $\mathscr{E}$ *to* 0 *if it converges uniformly to* 0 *on every compact, together with all its derivatives.*

We shall prove the following theorem.

Theorem. *The necessary and sufficient condition for a distribution to have a bounded support is that it should be a linear continuous functional on* $\mathscr{E}$.

The necessary condition is obvious. Let us show the sufficiency of the condition, namely prove that a linear continuous functional on $\mathscr{E}$, $L(\varphi)$ defines a distribution with bounded support. Indeed, every function $\varphi \in \mathscr{D}$ is in $\mathscr{E}$ and if a sequence $\{\varphi_j\}$ converges to 0 in $\mathscr{D}$ then it converges to 0 in $\mathscr{E}$. Consequently L is a linear continuous functional on $\mathscr{D}$. It follows that it defines a distribution $T \in \mathscr{D}'$, that is

$$L(\varphi) = \langle T, \varphi \rangle \text{ for every } \varphi \in \mathscr{D}. \tag{2.58}$$

It must still be shown that T has a bounded support. Let us suppose the contrary. Then there should be a sequence of functions $\varphi_n \in \mathscr{D}$ with the support in the complement of the set $|x| < n$, so that $\langle T, \varphi_n \rangle = 1$. But $\varphi_n \to 0$ in $\mathscr{E}$, consequently $L(\varphi_n) \to 0$ since L is continuous on $\mathscr{E}$ by hypothesis. The relationship $\langle T, \varphi_n \rangle = 1$ does not hold for any n and thus the theorem is proved completely.

Corollary 1. $L(\varphi) = \langle T, \varphi \rangle$ for every $\varphi \in \mathscr{E}$.

Indeed, let $\{\alpha_j\}$ be a sequence of functions from $\mathscr{D}$, defined as follows $\alpha_j(x) = 1$ for $|x| < j$ and $\alpha_j(x) = 0$ for $|x| \geqslant 2j$. Then obviously $\alpha_j\varphi \in \mathscr{D}$ and $\{\alpha_j \varphi\}$ converges in $\mathscr{E}$ to φ as $j \to \infty$.

It follows that $L(\alpha_j \varphi)$ tends to $L(\varphi)$. But the support K of the distribution T is included for j large enough in $|x| < j$. From relationship (2.57) it follows that we can write

$$\langle T, \varphi \rangle = \langle T, \alpha_j\varphi \rangle. \tag{2.59}$$

Since $\alpha_j\varphi \in \mathscr{D}$, it follows from formula (2.58) that

$$L(\alpha_j\varphi) = \langle T, \alpha_j\varphi \rangle \text{ for every } j. \tag{2.60}$$

From formulae (2.59) and (2.60) it follows that for j sufficiently large we have

$$\langle T, \varphi \rangle = L(\alpha_j\varphi)$$

and thus the corollary is proved completely.

Corollary 2. The distributions with bounded support define a vector space $\mathscr{E}'$, *the dual of the space* $\mathscr{E}$, *that is the space of the linear continuous functionals on* $\mathscr{E}$.

2.6 CONVOLUTION

2.6.1 Tensor product of distributions

Let X^m, Y^n be two euclidian spaces of dimension m and n respectively and let $x(x_1, \ldots, x_m) \in X^m$ and $y(y_1, \ldots, y_n) \in Y^n$. Let $Z^{m+n} = X^m \times Y^n$ that is the set $(x, y) = (x_1, \ldots, x_m, y_1, \ldots, y_n)$. Obviously Z^{m+n} is a euclidian space with $m + n$ dimensions.

Definition. Let $f(x)$ be a numerical function defined on X^m, and $g(y)$ a numerical function defined on Y^n. The function $h(x, y) = f(x)\, g(y)$ defined on Z^{m+n} is called the tensor product of the functions f and g and is denoted by $f(x) \otimes g(y)$.

Remark. If the functions f, g are locally summable on X^m and Y^n respectively, h is locally summable on Z^{m+n}.

From this remark we can see the possibility of the extension of the tensor product to distributions.

Notations. We denote by $(\mathfrak{D})_x$, $(\mathfrak{D})_y$, $(\mathfrak{D})_{x,y}$ the spaces $\mathfrak{D}$ of indefinitely differentiable functions with bounded support on X^m, Y^n, $X^m \times Y^n$ respectively, and let $(\mathfrak{D}')_x$, $(\mathfrak{D}')_y$, $(\mathfrak{D}')_{x,y}$ be the space $\mathfrak{D}'$ of the corresponding distributions.

Let us suppose that $\varphi(x, y) \in (\mathfrak{D})_{x,y}$ is a function of the form $u(x)\ v(y)$ with $u(x) \in (\mathfrak{D})_x$ and $v(y) \in (\mathfrak{D})_y$. Then

$$\langle f(x) \otimes g(y),\ u(x)\ v(y)\rangle = \int \cdots \int_{X^m \times Y^n} f(x)\ g(y)\ u(x)\ v(y)\ dx\ dy =$$

$$= \int_{X^m} F(x)\ u(x)\ dx \int_{Y^n} g(y)\ v(y)\ dy = \langle f, u\rangle\ \langle g, v\rangle. \tag{2.61}$$

If φ does not have this form then we apply Fubini's theorem and we obtain

$$\langle f(x) \otimes g(y),\ \varphi(x, y)\rangle = \int \cdots \int_{X^m \times Y^n} f(x)\ g(y)\ \varphi(x, y)\ dx\ dy =$$

$$= \int_{X^m} f(x)\ dx \int_{Y^n} g(y)\ \varphi(x, y)\ dy = \langle f(x), \langle g(y), \varphi(x, y)\rangle\rangle \tag{2.62}$$

and

$$\langle f(x) \otimes g(y),\ \varphi(x, y)\rangle = \langle g(y),\ \langle f(x),\ \varphi(x, y)\rangle\rangle. \tag{2.63}$$

The above formulae can be thus generalised.

Proposition. *If $S_x \in (\mathcal{D}')_x$ and $T_y \in (\mathcal{D}')_y$ then there is a uniquely determined distribution $W_{x,y} \in (\mathcal{D}')_{x,y}$, so that for every $\varphi \in (\mathcal{D})_{x,y}$ of the form $\varphi(x, y) = u(x)\, v(y)$ with $u \in (\mathcal{D})_x$ and $v \in (\mathcal{D})_y$ we have*

$$\langle W_{x,y},\ u(x)\, v(y)\rangle = \langle S_x,\ u\rangle\ \langle T_y,\ v\rangle. \tag{2.64}$$

Definition. *$W_{x,y}$ is called the tensor product of the distributions S_x and T_y and is denoted by $S_x \otimes T_y$.*

Remark. If $\varphi(x, y)$ does not have the form $u(x)\, v(y)$ then we proceed as for formulae (2.62) and (2.63), that is $\langle W_{x,y},\ \varphi\rangle$ is calculated with the aid of Fubini's theorem.

For x fixed, $\varphi(x, y) \in (\mathcal{D})_y$ and we write

$$\theta(x) = \langle T_y,\ \varphi(x, y)\rangle. \tag{2.65}$$

As $\theta(x) \in (\mathcal{D})_x$ we can calculate $\langle S_x,\ \theta\rangle$ and we get

$$\langle W_{x,y},\ \varphi\rangle = \langle S_x,\ \theta\rangle = \langle S_x,\ \langle T_y,\ \varphi(x, y)\rangle\rangle.$$

Similarly

$$\langle W_{x,y},\ \varphi\rangle = \langle T_y,\ \langle S_x,\ \varphi(x, y)\rangle\rangle.$$

Let us denote by A and B the supports of the distributions S_x, and T_y, respectively. What is the support of the distribution $W_{x,y}$. The answer is included in the following proposition.

Proposition. *The support of the distribution $W_{x,y}$ is the product $A \times B$ that is the set of the points (x, y) with $x \in A$, $y \in B$.*

Indeed, let us consider a function $\varphi(x, y) \in (\mathcal{D})_{x,y}$ with its support contained in $\complement(A \times B)$. Thus for $x \in A$ fixed, the function $\varphi(x, y)$ has the support in $\complement B$, that is $\theta(x) = 0$.

It follows that the support of the function $\theta(x)$ is contained in $\complement A$ hence $\langle S_x,\ \theta(x)\rangle = 0$ that is $\langle W_{x,y},\ \varphi\rangle = 0$, which shows that the support of $W_{x,y}$ is in $A \times B$. From equation (2.65) it follows immediately that every point from $A \times B$ belongs to the support of $W_{x,y}$ and thus the proposition is proved completely.

Example 1. Let D_x^p be a derivative with respect to x, D_y^q a derivative with respect to y, then

$$D_x^p D_y^q (S_x \otimes T_y) = (D_x^p S_x) \otimes (D_y^q \; T_y).$$

Definition. We say that a distribution is independent of x if it has the form $1_x \otimes T_y$, *and is written*

$$\langle 1_x \otimes T_y, \varphi \rangle = \int_{X^m} \langle T_y, \varphi(x, y) \rangle \, dx = \left\langle T_y, \int_{X^m} \varphi(x, y) \, dx \right\rangle.$$

The definition of the tensor product of two distributions can be extended to a finite number of distributions.

Definition. Let X^m, Y^n, Z^p *be euclidian spaces of dimensions m, n, p respectively, and let* R_x, S_y, T_z *be distributions on* X^m, Y^n, Z^p *respectively. Their tensor product* $R_x \otimes S_y \otimes T_z$ *is defined by the formula*

$$\langle R_x \otimes S_y \otimes Z_z, u(x) \; v(y) \; w(z) \rangle = \langle R_x, u \rangle \; \langle S_y, v \rangle \; \langle T_z, w \rangle.$$

Proposition. *The tensor product defined above is associative.*

Indeed, applying Fubini's theorem to the expression

$$\langle R_x \otimes S_y \otimes T_z, \varphi \rangle \quad \text{with} \quad \varphi \in (\mathcal{D})_{x, y, z}$$

we obtain

$$R_x \otimes S_y \otimes T_z = R_x \otimes (S_y \otimes T_z) = (R_x \otimes S_y) \otimes T_z.$$

Example. Let u be Heaviside's function of n, variables equal to 1 for $x_1 \geqslant 0$, $x_2 \geqslant 0, \ldots, x_n \geqslant 0$ and null in the rest. As it is equal to $u(x_1) \otimes \ldots \otimes u(x_n)$ and as

$$\frac{du(x_i)}{dx_i} = \delta_{x_i}$$

then

$$\frac{\partial^n u}{\partial x_1 \ldots \partial x_n} = \frac{\partial^n (u(x_1) \otimes u(x_2) \ldots \otimes u(x_n))}{\partial x_1 \ldots \partial x_n} = \frac{\partial u(x_1)}{\partial x_1} \otimes \frac{\partial u(x_2)}{\partial x_2} \ldots \otimes \frac{\partial u(x_n)}{\partial x_n} =$$

$$= \delta_{x_1} \otimes \partial_{x_2} \ldots \otimes \delta_{x_n} = \delta_{x_1, x_2 \ldots x_n}.$$

2.6.2 The convolution of two distributions

Definition. Let S, T *be two distributions on* R^n. *The distribution* $S*T$ *defined on* R^n *by formula*

$$\langle S*T,\ \varphi\rangle = \langle S_\xi \otimes T_\eta,\ \varphi(\xi+\eta)\rangle \qquad (2.66)$$

is called the convolution of the distributions S *and* T.

Let us consider formula (2.66).

Remark 1. If A and B are the supports of the distributions S_ξ, T_η, respectively, then the support of the distribution $S_\xi \otimes T_\eta$ is $A \times B$, as it follows from the last proposition of the previous Section.

Remark 2. The function $\varphi(\xi+\eta)$ has been obtained by replacing in $\varphi(x)$, x with $\xi+\eta$, $\xi, \eta \in R^n$. It is an indefinitely differentiable function of ξ and η.

Proposition. *Let* K *be the support of the function* $\varphi(x)$, *then the support of the function* $\varphi(\xi+\eta)$ *is not bounded.*

Indeed the support of the function $\varphi(\xi+\eta)$ is the set of the pairs (ξ, η) so that $\xi+\eta \in K$. This set is a strip parallel to the first bisector of equation $\xi+\eta=0$, that is the subspace with n dimensions of equations $\xi_1+\eta_1=0, \ldots, \xi_n+\eta_n=0$ of the euclidian space with $2n$ dimensions.

Remark 3. The right-hand side of relationship (2.66) makes sense if the intersection of the support of $S_\xi \otimes T_\eta$ with the support of $\varphi(\xi+\eta)$ is bounded, that is if for every K bounded the set of the points (ξ, η) which verify conditions $\xi \in A$, $\eta \in B$, $\xi+\eta \in K$ is bounded.

If this condition is fulfilled then $\langle S*T, \varphi\rangle$ is defined for every $\varphi \in \mathfrak{D}$.

Consequence. From the above proposition and from the three remarks it follows that the convolution $S*T$ defined by (2.66) makes sense if the supports A, B of S, and T respectively are such that for $\xi \in A$, $\eta \in B$, $\xi+\eta$ cannot remain bounded unless ξ and η remain bounded.

Proposition. *If the convolution* $S*T$ *exists, then the product* $T*S$ *exists and*

$$S*T = T*S.$$

Proposition. *If one of the two distributions* S, T *has a bounded support then the convolution* $S*T$ *exists.*

Indeed let us suppose that the set A is bounded then $\xi \in A$ is bounded. If $\xi + \eta$ is bounded, then $\eta = (\xi + \eta) - \xi$ is bounded.

Let S be a distribution on the real line R.

Definition. *We say that the distribution S has the support bounded to the left if it is included in the halfline (a, ∞).*

Proposition. *Let S, T be two distributions on R. If S, T have supports bounded to the left then the convolution $S * T$ exists.*

Indeed, let A, B be the supports of S, and T respectively, then $A \subset (a, \infty)$, $B \subset (a, \infty)$, an dif $\xi \in A$, $\eta \in B$ we have $\xi \geqslant a$, $\eta \geqslant a$. If the set $\xi + \eta$ is bounded then there is a constant c so that $\xi + \eta \leqslant c$. Since $\xi \geqslant a$ we have

$$\eta \leqslant c - \xi \leqslant c - a.$$

Similarly

$$\xi \leqslant c - \eta \leqslant c - a$$

because $\eta \geqslant a$. Hence ξ and η remain bounded and hence from the above corollary it follows that $S * T$ exists.

Remark 4. The above proposition remains valid if the supports of the distributions S and T are both bounded to the right.

Remark. The proposition is not true if one of the supports is bounded to the left and the another one to the right.

2.6.3 Properties of the convolution

We shall give a few properties of the convolution.

Proposition 1. *If S, T are two locally summable functions f, g, defined almost everywhere, then $S * T$ is a function h locally summable defined almost everywhere by the formula*

$$h(x) = \int_{R^n} f(x-t)\, g(t)\, dt = \int_{R^n} f(t) g(x-t)\, dt. \tag{2.67}$$

Indeed if $w = f * g$ then

$$\langle w, \varphi \rangle = \langle f_\xi\, g_\eta, \ \varphi(\xi + \eta) \rangle = \iint_{R^n \times R^n} f(\xi)\, g(\eta)\, \varphi(\xi + \eta)\, d\xi\, d\eta. \quad (2.68)$$

Since the supports A and B of S and T respectively are bounded, it follows that $f(\xi)\, g(\eta)\, \varphi(\xi + \eta)$ has the support bounded. But f and g are locally summable and hence the function $f(\xi)\, g(\eta)\, \varphi(\xi + \eta)$ is locally summable and as its support is bounded it follows that it is summable. Let us consider the change of variables $x = \xi + \eta$, $t = \eta$ whose jacobian is 1. Then

$$d\xi\, d\eta = dx\, dt$$

and hence

$$\langle w, \varphi \rangle = \iint_{R^n \times R^n} f(x - t)\, g(t)\, \varphi(x)\, dx\, dt =$$

$$= \int_{R^n} dx \int_{R^n} \varphi(x)\, f(x - t)\, g(t)\, dt. \quad (2.69)$$

Let us show that

$$h(x) = \int_{R^n} f(x - t)\, g(t)\, dt \quad (2.70)$$

makes sense for almost every x and hence the function $h(x)$ is defined almost everywhere.

Indeed from Fubini's theorem it follows that the last integral of relationship (2.69) makes sense for almost every value of x. It follows that the integral in relationship (2.70) makes sense for almost every value of x for which $\varphi(x) \neq 0$. As this is true for every $\varphi(x) \in (\mathfrak{D})_x$, it follows that the integral in relationship (2.70) makes sense for almost every value of x.

Also from Fubini's theorem it follows that $\varphi(x)\, h(x)$ is summable, hence h is locally summable and

$$\langle w, \varphi \rangle = \int_{R^n} h(x)\, \varphi(x)\, dx = \langle h, \varphi \rangle$$

hence $w = h$. The right-hand side of relationship (2.67) is obtained in the same way as above if we make the change of variables $\xi = t$, $\xi + \eta = x$.

Corollary 1. *If f and g are continuous then $h = f * g$ is continuous.*

Corollary 2. *If one of the functions f, g is continuous or locally bounded and the other is locally summable then $h = f * g$ is continuous.*

Corollary 3. *Let f be a summable function on R^n and g a function bounded on R^n. Then $h = f * g$ makes sense for ev ry value of x and*

$$\| h \|_{L^\infty} \leqslant \| f \|_{L^1} \| g \|_{L^\infty} . \tag{2.71}$$

Indeed we have

$$| h(x) | \leqslant \left| \int_{R^n} f(x) \, dx \right| \sup_{x \in R^n} | g(x) |$$

whence inequality (2.71) follows.

Corollary 4. *If f, $g \in L^1$ then $f * g$ exists almost everywhere, $f * g \in L^1$, and*

$$\| f * g \|_{L^1} \leqslant \| f \|_{L^1} \| g \|_{L^1} .$$

Remark. Corollary 4 has been directly proved in [Vol. 1, Section 6.2.3].

Example. We have

$$u_{(\lambda)}^{(\alpha)} * u_{(\lambda)}^{(\beta)} = u_{(\lambda)}^{(\alpha+\beta)} \tag{2.72}$$

where

$u_{(\lambda)}^{(\alpha)} = u(x) \dfrac{x^{\alpha-1}}{\Gamma(\alpha)} e^{\lambda x}$, $\alpha > 0$, λ is a complex number, and u is Heaviside's function.

Indeed $u_{(\lambda)}^{(\alpha)} * u_{(\lambda)}^{(\beta)} = 0$ for $x \leqslant 0$ and for $x \geqslant 0$ we have

$$u_{(\lambda)}^{(\alpha)} * u_{(\lambda)}^{(\beta)} = \int_0^x \frac{(x-t)^{\alpha-1} t^{\beta-1}}{\Gamma(\alpha)\,\Gamma(\beta)} e^{\lambda(x-t)} e^{\lambda t} \, dt = \frac{x^{\alpha+\beta-1} e^{\lambda x}}{\Gamma(\alpha)\,\Gamma(\beta)} \int_0^1 (1-v)^{\alpha-1} v^{\beta-1} \, dv.$$

In the first integral above we changed the variables $t = xv$. But from Section 1.1.3. Consequence 5, we know that

$$B(\alpha, \beta) = \frac{\Gamma(\alpha)\,\Gamma(\beta)}{\Gamma(\alpha+\beta)} = \int_0^1 (1-v)^{\alpha-1} v^{\beta-1} \, dv$$

whence formula (2.72) follows.

In particular

$$u_{(\lambda)}^{(n)}(x) = u(x) \frac{x^{n-1}}{(n-1)!} e^{\lambda x}$$

is the convolution of the function $u(x)e^{\lambda x}$ by itself, n times.

Proposition 2. *If α is an indefinitely differentiable function and T is a distribution then $T*\alpha$ is an indefinitely differentiable function and*

$$(T*\alpha)(x) = \langle T_t, \alpha(x-t)\rangle. \tag{2.73}$$

Indeed, if we fix x then $\alpha(x-t)$ considered as a function of t, belongs to the space $\mathfrak{D}$. We have

$$h(x) = \langle T_t, \quad \alpha(x-t)\rangle,$$

where $h(x)$ is indefinitely differentiable. From the definition of the convolution we have

$$\langle T*\alpha, \varphi\rangle = \langle T_\xi \otimes \alpha(\eta), \varphi(\xi+\eta)\rangle = \langle T_\xi, \langle \alpha(\eta), \varphi(\xi+\eta)\rangle\rangle =$$

$$= \left\langle T_\xi, \int_{R^n} \alpha(\eta)\ \varphi(\xi+\eta)\ d\eta \right\rangle = \left\langle T_\xi, \int_{R^n} \alpha(x-\xi)\ \varphi(x)\ dx \right\rangle =$$

$$= \langle T_\xi, \langle \varphi(x), \alpha(x-\xi)\rangle\rangle = \langle T_\xi \otimes \varphi(x), \alpha(x-\xi)\rangle =$$

$$= \langle \varphi(x), \langle T_\xi, \alpha(x-\xi)\rangle\rangle = \int_{R^n} \langle T_\xi, \alpha(x-\xi)\rangle\ \varphi(x)\ dx =$$

$$= \int_{R^n} h(x)\ \varphi(x)\ dx = \langle h, \varphi\rangle$$

and thus the proposition is proved completely.

Particular case. Formula (2.73) is reduced to formula (2.67) if T is a function f.

Example 1. $T * 1 = \langle T, 1\rangle$.

Let T be a distribution with bounded support on R.

Definition. We call Laplace transform of T, the holomorphic function of complex variable λ, defined by

$$L(\lambda) = \langle T_x, e^{-\lambda x}\rangle.$$

Proposition 3. *Given two distributions S, T, the Laplace transform of their convolution is equal to the usual product of the Laplace transforms of the distributions S and T.*

Indeed, let there be

$$U(\lambda) = \langle S_\xi, e^{-\lambda\xi} \rangle$$

and

$$L(\lambda) = \langle T_\eta, e^{-\lambda\eta} \rangle$$

then

$$\langle S * T, e^{-\lambda x} \rangle = \langle S_\xi \otimes T_\eta, e^{-(\xi+\eta)} \rangle =$$
$$= \langle S_\xi \otimes T_\eta, e^{-\lambda\xi} e^{-\lambda\eta} \rangle = \langle S_\xi, e^{-\lambda\xi} \rangle \langle T_\eta, e^{-\lambda\eta} \rangle = U(\lambda) L(\lambda).$$

Proposition 4. *δ_0 represents the unit with respect to the convolution that is*

$$\delta_0 * T = T. \tag{2.74}$$

Indeed

$$\langle \delta_0 * T, \varphi \rangle = \langle \delta_{0\xi} \otimes T_\eta, \varphi(\xi + \eta) \rangle = \langle T_\eta, \langle \delta_{0\xi}, \varphi(\xi + \eta) \rangle\rangle =$$
$$= \langle T_\eta, \varphi(\eta) \rangle = \langle T, \varphi \rangle.$$

Remark. The physicists write formula (2.74) thus:

$$\int f(x - t)\, \delta(t)\, \mathrm{d}t = \int f(t)\, \delta(x - t)\, \mathrm{d}t = f(x).$$

Proposition 5. *Let T be a distribution and $\tau_a T$ the translate of T through the translation a, defined by*

$$\langle \tau_a T, \varphi \rangle = \langle T_x, \varphi(x + a) \rangle.$$

Then the convolution of δ_a by T is equal to the translate of T through a, that is

$$\delta_a * T = \tau_a T. \tag{2.75}$$

Indeed

$$\langle \delta_a * T, \varphi \rangle = \langle \delta_{a\xi} \otimes T_\eta, \varphi(\xi + \eta) \rangle = \langle T_\eta, \langle \delta_{a\xi}, \varphi(\xi + \eta) \rangle\rangle =$$
$$= \langle T_\eta, \varphi(a + \eta) \rangle = \langle \tau_a T, \varphi \rangle.$$

Particular case. If in formula (2.75) we take $T = \delta_b$ we obtain

$$\delta_a * \delta_b = \delta_{a+b}. \tag{2.76}$$

Proposition 6. *The convolution of δ_0' by a distribution T is equal to the derivative of the distribution, that is*

$$\delta_0' * T = T'. \tag{2.77}$$

Indeed,

$$\langle \delta_0' * T, \varphi \rangle = \langle \delta_{0\xi}' \otimes T_\eta, \varphi(\xi + \eta) \rangle = \langle T_\eta, \langle \delta_{0\xi}', \varphi(\xi + \eta) \rangle\rangle =$$
$$= \langle T_\eta, -\varphi'(\eta) \rangle = -\langle T, \varphi' \rangle = \langle T', \varphi \rangle.$$

Remark. In quantum physics formula (2.77) is often written thus:

$$\int \delta'(x - t) \, f(t) \, \mathrm{d}t = f'(x).$$

Corollary 1. From proposition 6 *we have immediately*

$$\delta_0^{(m)} * T = T^{(m)}.$$

Corollary 2. If D *is a differential operator with constant coefficients in* R^n, *then we have from proposition* 6 *and corollary* 1

$$\mathrm{D}\delta_0 * T = \mathrm{D}T.$$

Particular case: (1) $\mathrm{D} = \Delta = \sum_{i=1}^{n} \frac{\partial^2}{\partial x_i^2}$, then

$$\Delta\delta_0 * T = \Delta T.$$

(2) $\mathrm{D} = \Box = \frac{1}{v^2} \cdot \frac{\partial^2}{\partial t^2} - \frac{\partial^2}{\partial x^2} - \frac{\partial^2}{\partial y^2} - \frac{\partial^2}{\partial z^2} = \frac{1}{v^2}\frac{\partial^2}{\partial t^2} - \Delta$

then

$$\Box\,\delta_0 * T = \Box\, T.$$

The operator $\Box$ is called *the dalembertian.*

We shall give without proof a continuity property of the convolution

Theorem. *Let T_j and S_j be two sequences of distributions depending on an integer index j and so that S_j and T_j should tend in $\mathcal{D}'$ to S and T respectively. Let us suppose that all the distributions S_j and T_j have the supports included in the closed fixed sets A and B from R^n and that for $\xi \in A$ and $\eta \in B$, $\xi + \eta$ cannot remain bounded unless both ξ and η remain bounded. In this case the sequence of distributions $S_j * T_j$ converges as $j \to \infty$ to the distribution $S * T$.*

From the above theorem we have, taking into account that δ_0 is a limit in $\mathfrak{D}'$ of indefinitely differentiable functions, the following corollary.

Corollary. Every distribution is a limit in $\mathfrak{D}'$ of a sequence of polynomials.

Indeed, it can be shown that there is a sequence of polynomials α converging to δ_0. Therefore if T has a bounded support, $T * \delta_0 = T$ is the limit of the sequence of polynomials $T * \alpha$.

2.6.4 The support of the convolution of two distributions

Proposition. *Let S and T be two distributions on R^n, and let A and B respectively be their supports. Then the support of the distribution $S * T$ is included in $\overline{A \times B}$ that is in the adherence of the set $\xi + \eta$ $(\xi \in A, \eta \in B)$.*

Indeed, in order to prove the proposition it is sufficient to show that if the support K of the function φ is in the complement Ω of the set $\overline{A \times B}$ then

$$\langle S * T, \varphi \rangle = 0.$$

But we have

$$\langle S * T, \varphi \rangle = \langle S_\xi \otimes T_\eta, \varphi(\xi + \eta) \rangle$$

and the support of the product $S_\xi \otimes T_\eta$ is contained in $A \times B$, that is in the set of the pairs (ξ, η) with $\xi \in A$, $\eta \in B$ and the support of $\varphi(\xi + \eta)$ is the set of the pairs (ξ, η) so that $\xi + \eta \in K$.

Let us show that the intersection of two supports is the empty set. Indeed, $\xi \in A$, $\eta \in B$ implies $\xi + \eta \in A \times B$. But $A \times B$ and K are disjoint hence:

$$\langle S * T, \varphi \rangle = 0.$$

But $\Omega = \complement(\overline{A \times B})$, hence it is open. It follows that $S * T = 0$ in Ω which means that the support of $S * T$ is included in $\overline{A \times B}$ and thus the proposition is proved completely.

Corollary 1. If the distributions S and T have a bounded support then the distribution $S * T$ has a bounded support.

Corollary 2. If the support of S is included in (a, ∞) and the support of T is included in (b, ∞) then the support of $S * T$ is included in $(a + b, \infty)$.

2.6.5 The convolution of several distributions

Definition. *Let R, S, T be three distributions in R^n and let A, B and C be their supports. The convolution $R * S * T$ is defined by the formula*

$$\langle R * S * T, \varphi \rangle = \langle R_\xi \otimes S_\eta \otimes T_\zeta, \varphi(\xi + \eta + \zeta) \rangle \tag{2.78}$$

and makes sense if for $\xi \in A$, $\eta \in B$, $\zeta \in C$ the sum $\xi + \eta + \zeta$ remains bounded only if ξ, η, ζ remain bounded.

Proposition 1. *The convolution is associative i. e.*

$$R * S * T = (R * S) * T = R * (S * T). \tag{2.79}$$

This property follows immediately from the property of associativity of the tensor product.

Proposition 2. *The convolution of several distributions makes sense and hence is associative and commutative if all the distributions except one at the most have bounded support.*

Proposition 3. *The convolution of several distributions on R makes sense and hence is associative and commutative, if all the distributions have the supports bounded to the left (or right respectively).*

2.6.6 The differentiation of a convolution

Proposition. *In order to differentiate a convolution one of the factors is differentiated, that is*

$$(S * T)' = S' * T = S * T'.$$

Indeed, from proposition 6, equation (2.15) and from the property of associativity of the convolution we have:

$$(S * T)' = \delta_0' * (S * T) = \delta_0' * S * T = (\delta_0' * S) * T = S' * T = S * \delta_0' * T =$$

$$= S * (\delta_0' * T) = S * T'.$$

Remark. The above proposition remains true if the differentiation is replaced by the translation.

Corollary. *We have*

$$\mathrm{D}(S * T) = \mathrm{D}S * T = S * \mathrm{D}T, \tag{2.80}$$

where D *is a differential operator with constant coefficients.*

Application. *Poisson's formula.* Let $U = T * \dfrac{1}{r}$, where $r = \sqrt{x^2 + y^2 + z^2}$, be the potential of a distribution T. Then

$$\Delta U = -4\pi T. \tag{2.81}$$

Formula (2.81) follows from the above corollary, from the particular case (2), Section 2.2.4 and from proposition 4, Section 2.6.3. Indeed, we have

$$\Delta U = T * \Delta \frac{1}{r} = T * (-4\pi\delta_0) = -4\pi T.$$

2.7 CONVOLUTION EQUATIONS

2.7.1 Convolution algebras

Definition. *We say that a vector subspace* $\mathscr{A}'$ *of the space* $\mathfrak{D}'$ *forms a convolution algebra if it has the following properties:*

(1) $\delta_0 \in \mathscr{A}'$.

(2) The convolution of two or more distributions from $\mathscr{A}'$ is defined and belongs to $\mathscr{A}'$.

(3) The convolution is associative and commutative.

Example 1. $\mathscr{A}' = \mathscr{E}'$ is the convolution algebra of the distributions with bounded support in R^n.

Example 2. $\mathfrak{D}'(\Gamma)$ is the convolution algebra of all the distributions on the circle.

Example 3. $\mathscr{A}' = \mathfrak{D}'_+$ is the convolution algebra of the distributions with the support included in the half-line $x \geqslant 0$.

2.7.2 Convolution equations

Let $\mathscr{A}'$ be a convolution algebra and let A, $B \in \mathscr{A}'$.

Definition. *An equation of the form*

$$A * X = B \tag{2.82}$$

where the unknown is $X \in \mathscr{A}'$, is called a convolution equation.

Definition. *We say that an element $A \in \mathscr{A}'$ has an inverse in $\mathscr{A}'$, if there is an element $A^{-1} \in \mathscr{A}'$ so that*

$$A * A^{-1} = A^{-1} * A = \delta_0. \tag{2.83}$$

Proposition. *The necessary and sufficient condition for the equation* (2.82) *to have a solution for arbitrary B is that A should have an inverse in $\mathscr{A}'$. If A^{-1} exists, then it is unique and the solution of equation* (2.82) *is given by the formula*

$$X = A^{-1} * B. \tag{2.84}$$

Indeed, if the convolution equation (2.82) has a solution for every B, it follows, in particular, that it has a solution for $B = \delta_0$ hence A has an inverse A^{-1}. Conversely if A has an inverse we take the convolution of the two terms of equation (2.82) by A^{-1} and we obtain

$$A^{-1} * A * X = A^{-1} * B,$$

that is, equation (2.84). Analogously if we take the convolution of relationship (2.84) by A, we obtain equation (2.82) which means the two relationships (2.82) and (84) are equivalent. Consequently equation (2.82) admits as unique solution $X \in \mathscr{A}'$ given by equation (2.84). Since A^{-1} is a solution of equation (2.82) for $B = \delta_0$ it follows that it is unique and thus the proposition is proved completely.

Definition. *The inverse A^{-1} of an element $A \in \mathscr{A}'$ is called the elementary solution of the convolution equation* (2.82).

Remark 1. If A has no inverse then equation (2.82) has no solution for $B = \delta_0$. For certain values of B it has solutions, that is it can have a solution or an infinity of solutions, the difference between them being a solution of the homogeneous equation $A * X = 0$.

Example. In the algebra $\mathscr{D}'(\Gamma)$, equation (2.82) can have a solution or an infinity of solutions.

In $\mathscr{E}'$ and D'_+ equation (2.82) has never more than one solution.

Proposition. *If A is an indefinitely differentiable function then equation (2.82) has no elementary solution.*

Indeed for equation (2.82) to have an elementary solution it is necessary that A should have an inverse. But from proposition 2, Section 2.6.3 it follows that $A * A^{-1}$ is an indefinitely differentiable function, consequently it can not be equal to δ_0.

Example. Let us take in R^3, $A = \Delta\delta_0$. Then from the particular case (2), Section 2.2.4 we have $A^{-1} = \dfrac{-1}{4\pi r}$. But A^{-1} has as support the whole space and $\mathfrak{D}'$ is not a convolution algebra.

Equations (2.82) and (2.84) are not equivalent since $A^{-1} * A * X$ makes no sense if X has no bounded support. We can only say that if X has a bounded support and verifies equation (2.82) then X is given by formula (2.84). We notice that in this case the solution of equation (2.82) is not unique because the homogeneous equation

$$\Delta\delta_0 * X = 0 \text{ or } \Delta X = 0$$

has an infinity of solutions, namely 'the harmonic distributions' and in particular the usual harmonic functions. By analogy, relationship (2.84) makes sense provided that B has a bounded support, whence follows equation (2.82).

Consequently a particular solution of the equation

$$\Delta\delta_0 * X = \Delta X = B,$$

when B has a bounded support is

$$X = -\frac{1}{4\pi r} * B$$

and the general solution is obtained by adding to this solution a harmonic distribution.

Let D be a differential operator of order m with constant coefficients

$$\mathrm{D} = \frac{\mathrm{d}^m}{\mathrm{d}x^m} + a_1 \frac{\mathrm{d}^{m-1}}{\mathrm{d}x^{m-1}} + \ldots + a_{m-1}\frac{\mathrm{d}}{\mathrm{d}x} + a_m. \tag{2.85}$$

Then

Proposition. D δ *is invertible in* $\mathfrak{D}'_+$ *and its inverse is the product of Heaviside's function u and the solution Z of the homogeneous equation* $\mathrm{D}Z = 0$ *which verifies the initial conditions*

$$Z(0) = Z'(0) = \ldots = Z^{(m-2)}(0) = 0,\ Z^{(m-1)}(0) = 1. \tag{2.86}$$

Indeed, since Z is indefinitely differentiable we can differentiate uZ as a product (proposition 2, Section 2.3.1) and we obtain

$$\begin{aligned}
(uZ)' &= uZ' + \delta_0 Z(0),\\
(uZ)'' &= uZ'' + \delta_0 Z'(0) + \delta_0' Z(0),\\
&\dots\dots\dots\dots\dots\dots\\
(uZ)^{(m-1)} &= uZ^{(m-1)} + \delta_0 Z^{(m-2)}(0) + \dots + \delta_0^{(m-2)} Z(0),\\
(uZ)^{(m)} &= uZ^{(m)} + \delta_0 Z^{(m-1)}(0) + \dots + \delta_0^{(m-1)} Z(0).
\end{aligned} \tag{2.87}$$

Taking into account the initial conditions (2.86) we obtain

$$(uZ)^{(k)} = uZ^{(k)} \text{ for } k \leqslant m - 1,$$

$$(uZ)^{(m)} = uZ^{(m)} + \delta_0.$$

Hence we have

$$\mathrm{D}\delta_0 * (uZ) = \mathrm{D}(uZ) = u\mathrm{D}Z + \delta_0$$

and since Z is the solution of the homogeneous equation we obtain

$$\mathrm{D}(uZ) = \delta_0$$

and thus the proposition is proved completely.

Example 1. The elementary solution of $\mathrm{D} = \dfrac{\mathrm{d}}{\mathrm{d}x} - \lambda$ where λ is a complex number is

$$(\delta_0' - \lambda\delta_0)^{-1} = u(x)\, e^{\lambda x}. \tag{2.88}$$

The elementary solution of $\mathrm{D} = \left(\dfrac{\mathrm{d}}{\mathrm{d}x} - \lambda\right)^m$ is

$$(\delta_0' - \lambda\delta_0)^{-m} = u(x)\, e^{\lambda x} \frac{x^{m-1}}{(m-1)!} \tag{2.89}$$

Example 2. Determine the solution of the equation

$$\mathrm{D}z = f \tag{2.90}$$

with the initial conditions

$$z^{(k)}(0) = z_k,\ k \leqslant m - 1, \tag{2.91}$$

where D is the operator (2.85) and f is a given function.

We have to find the function uz. But $uz \in \mathfrak{D}'_+$ and verifies the relationship (2.87). Hence we have

$$\mathrm{D}(uz) = u\mathrm{D}z + \sum_{k=0}^{m-1} e_k \, \delta_0^{(k)}$$

with

$$\mathrm{D}z = f,$$

$$e_k = z_{m-1-k} + a_1 z_{m-2-k} + \ldots + a_{m-k-1} z_0.$$

If we denote by Z the particular solution of the equation $\mathrm{D}Z = 0$ with the conditions shown in equation (2.91) and we take $(\mathrm{D}\ \delta_0)^{-1} = uZ$ then we obtain

$$uz = uZ * \left(uf + \sum_{k=0}^{m-1} e_k \, \delta_0^{(k)} \right)$$

that is, for $x \geqslant 0$

$$z(x) = \int_0^x Z(x-t)\, f(t)\, \mathrm{d}t + \sum_{k=0}^{m-1} e_k \, Z^{(k)}. \tag{2.92}$$

In the case $x \leqslant 0$ in order to determine z we repeat the reasoning in the algebra $\mathfrak{D}'_-$ of the distributions having the support in $(-\infty, 0)$ and we obtain the same formula (2.92).

Let us prove the following proposition.

Proposition. *If A_1, $A_2 \in \mathfrak{D}'_+$ are invertible then $A_1 * A_2$ is invertible and*

$$(A_1 * A_2)^{-1} = A_1^{-1} * A_2^{-1}.$$

Indeed

$$(A_1 * A_2) * (A_1^{-1} * A_2^{-1}) = (A_1 * A_1^{-1}) * (A_2 * A_2^{-1}) = \delta_0 * \delta_0 = \delta_0.$$

Example. Let D be a differential operator with constant coefficients of the form (2.85) which we decompose thus

$$\mathrm{D} = \left(\frac{\mathrm{d}}{\mathrm{d}x} - z_1 \right) \left(\frac{\mathrm{d}}{\mathrm{d}x} - z_2 \right) \ldots \left(\frac{\mathrm{d}}{\mathrm{d}x} - z_m \right).$$

Then

$$\mathrm{D}\delta_0 = (\delta_0' - z_1\delta_0) * (\delta'_0 - z_2\delta_0) * \ldots * (\delta_0' - z_m\delta_0).$$

From the previous proposition $(D\delta_0)^{-1}$ exists and taking example 1 into account we have

$$(D\delta_0)^{-1} = u(x)e^{z_1 x} * u(x)e^{z_2 x} * \dots * u(x)e^{z_m x}.$$

In particular

$$(\delta_0' - \lambda\delta_0)^m = u(x)e^{\lambda x} * u(x)e^{\lambda x} * \dots * u(x)e^{\lambda x} = u(x)e^{\lambda x}\frac{x^{m-1}}{(m-1)!}$$

and we find thus formula (2.89).

2.7.3 Application to a system of convolution equations

Definition. *A system of the following form is called a system of convolution equations with n unknown distributions* X_i.

$$\begin{aligned} A_{11} * X_1 + A_{12} * X_2 + \dots + A_{1n} * X_n &= B_1, \\ \dots\dots\dots\dots\dots\dots\dots\dots\dots\dots \\ A_{i1} * X_1 + A_{i2} * X_2 + \dots + A_{in} * X_n &= B_i, \\ \dots\dots\dots\dots\dots\dots\dots\dots\dots\dots \\ A_{n1} * X_1 + A_{n2} * X_2 + \dots + A_{nn} * X_n &= B_n, \end{aligned} \tag{2.93}$$

where $X_i, B_i, A_{ij} \in \mathscr{A}'$.

We shall prove the following theorem.

Theorem. *The necessary and sufficient condition for the system of equations (2.93) to have a solution is that the determinant of the coeficients* A_{ij}, $|A| = |A_{ij}|$ *should have an inverse in* $\mathscr{A}'$. *If it exists, the inverse of A is unique and the system (2.93) has a unique solution given by:*

$$\begin{aligned} X_1 &= C_{11} * B_1 + C_{12} * B_2 + \dots + C_{1n} * B_n, \\ X_2 &= C_{21} * B_1 + C_{22} * B_2 + \dots + C_{2n} * B_n, \\ &\dots\dots\dots\dots\dots\dots\dots\dots\dots\dots \\ X_n &= C_{n1} * B_1 + C_{n2} * B_2 + \dots + C_{nn} * B_n \end{aligned} \tag{2.94}$$

where the matrix $C = \| C_{ij} \|$ *is the inverse of the matrix* * $A = \| A_{ij} \|$.

---

* As in Vol. 1, Chapter 3 we denote by $A = \| A_{ij} \|$ the matrix formed with the elements A_{ij} and by $|A| = |A_{ij}|$ its determinant.

Let us note first that the products which appear in the calculus of the determinants are convolutions. The system (2.93) can be written in the matrix form

$$A * X = B, \tag{2.95}$$

$$X = \begin{Vmatrix} X_1 \\ X_2 \\ \cdot \\ \cdot \\ \cdot \\ X_n \end{Vmatrix}, \qquad B = \begin{Vmatrix} B_1 \\ B_2 \\ \cdot \\ \cdot \\ \cdot \\ B_n \end{Vmatrix}.$$

Let us first prove that this condition is necessary. We take for B the system $B_j = \delta_0$, $B_k = 0$ for $k \neq j$. Let $X_i = C_{ij}$, $i = 1, \ldots, n$, be the corresponding solution. Then we take j equal to 1, 2, . . ., n. We have

$$\sum_{k=1}^{n} A_{ik} * C_{kj} = \begin{cases} 0, & \text{if } i \neq j, \\ \delta_0, & \text{if } i = j. \end{cases}$$

The above relationship can be written in the matrix form:

$$A * C = \delta_0 I, \tag{2.96}$$

where $\delta_0 I$ is a square diagonal matrix whose elements are δ_0 on the main diagonal and 0 on the rest.

From relationship (2.96) we deduce the following relationship between the determinants of the above matrices

$$|A| * |C| = |\delta_0 I| = \delta_0, \tag{2.97}$$

whence it follows that the determinant $|A|$ is invertible in the algebra $\mathscr{A}'$ and thus the necessity of the condition has been proved.

Let us prove that the condition is sufficient. We suppose that $|A|$ is invertible and we denote by α_{ij} the minor of the element A_{ij}. Let C be the matrix whose elements are

$$C_{ij} = A^{-1} * \alpha_{ij}$$

We have (Vol. 1, Section 3.2)

$$A * C = C * A = \delta_0 I. \tag{2.98}$$

Let X be a solution of equation (2.95). Taking the convolution to the left of relationship (2.95) with C, and taking into account equation (2.98) we obtain

$$C * B = C * A * X = \delta_0 I * X = X \tag{2.99}$$

that is

$$X = C * B.$$

Conversely, if X verifies relationship (2.99) then by taking the convolution to the left of relationship (2.99) with A we obtain equation (2.95). Consequently equations (2.95) and (2.99) are equivalent. In other words equation (2.95) has as unique solution X given by equation (2.99), that is the solution of system (2.93) is given by equation (2.94).

Moreover we have to show that the matrix C the inverse of matrix A is unique. Indeed, by taking the convolution to the right of the relationship

$$A^{-1} * A = \delta_0 I$$

with C we obtain

$$A^{-1} * A * C = A^{-1},$$

$$A^{-1} * A * C = A^{-1} * \delta_0 I = A^{-1} = \delta_0 I * C = C.$$

2.7.4 Application of the convolution to the integral equations

Definition. An equation of the form

$$f(x) + \int_0^x K(x - t)\, f(t)\, \mathrm{d}t = g(x), \quad x \geqslant 0, \tag{2.100}$$

where g and K are given locally summable functions and f is the unknown function is called the integral equation of second species and the function K is called the kernel of the equation.

Remark. If we extend the functions f, g, K by 0 for $x < 0$ we obtain a convolution equation in $\mathfrak{D}'_+$ of the type shown in equation (2.82) where $A = \delta_0 + K$, $X = f$, $B = g$.

Let us denote symbolically, $\delta_0 = 1$, $K = g$ and prove the following proposition.

Proposition. *If the kernel of equation* (2.100) *is bounded on every finite interval, then* $A = \delta_0 + K$ *is invertible in* $\mathfrak{D}'_+$ *for every* $K \in \mathfrak{D}'_+$ *and its inverse has the form* $\delta_0 + H$, *where* $H \in \mathfrak{D}'_+$.

With the above symbolic notation it follows that we must invert $1 + q$. But we shall take as the inverse of $1 + q$ the series $1 - q + q^2 + \ldots + (-1)^n q^n + \ldots$ if it is convergent. This means that we have to calculate

$$E = \delta_0 - K + K^{*2} + \ldots + (-1)^n K^{*n} + \ldots \tag{2.101}$$

where we denoted by K^{*n} the convolution of n functions equal to K. Let us show that the series (2.101) is convergent in $\mathfrak{D}'_+$. Leaving out the first term it will be sufficient to show that K^{*n} is dominated in the modulus on every finite interval $(0, a)$ by the general term of a convergent numerical series. We shall prove this property through recurrence on n. For $0 \leqslant x \leqslant a$ we have

$$| K^{*2}(x) | = \left| \int_0^x K(x-t)\, K(t)\, \mathrm{d}t \right| \leqslant x M_a^2$$

where

$$M_a = \max_{0 \leqslant x \leqslant a} | K(x) |.$$

Let us suppose that for $x \in [0, a]$

$$| K^{*(n-1)}(x) | \leqslant \frac{x^{n-2}}{(n-2)!} M_a^{n-1}$$

then

$$| K^{*n}(x) | = \left| \int_0^x K^{*(n-1)}(t)\, K(x-t)\, \mathrm{d}t \right| \leqslant M_a^n \int_0^x \frac{t^{n-2}}{(n-2)!}\, \mathrm{d}t = \frac{x^{n-1}}{(n-1)!} M_a^n.$$

The series (2.101) is dominated term by term in $[0, a]$ by the convergent numerical series

$$\sum_{n=1}^{\infty} \frac{a^{n-1}}{(n-1)!} M_a^n.$$

Let us show that the distribution E, sum of the series (2.101), verifies the equation

$$A * E = \delta_0.$$

Since from the theorem of Section 2.6.3 it follows that the convolution can be calculated term by term, and keeping the above notation we have only to verify that

$$(1+q)(1-q+q^2-\ldots+(-1)^n q^n+\ldots)=1$$

which is self evident. If we write

$$-H=K-K^{*2}+K^{*3}-\ldots+(-1)^{n-1}K^{*n}+\ldots$$

then

$$A^{-1}=\delta_0+H.$$

Let us notice that H is null for $x<0$. If we suppose that K is continuous then H is a continuous function as the sum of a uniformly convergent series on every interval $[0, a]$ of continuous functions.

Consequently the solution of equation (2.100) is given by

$$f=A^{-1}*g=(\delta_0+H)*g$$

or, in another way written

$$f(x)=g(x)+\int_0^x H(x-t)\,g(t)\,\mathrm{d}t.$$

Remark. We can interchange f and g.

Definition. We call the integral equation of first species an equation of the form

$$\int_0^x K(x-t)\,f(t)\,\mathrm{d}t=g(x).$$

Such an equation is a convolution equation where $A=K$. For such an equation two cases can appear:

(1) The inverse A^{-1} does not exist. For instance this is the case where K is an indefinitely differentiable function because from proposition 2, Section 2.6.3 it follows that for arbitrary $E\in\mathscr{D}'_+$, $K*E$ is an indefinitely differentiable function, hence it cannot be equal to δ_0.

(2) If A^{-1} exists, then it is a distribution which can never be a function because, if E is a function, $K*E$ is a function, hence it cannot be equal to δ_0.

Remark. There are integral equations which are not convolution equations, for instance, the integral equation of second general species

$$f(x) + \int_0^x K(x, \xi)\, f(\xi)\, d\xi = g(x),$$

where K is a continuous function of two variables for $x \geqslant 0$ and $0 \leqslant \xi \leqslant x$.

2.8 EXERCISES

1. Calculate, in the sense of the theory of distributions, the successive derivatives of $|x|$.

2. Show that the multiplication by a function $\alpha \in \mathscr{E}$ is a linear continuous mapping in $\mathscr{D}'$. Find the limits in $\mathscr{D}'$, as $a \to 0$, of

$$\frac{a}{x^2 + a^2} \text{ and } \frac{ax}{x^2 + q^2}, \quad a > 0.$$

3. Determine the limits in $\mathscr{D}$, when $h \to 0$ of the following distributions

(a) $$\frac{\delta(h) - \delta(-h)}{2h},$$

(b) $$\frac{\delta(2h) + \delta(-2h) - 2\delta(0)}{4h^2}.$$

4. (a) Calculate the successive convolution powers $(f)^{*n}$ of

$$f(x) = \begin{cases} 1 \text{ for } -1 < x < 1, \\ 0 \text{ in the rest.} \end{cases}$$

(*b*) How can one deduce the result for $f(f')^{*n}$?
(c) Find is the support of $(f)^{*n}$.

3

Elements of Differential Calculus in Affine Spaces

3.1 AFFINE SPACES

3.1.1 Definition of affine space. Affine manifolds

In Volume 1, Chapter 3, the notion of vector space and some properties of these spaces have been studied. Starting from these spaces we shall introduce here the space of the elementary geometry which is an affine euclidian space. We shall use here a slightly different notation than the one used in Volume 1 and we shall denote the elements of a vector space by $\bar{a}, \bar{b}, \bar{c}, \ldots$ and call them vectors. Let K be the field of real or complex numbers.

Definition. *We call affine space E in the field K a non-empty set whose elements we shall call points to which are associated:*

(1) *A vector space $\bar{E}$ in the field K.*

(2) *A mapping of $E \times E$ in $\bar{E}$, which establishes a correspondence between the couple $(a, b) \in E \times E$ and a vector from $\bar{E}$ which we shall denote by $\overline{ab}$ called a vector with its origin in a and its extremity in b, with the following properties:*

(a) *For every, $a, b, c \in E$, $\overline{ab} + \overline{bc} + \overline{ca} = \bar{0}$ the null element of $\bar{E}$. The above relation is, in fact, Chasles' relation.*

(b) *For every fixed point a, the mapping $x \to \overline{ax}$ is a one to one mapping of E on $\bar{E}$.*

Remark 1. From Chasles' relation we deduce that for every a, the vector $\overline{aa}$ is the null vector.

Remark 2. If instead of $\overline{ab}$ we write $\overline{b - a}$ then Chasles' relation becomes $\overline{b - a} + \overline{c - b} + \overline{a - c} = \bar{0}$.

Remark 3. From (b) it follows that if $a \in E$ and $\bar{h} \in \bar{E}$, there is a point $b \in E$, and only one so that $\overline{b - a} = \bar{h}$. Sometimes it is simpler to denote by $a + \bar{h}$ this point.

Remark 4. From Chasles' relation we have $a + (\bar{h} + \bar{k}) = (a + \bar{h}) + \bar{k}$, consequently this notation is compatible with the usual properties of the addition.

Definition. *The dimension of the associate space $\bar{E}$ is called the dimension of the affine space E.*

Examples of affine spaces. (1) A vector space is an affine space. Indeed, it is sufficient to consider that the vector space and the affine space coincide and to make the correspondence between two arbitrary elements $\bar{a}, \bar{b} \in E$ and the vector $\overline{ab} = \bar{b} - \bar{a}$.

(2) As a particular case of (1), the field of the scalars K is an affine space of dimension 1.

Definition. The system formed by the origin 0 *of* E *and by a basis* $(\bar{e}_i)_{i \in I}$ *of the associated vector space* $\bar{E}$ *is called the frame of reference of an affine space of finite dimension.*

In this case if we consider a point $x \in E$, the vector $x - \bar{0}$ has the coordinates $(x_i)_{i \in I}$ with respect to the basis chosen in E.

Definition. The coordinates $(x_i)_{i \in I}$ *are called the coordinates of* x *with respect to the considered frame of reference and we write*

$$x = 0 + \sum_{i \in I} x_i \, \bar{e}_i.$$

Let F be a non-empty subset of an affine space E.

Definition. If there is a vector subspace $\bar{F}$ *of* $\bar{E}$ *so that for every couple* $(a, b) \in F \times F$, *the vector* $\overline{b - a} \in \bar{F}$, *and for every couple* $(a, \bar{h}) \in F \times \bar{F}$ *the point* $a + \bar{h} \in F$, *then obviously such a space is unique being exactly the following set*

$$\{\overline{b - a} \mid (a, b) \in F \times F\}.$$

In this case F *is called the affine subspace or affine manifold of* E.

Remark 1. F is an affine space, $\bar{F}$ is an associated vector space and the mapping $F \times F$ in $\bar{F}$ is the restriction of the given mapping of $E \times E$ in $\bar{E}$.

Remark 2. E is itself an affine manifold and each of its points is an affine manifold of zero dimension.

Definition. An affine manifold of dimension 1 *is called a line, and an affine manifold of dimension* 2 *is called a plane.*

Definition. A vector subspace $\bar{F}$ *of* $\bar{E}$ *whose supplimentary subspaces are of dimension* 1 *is called hyperplane. An affine manifold* F *of* E *is a hyperplane if its associated vector subspace is a hyperplane.*

Remark. If E is of finite dimension n, a hyperplane is an affine manifold of dimension $n - 1$.

Proposition. *A cartesian product of two affine spaces E_1, E_2 is an affine space $E_1 \times E_2$; the vector space associated to $E_1 \times E_2$ is $\overline{E}_1 \times \overline{E}_2$.*

Indeed the proposition is obvious if we take

$$\overline{(b_1, b_2) - (a_1, a_2)} = (\overline{b_1 - a_1}, \overline{b_2 - a_2}).$$

3.1.2 Linear mappings. Affine mappings

Let E and F be two affine spaces.

Definition. *We say that a mapping u of E in F is an affine mapping if there is a linear mapping $\overline{u}$ of $\overline{E}$ in $\overline{F}$ so that*

$$\overline{u(b) - u(a)} = \overline{u}(\overline{b - a}).$$

Example 1. Let E and F be affine spaces of finite dimension. Let a be the origin in E, $(\overline{e}_j)_{j \in \mathfrak{J}}$ be a basis in $\overline{E}$, b the origin in F and $(\overline{f}_i)_{i \in I}$ a basis in $\overline{F}$ then an affine mapping u is completely determined if we know the coordinates $(c_i)_{i \in I}$ of the point $u(a) \in F$ and the coordinates $(u_{ij})_{i \in I}$ of each vector $\overline{u}(\overline{e}_j)_{j \in \mathfrak{J}}$ from the vector space $\overline{F}$. In this case we have

$$u(a) = b + \sum_{i \in I} c_i \overline{f}_i,$$

$$\overline{u}(\overline{e}_j) = \sum_{i \in I} u_{ij} \overline{f}_i. \tag{3.1}$$

It can also be said that the mapping u establishes the correspondence between the point $x = a + \sum_{j \in \mathfrak{J}} x_j \overline{e}_j \in E$ of coordinates $(x_j)_{i \in \mathfrak{J}}$ and the point $y = b + \sum_{i \in I} y_i \overline{f}_i \in F$, of coordinates $(y_i)_{i \in I}$ according to the formula

$$y_i = c_i + \sum_{j \in \mathfrak{J}} u_{ij} \, x_j, \; i \in I. \tag{3.2}$$

Example 2. Let F be the field of the real or complex numbers where we take the origin and the unit vector as frame of reference. In this case u will be a real or complex affine function which will establish the correspondence between the point x of coordinates $(x_j)_{j \in \mathfrak{J}}$ and the real or complex number

$$u(x) = c + \sum_{j \in \mathfrak{J}} u_j x_j, \quad c = u(a), \quad u_j = \overline{u}(\overline{e}_j).$$

Remark. It is seen that what we usually call a linear function is in fact an affine function and what we call a homogeneous linear function could be called an associate linear function.

Example 3. $y = ax + b$ is an affine function, $y = ax$ is its associate linear function.

3.1.3 Normed affine spaces

Definition. *We say that an affine space is normed if its associate vector space is normed.*

Remark 1. The norm is a function defined in the associate vector space and not in the affine space, in other words we can talk about the norm of a vector but not about the norm of a point.

Remark 2. A normed affine space possesses a metric defined by $d(x, y) = \|\overline{x - y}\|$ which is invariant to translations.

Remark 3. It can be easily seen that if E and F are normed affine spaces and E is of finite dimension every affine mapping of E in F is continuous.

Remark 4. If E and F are of infinite dimensions then an affine mapping of E in F might be not continuous.

Proposition 1. *A normed affine space is complete if, and only if, the associate vector space is complete.*

Indeed let a be the origin chosen in E. The mapping $\bar{x} \to a + \bar{x}$ of $\bar{E}$ in E is one to one and preserves the distances. Consequently $\bar{E}$ is complete.

Proposition 2. *An affine subspace F of E is closed if and only if its associate vector subspace $\bar{F}$ is closed in $\bar{E}$. In particular an affine subspace of finite dimension is closed.*

Indeed let $a \in F$. Then $\bar{x} \to a + \bar{x}$ is a homeomorphism* of $\bar{E}$ on E and the image of $\bar{F}$ is F. Consequently F is closed in E if and only if $\bar{F}$ is closed in $\bar{E}$.

---

* A one to one mapping of a metric space E on a metric space F is called homeomorphism if it is continuous and if its inverse is continuous.

Proposition 3. *The necessary and sufficient condition for an affine mapping u, of an affine normed space E in a normed affine space F, to be continuous is that the linear associate mapping be continuous. In this case, u is uniformly continuous.*

Indeed if u is a continuous affine mapping of E in F the linear associate mapping $\bar{u}$ is

$$\bar{u}(\bar{x}) = \overline{u(a + \bar{x}) - u(a)}$$

with a fixed and hence obviously continuous. Conversely, if $\bar{u}$ is continuous, the inequality

$$\| \overline{u(x) - u(y)} \| = \| \bar{u}(\bar{x} - \bar{y}) \| \leqslant \| \bar{u} \| \; \| \bar{x} - \bar{y} \|$$

shows that u is uniformly continuous.

3.2 THE DERIVATIVE OF THE MAPPING OF AN AFFINE SPACE IN ANOTHER AFFINE SPACE

3.2.1 The derivative vector of a function of scalar variable

Let Ω be an open set of the field K, F a normed affine space and $f: \Omega \rightarrow F$.

Definition. *If for $a \in \Omega$, $h \neq 0$, $a + h \in \Omega$ the ratio $\dfrac{\overline{f(a + h) - f(a)}}{h}$ has a limit as $h \rightarrow 0$ it is said that f is differentiable at the point a and we write*

$$\overline{f'(a)} = \lim_{h \neq 0,\ h \rightarrow 0,\ a + h \in \Omega} \frac{\overline{f(a+h) - f(a)},.}{h} \tag{3.3}$$

$\overline{f'(a)}$ is called the derivative vector or the derivative of the function f at the point a.

Remark. The function $\overline{f'}$ is a mapping of Ω in the normed vector space $\bar{F}$. Similarly we can consider as for the real functions $\overline{f''}, \overline{f'''}, \ldots$

Notation. We denote by C^1, $C^2, \ldots, C^m, \ldots$, C^∞ the class of the functions once, twice, ..., m times, ... indefinitely differentiable with values in F.

Proposition. *A differentiable function is continuous.*

Example. If F is of finite dimension and if we consider in F a frame of reference formed by an origin b and a basis $(\overline{f_i})_{i \in I}$ of $\overline{F}$ then any point of F is represented by its coordinates $(y_i)_{i \in I}$ and the function , defined on $\Omega \subset K$ with values in F is expressed by means of the scalar functions $(F_i)_{i \in I}$ by the formula

$$f(x) = b + \sum_{i \in I} F_i(x)\overline{f_i}, \text{ with } y_i = F_i(x).$$

In these conditions the derivative function is given by the formula

$$\overline{f'(x)} = \sum_{i \in I} F_i'(x)\overline{f_i}.$$

From the above example we have the following proposition.

Proposition. *For a function defined in $\Omega \subset K$ with values in a normed affine space of finite dimension to be differentiable it is necessary and sufficient that its components with respect to an arbitrary frame of reference should be differentiable scalar functions. In this case the components of the derivatives are exactly the derivatives of the components.*

3.2.2 The general case of a derivative along a vector of Gâteaux's derivative

Let E, F be two affine spaces, $\Omega \subset E$ an open set and f: $E \to F$. Let us consider a point $a \in \Omega$ and define the mapping $t \to f(a + t\,\overline{x})$ where t belongs to the field of scalars K. Let us denote by $K_{a\overline{x}}$ the set t of the elements of K for which $a + t\overline{x} \in \Omega$. The set $K_{a\overline{x}}$ is the reciprocal image of the open set Ω through the continuous mapping $t \to a + t\overline{x}$ of K in E.

Consequently it is an open set in K containing the origin. We can therefore speak of the derivative of the mapping $t \to f(a + t\overline{x})$ at the origin.

Definition. If the mapping $t \to f(a + t\overline{x})$ has a derivative at the origin, that is if

$$\lim_{\substack{t \neq 0,\, t \to 0 \\ a + t\overline{x} \in \Omega}} \frac{\overline{f(a + t\overline{x}) - f(a)}}{t} = \left[\frac{\mathrm{d}}{\mathrm{d}t} f(a + t\overline{x})\right]_{t=0} \tag{3.4}$$

exists, then we shall say that at the point a, f has a derivative along the vector $\overline{x}$ and we shall denote it by $\overline{\mathrm{D}_{\overline{x}}\, f(a)} \in \overline{F}$.

Consequently

$$\overline{D_{\bar{x}}f(a)} = \left[\frac{d}{dt} f(a + t\bar{x})\right]_{t=0}.$$

For the derivative along a vector, the differentiation rules of a sum, a product, and a quotient are the same as for the derivative of a function of a scalar variable, since from formula (3.4) it follows that the differentiation along a vector is reduced to the differentiation at $t = 0$ of a function of a scalar variable t.

Example 1. If E is of finite dimension and S a regular hypersurface contained in Ω, $a \in S$ and $\bar{\nu}$ is the normal of S at a, then $\frac{df}{d\bar{\nu}}$ at the point a is exactly the derivative after the unit vector of $\bar{\nu}$.

Example 2. If E is the field of the scalars and if x is the element 1 of this field, the derivative along x is exactly the derivative $f'(a)$ given in the expression (3.3), that is $\overline{D_{\bar{x}}f(a)} = \overline{f'(a)}$.

If $\overline{D_{\bar{x}}f(x)}$ exists for every x, then the function $\overline{D_{\bar{x}}f}:\ x \to \overline{D_{\bar{x}}f(x)}$ represents a mapping of Ω in $\bar{F}$ and is called derivative function of f along $\bar{x}$.

If $\overline{D_{\bar{x}}f}$ admits in turn a derivative along a vector $\bar{y}$, we shall denote by $\overline{D_{\bar{y}}D_{\bar{x}}f(a}$ this derivative a.s.o.

3.2.3 The derivative matrix

Let us consider the following cases:

(1) F is an affine space of finite dimension, b is one of its origins and $\overline{(f_i)_{i \in I}}$ a basis in $\bar{F}$; consequently

$$f(x) = b + \sum_{i \in I} F_i(x)\bar{f}_i,$$

$$\overline{D_{\bar{x}}f(x)} = \sum_{i \in I} D_{\bar{x}}F_i(x)\bar{f}_i.$$

(2) If F has a finite dimension and if a, $(\bar{e}_j)_{j \in \mathfrak{J}}$ is a frame of reference in E, then the derivatives along the vector $\bar{e}_i$ of the basis in $\bar{E}$ are what we generally call the partial derivatives of f, in other words,

$$\overline{\partial_j f(x)} = \overline{\frac{\partial f}{\partial x_j}} = \overline{De_j f(x)} = \lim_{t \to 0} \overline{\frac{f(x + t\bar{e}) - f(x)}{t}}.$$

(3) Both spaces E and F are of finite dimension, then the derivative $\overline{\partial_j f} = \overline{\frac{\partial f}{\partial x_j}}$ is exactly

$$\overline{\frac{\partial f}{\partial x_j}} = \sum_{i \in I} \frac{\partial F_i}{\partial x_j} \bar{f}_i.$$

Definition. If $I = \{1, 2, \ldots, m\}$ and $\mathcal{J} = \{1, 2, \ldots, n\}$ the matrix

$$\left\| \begin{matrix} \frac{\partial F_1}{\partial x_1}(x) & \frac{\partial F_1}{\partial x_2}(x) & \ldots & \frac{\partial F_1}{\partial x_n}(x) \\ \ldots & \ldots & \ldots & \ldots \\ \frac{\partial F_m}{\partial x_1}(x) & \frac{\partial F_m}{\partial x_2}(x) & \ldots & \frac{\partial F_m}{\partial x_n}(x) \end{matrix} \right\|$$

is called derivative matrix of f at the point $x \in \Omega$. If $m = n$ its determinant is called the Jacobean of f at the point x with respect to the frame of reference we considered.

Notation. The Jacobean of the function $y = f(x)$ defined in the considered frame of reference by the scalar functions $y_i = F_i(x) = F_i(x_1, \ldots, x_n)$, $i = 1, 2, \ldots, n$, is denoted by $\frac{D(y_1, \ldots, y_n)}{D(x_1, \ldots, x_n)}$.

If $n = 1$, the Jacobean is reduced to the usual derivative.

3.2.4 The total derivative or Fréchet's differential

The notion of Gâteaux's derivative is not sufficient. We shall give an example of Gâteaux differentiable function which is not continuous.

Example. Let

$$f(x, y) = \begin{cases} 0 \text{ for } (x, y) = (0, 0), \\ \dfrac{x^5}{(y - x^2)^2 + x^8} \text{ for } (x, y) \neq (0, 0). \end{cases}$$

For $Y \neq 0$,

$$f(tX, tY) = \frac{t^3 X^5}{Y^2 + \ldots} = \frac{X^5}{Y^2} t^3 + \ldots$$

But $f = 0$ in the origin, hence its derivative along the vector (X, Y) at the origin is zero. If $Y = 0$, $X \neq 0$, $t \neq 0$ then

$$f(tX, tY) = \frac{t^5X^5}{t^4X^4 + \ldots} = tX + \ldots$$

Thus in this case its derivative along the vector (X, Y) is X. Along the vector $(0, 0)$ the derivative is always zero. Consequently f is differentiable along every vector at the origin. But at the origin f is obviously discontinuous.

We shall now give the definition of the total derivative or of Fréchet's derivative.

Let f be a mapping of an open set of a normed affine space E in a normed affine space F.

Definition. *We say that at the point $a \in \Omega$, f admits a total derivative or a differential in Fréchet's sense, if there is a continuous linear mapping of $\bar{E}$ in $\bar{F}$ so that for $a + \bar{h} \in \Omega$*

$$f(a + \bar{h}) = f(a) + L \cdot \bar{h} + \varphi(\bar{h}) \; \|\bar{h}\| \tag{3.5}$$

where $\varphi(h)$ tends to $\bar{0}$ as $\bar{h} \to \bar{0}$.*

Consequently, when we say that f has L as a derivative at a, we mean that

$$\varphi(\bar{h}) = \overline{\frac{f(a + \bar{h}) - f(a) - L(\bar{h})}{\|\bar{h}\|}}$$

tends to $\bar{0}$ as $\bar{h} \to \bar{0}$.

Proposition 1. *If the mapping f admits a total derivative L at the point a then f is continuous at this point.*

Indeed if $\bar{h} \to \bar{0}$, $L(\bar{h}) \to \bar{0}$ since the mapping L is supposed to be continuous.

Similarly $\varphi(\bar{h}) \|\bar{h}\| \to 0$ as $\bar{h} \to 0$, hence taking (3.5) into account we have the continuity of the function f at a.

* We note $L\bar{h}$ or $L(\bar{h})$.

Proposition 2. *If the mapping f has a total derivative at the point a then at a, f has a derivative along every vector* $\bar{x} \in \bar{E}$ *and the mapping* $\bar{x} \to \overline{D_{\bar{x}} f(a)}$ *is a linear continuous mapping from* $\bar{E}$ *in* $\bar{F}$ *equal to L.*

Indeed, let $\bar{x} \to \bar{E}$. Then replacing $\bar{h}$ by $t\bar{x}$ in (3.5) and noticing that $a + t\bar{x} \in \Omega$ for $|t|$ small enough, we get the formula

$$\frac{\overline{f(a + t\bar{x}) - f(a)}}{t} = L(\bar{x}) + \frac{|t|}{t} \varphi(t\bar{x}) \|\bar{x}\| \qquad (3.6)$$

which proves the proposition.

Remark. If f is the mapping considered above we can denote by $f'(a)$ or $\frac{df}{dx}(a)$ the total derivative of f at a. Obviously $f'(a) \in \mathfrak{L}(\bar{E}, \bar{F})$ (see Vol. I, Section 3.2.7).

If $\bar{x} \in \bar{E}$ we denote by $f'(a)(\bar{x})$ or for simplicity's sake $f'(a)\bar{x}$ the value of this mapping for the vector $\bar{x}$. We have from proposition 2

$$\overline{D_{\bar{x}} f(a)} = f'(a)\bar{x} \in \bar{F}.$$

Example 1. Let E be a space of finite dimension, let us suppose we chose a frame of reference formed by $l \in E$ and a basis $(\bar{e}_j)_{j \in \mathfrak{J}}$ then the derivative $f(a) \in \mathfrak{L}(\bar{E}, \bar{F})$ is linked with the partial derivatives $f'(a)(\bar{e}_j) = \frac{\overline{\partial f}}{\partial x_j}(a)$ by the formula

$$f'(a) = f'(x)\left(\sum_{j \in \mathfrak{J}} x_j \bar{e}_j\right) = \sum_{j \in \mathfrak{J}} x_j f'(x)(\bar{e}_j) = \sum_{j \in \mathfrak{J}} x_j \frac{\overline{\partial f}}{\partial x_j}(x). \qquad (3.7)$$

Let us suppose that E and F are of finite dimension and let l, $(\bar{e}_j)_{j \in \mathfrak{J}}$ and b, $(\bar{f}_i)_{i \in I}$ be the frames of reference in E and F respectively. In this case every point in E and F is completely determined by its coordinates and the mapping $f: \Omega \to F$ can be defined by a system of m functions of n variables, namely $y_i = F_i(x_j)_{j \in \mathfrak{J}}$.

If $I = \{1, 2, \ldots, m\}$ and $\mathfrak{J} = \{1, 2, \ldots, n\}$ then $y_i = F_i(x_1, \ldots, x_n)$ for $i = 1, \ldots, m$.

Let us see how the total derivative is defined, if it exists, at the point x. Let $\bar{X} = \sum_{j \in \mathfrak{J}} X_j \bar{e}_j \in \bar{E}$ and let $\bar{Y} = \sum_{i \in I} Y_i \bar{e}_i$ be its image through the total derivative. Then

$$\sum_{i \in I} Y_i \bar{f}_i = f'(x)(\bar{X}) = \sum_{j \in \mathfrak{J}} \frac{\overline{\partial f}}{\partial x_j}(x) X_j = \sum_{i,\, j} \frac{\partial F_i}{\partial x_j}(x) X_j \bar{f}_i,$$

where

$$Y_i = \sum_{j \in \mathfrak{J}} \frac{\partial F_i}{\partial x_j}(x)\, X_j, \qquad i \in I.$$

We obtained above the following proposition.

Proposition. *If E and F are of finite dimension and if $f: \Omega \to F$ then the matrix of the total derivative $f'(x)$ with respect to the considered frame of reference is what we called in Section 3.2.3 the derivative matrix.*

3.2.5 The case where F or E are products of affine spaces

(1) Let us first consider the case where $F = F_1 \times F_2 \times \ldots \times F_m$ where $F_1, \ldots, F_m$ are normed affine spaces $f: \Omega \to F$ and Ω is an open set in E. Then f is defined by the mappings

$$f_i: E \to F_i, \; i = 1, \ldots, m.$$

We shall prove the following proposition.

Proposition 1. *The necessary and sufficient condition for f to be differentiable at the point $a \in \Omega$ is that f_i be differentiable at a. In this case $f'(a)$ is the linear continuous mapping of $\overline{E}$ in $\overline{F} = \overline{F}_1 \times \overline{F}_2 \times \ldots \times \overline{F}_m$ defined by the linear continuous mappings $f_i'(a)$ of $\overline{E}$ in $\overline{F}_i$. In other words the components of the derivatives are equal to the derivatives of the components and*

$$f'(a) = (f_1'(a), f_2'(a), \ldots, f_m'(a)); \; \overline{\mathrm{d}f} = (\overline{\mathrm{d}f_1}, \overline{\mathrm{d}f_2}, \ldots, \overline{\mathrm{d}f_m}).$$

Indeed let us suppose that f_i are differentiable at a, then if we take $y_i = f_i(x)$ we have

$$\Delta y_i = f_i'(a)\,\overline{\mathrm{d}x} + \bar{\alpha}_i \|\overline{\mathrm{d}x}\|$$

where $\bar{\alpha}_i \to \bar{0}$ at the same time as $\overline{\mathrm{d}x}$ tends to $\bar{0}$. We have

$$\overline{\Delta(y_1, y_2, \ldots, y_m)} = (\overline{\Delta y_1}, \ldots, \overline{\Delta y_m}) = (f_1'(a)\,\overline{\mathrm{d}x}, f_2'(a)\,\overline{\mathrm{d}x}, \ldots, f_m'(a)\,\overline{\mathrm{d}x}) +$$
$$+ (\bar{\alpha}_1 \|\overline{\mathrm{d}x}\|, \bar{\alpha}_2 \|\overline{\mathrm{d}x}\|, \ldots, \bar{\alpha}_m \|\overline{\mathrm{d}x}\|) =$$
$$= (f_1'(a), f_2'(a), \ldots, f_m'(a))\,\overline{\mathrm{d}x} + (\bar{\alpha}_1, \bar{\alpha}_2, \ldots, \bar{\alpha}_m) \|\overline{\mathrm{d}x}\|,$$

where $(\bar{\alpha}_1, \bar{\alpha}_2, \ldots, \bar{\alpha}_m) \to \bar{0}$ in $\overline{F}_1 \times \overline{F}_2 \times \ldots \times \overline{F}_m$ as $\overline{\mathrm{d}x} \to 0$. It follows that f is differentiable at a and that $f'(a) = (f_1'(a), f_2'(a), \ldots, f_m'(a))$.

The converse is proved in the same way.

(2) Let us now suppose that $E = E_1 \times E_2$ and let $f\colon \Omega \to F$ where $\Omega \subset E$. Then we can write $y = f(x_1, x_2)$, $(x_1 \in E_1, x_2 \in E_2)$.

If $x_1 = a_1 \in E_1$ is fixed and we consider the mapping $f_{a_1} : x_2 \to f(a_1, x_2)$ then if this mapping has a Fréchet derivative at a_2, it will be a mapping from E_2 in F.

Definition. *The above derivative is called a partial derivative or a partial differential mapping with respect to x_2 of f at the point (a_1, a_2) and will be denoted by $\partial_2 f(a_1, a_2)$ or $\frac{\partial f}{\partial x_2}(a_1, a_2)$.*

Proposition 2. *If $E = E_1 \times E_2$ and the total derivative of f at the point $a = (a_1, a_2) \in \Omega$ is $f'(a_1, a_2)$ then f has partial derivative mappings at this point and we can write*

$$f'(a_1, a_2)(\bar{x}_1, \bar{x}_2) = \frac{\partial f}{\partial x_1}(a_1, a_2)\,\bar{x}_1 + \frac{\partial f}{\partial x_2}(a_1, a_2)\,\bar{x}_2. \tag{3.8}$$

The proof of this proposition is immediate and we leave it to the reader.

3.2.6 Continuously differentiable functions. Spaces of differentiable functions

Definition 1. *If f has a derivative $f'(x)$ at every point $x \in \Omega$, then we say that f is differentiable in Ω. The mapping $f'\colon x \to f'(x)$ is a mapping of Ω in $\mathfrak{L}(\bar{E}, \bar{F})$.*

Definition 2. *We say that the function f is continuously differentiable or of class C^1 if the mapping $f'\colon \Omega \to \mathfrak{L}(\bar{E}, \bar{F})$ is continuous.*

We shall give a proposition whose proof is found in Ref. 50.

Proposition 3. *If f is a continuously differentiable mapping of Ω in F, then the mapping $(x, \bar{X}) \to f'(x)\,\bar{X}$ of $\Omega \times \bar{E}$ in $\bar{F}$ is continuous.*

Conversely, if this mapping is continuous and E is of finite dimension it follows that f is of class C^1.

Notation. We shall denote by $\bar{F}^E$ the set of the mappings of a set E in the vector space $\bar{F}$. It is easily seen that $\bar{F}^E$ is a vector space. Let us denote by $(\bar{F}^E)_{cb}$ the set of the continuous and bounded mappings of E in $\bar{F}$ where E is a topological space and $\bar{F}$ a vector space.

If F is a normed affine space, then $(F^E)_{cb}$ is a normed affine space whose associate vector space is $(\overline{F}^E)_{cb}$.

Let E, F be two normed affine spaces, $\Omega \subset E$ an open set and $(F^\Omega)_{b;1}$ ($(F^\Omega)_{cb;1}$ respectively) the space of the functions defined on Ω with values in F, differentiable, bounded and with a bounded derivative (continuously differentiable, bounded and with a bounded derivative respectively).

Proposition 4. *The spaces $(F^\Omega)_{b;1}$, $(F^\Omega)_{cb;1}$ are affine spaces of associate vector spaces $(\overline{F}^\Omega)_{b;1}$ and $(\overline{F}^\Omega)_{cb;1}$ respectively. They become normed affine spaces if we take in the vector spaces the associate norm*

$$|||\overline{f}|||_1 = \max_{x\in\Omega} (\|\overline{f}(x)\|, \|f'(x)\|) = \max (|||\overline{f}|||, |||f'|||).$$

Remark. $\|\overline{f}(x)\|$ is the norm in $\overline{F}$ and $\|f'(x)\|$ the norm in $\mathfrak{L}(\overline{E}, \overline{F})$.

Proposition. *The spaces $(F^\Omega)_{b;1}$ and $(F^\Omega)_{cb;1}$ are complete if F is complete.*

We do not give the proof of this proposition here. It can be found in Ref. 50.

Remark. A sequence f_n converges to f in $(F^\Omega)_{b;1}$ if f_n converges uniformly to f and f_n' converges uniformly to f'.

Let us denote by $E = (F^{[a,b]})_{cd}$ the set of the continuously differentiable functions defined on $[a, b]$ with values in F. One can easily prove the following proposition.

Proposition 5. *$(F^{[a,b]})_{cd}$ is a normed affine space, $(\overline{F}^{[a,b]})_{cd}$ is an associate vector space and the distance between two elements $f, g \in E$ is given by the formula*

$$|||\overline{f-g}|||_1 = \sup_{a \leqslant x \leqslant b} (\|\overline{f(x)-g(x)}\|, \|\overline{f'(x)-g'(x)}\|).$$

We shall now give without proof a proposition concerning the differentiation of the composed functions.

Proposition 6. *Let E, F, G be normed affine spaces, Ω, Ω' open sets in E and F respectively, $f: \Omega \to \Omega'$ and $g: \Omega' \to G$. If f admits a derivative $f'(a) \in \mathfrak{L}(\overline{E}, \overline{F})$ at a point $a \in \Omega$ and g admits a derivative $g'(b) \in \mathfrak{L}(\overline{F}, \overline{G})$ at the point $b = f(a) \in \Omega'$, then the composed mapping $h = g \circ f$ admits a derivative at point a and*

$$h'(a) = g'(b) \circ f'(a) = g'(f(a)) \circ f'(a). \tag{3.9}$$

Proposition 7. *If $\bar{u}_1$ and $\bar{u}_2$ are differentiable mappings (respectively continuously differentiable mappings) of an open set Ω from a normed affine space E into a normed affine space $\overline{F}_1$ and $\overline{F}_2$ respectively and if $B: \overline{F}_1 \times \overline{F}_2 \to \overline{G}$ is a continuous bilinear mapping, then the mapping $B(\bar{u}_1, \bar{u}_2): x \to$ $\to B(u_1(x), u_2(x))$ of E in $\overline{G}$ is differentiable (respectively continuously differentiable) and its Fréchet derivative is given by the formula*

$$(B(\bar{u}_1, \bar{u}_2))'(a)\overline{X} = B(u_1'(a)\overline{X}, \bar{u}_2(a) + B(\bar{u}_1(a), u_2'(a)\overline{X})$$

or written differently

$$D_{\overline{X}}\, B(\bar{u}_1, \bar{u}_2)\ (a) = B(\overline{D_{\overline{X}}\, \bar{u}_1\ (a)}, \bar{u}_2(a)) + B(\bar{u}_1(a), \overline{D_{\overline{X}}\ \bar{u}(a)}),$$

where $\overline{X} \in \overline{E}$.

From the above proposition it is seen that the classical formula of a product of functions $(uv)' = u'v + uv'$ is a particular case of the derivative of a bilinear continuous mapping. The proof of proposition 7 is found in Ref. 50.

3.2.7 The mean-value formula

Lemma. *Let $f: [0, 1] \to F$ where F is a normed affine space and $g: [0, 1] \to$ $\to R$. If f and g are continuous on* [0, 1] *and differentiable on the open interval* (0, 1) *and if*

$$\| f'(x) \| \leqslant g'(x), \text{ for } 0 < x < 1,$$

then

$$\| \overline{f(1) - f(0)} \| \leqslant g(1) - g(0). \tag{3.10}$$

Indeed let $\varepsilon > 0$ and let A_ε be the set of the points $x \in [0, 1]$ so that

$$\| \overline{f(x) - f(0)} \| \leqslant g(x) - g(0) + \varepsilon x + \varepsilon. \tag{3.11}$$

The function

$$x \to \| \overline{f(x) - f(0)} \| - g(x) + g(0) - \varepsilon x - \varepsilon \tag{3.12}$$

is continuous and A_ε is exactly the set where this function is $\leqslant 0$, hence A_ε is a closed set. In particular A_ε contains its upper bound which is a maximum β. Let us notice that β cannot be 0. Indeed as f and g are continuous we have $\|\overline{f(x)-f(0)}\| \leqslant \varepsilon$ for x small enough, hence formula (3.11). Thus A_ε contains a neighbourhood of 0.

Let us now show that $\beta \notin (0, 1)$. Indeed if $0 < \beta < 1$ there would exist $\delta > 0$ so that

$$\|\overline{f(\beta+\delta)-f(\beta)}\| \leqslant \|\overline{f'(\beta)}\|\,\delta + \frac{\varepsilon}{2}\,\delta \leqslant g'(\beta)\,\delta + \frac{\varepsilon}{2}\,\delta,$$

$$g(\beta+\delta) - g(\beta) \geqslant g'(\beta)\,\delta - \frac{\varepsilon}{2}\,\delta$$

hence we would have

$$\|\overline{f(\beta+\delta)-f(\beta)}\| \leqslant g(\beta+\delta) - g(\beta) + \varepsilon\delta \tag{3.13}$$

but since $\beta \in A_\varepsilon$ we have

$$\|\overline{f(\beta)-f(0)}\| \leqslant g(\beta) - g(0) + \varepsilon\beta + \varepsilon. \tag{3.14}$$

Taking the sum of (3.13) and (3.14) we obtain

$$\|\overline{f(\beta+\delta)-f(0)}\| \leqslant g(\beta+\delta) - g(0) + \varepsilon(\beta+\delta) + \varepsilon. \tag{3.15}$$

From formula (3.15) we have $\beta + \delta \in A_\varepsilon$ which is absurd since β is the maximum of A_ε. Consequently $\beta = 1$ and formula (3.14) becomes

$$\|\overline{f(1)-f(0)}\| \leqslant g(1) - g(0) + 2\varepsilon$$

hence, as ε is arbitrary, we have formula (3.10).

We can now prove the following mean-value theorem.

Theorem. *Let f be a continuous mapping of an open set Ω of a normed affine space E into a normed affine space F. Then, if $[x, x+\overline{h}] \subset \Omega$ and f has a Fréchet derivative at every point of the open interval $(x, x+\overline{h})$ with the norm $\leqslant M$, it follows that*

$$\|\overline{f(x+\overline{h})-f(x)}\| \leqslant M\,\|\overline{h}\|. \tag{3.16}$$

Indeed, let $\Phi : t \to f(x + t\overline{h})$ be the mapping of the interval [0, 1] into F. Its derivative is

$$\Phi'(t) = f'(x + t\overline{h})\,(\overline{h}).$$

From the hypothesis we have

$$\|\overline{\Phi'(t)}\| \leqslant M\,\|\overline{h}\|.$$

If we consider the function $f: t \to M \| \overline{h} \|$, we find ourselves in the conditions of the lemma where Φ stands for the function f of the lemma. Then applying the lemma, we have formula (3.16).

Corollary. *In the conditions of the theorem, if L is a continuous linear mapping of $\overline{E}$ in $\overline{F}$ then*

$$\| \overline{f(x+\overline{h}) - f(x) - L\overline{h}} \| \leqslant \omega \| \overline{h} \|$$

where $\omega = \sup_{\xi \in (x, x+h)} \| f'(\xi) - L \|$. *We can take* $L = f'(x)$.

Indeed we can apply the mean-value theorem to the function $\xi \to f(\xi) - L(\overline{\xi - x})$ whose derivative at ξ is $f'(\xi) - L$.

3.3 MAXIMA AND MINIMA OF A REAL FUNCTION DEFINED IN AN AFFINE SPACE

3.3.1 Definitions. The necessary conditions of extremum

Let E be a topological space and $f: E \to R$.

Definition. *We say that f admits a relative maximum at a point* $a \in E$ *if there is a neighbourhood* $\mathcal{V} \ni a$ *so that for every* $x \in \mathcal{V}$, $f(x) \leqslant f(a)$. *The relative maximum is said to be strict if for every* $x \in \mathcal{V}$, $x \neq a$, $f(x) < f(a)$. *By analogy we define a relative minimum.*

Theorem. *Let f be a real function defined on an open set* Ω *of an affine space E. If f is differentiable in* Ω *then a necessary condition for the function to admit at a point* $a \in \Omega$ *a relative extremum (that is a relative maximum or a relative minimum) is that the mapping* $f'(a) \in \mathcal{L}(\overline{E}, R)$ *be null.*

Indeed if f admits a relative maximum or minimum at a, the function $t \to f(a + t\overline{x})$ with $\overline{x} \in \overline{E}$ fixed and t sufficiently small such that $a + t\overline{x}$ belongs to Ω, admits at the point $t = 0$ a relative maximum or minimum. It then follows that its derivative with respect to t is null in $t = 0$. But this derivative is $f'(a)(\overline{x})$. Hence

$$f'(a)(\overline{x}) = 0 \text{ for every } \overline{x} \in \overline{E} \text{ that is}$$

$$f'(a) = 0.$$

Particular case. If E is an affine space of finite dimension and $0, (\bar{e}_i)_{i \in I}$ a frame of reference in E, the condition of the theorem comes to the well-known theorem in classical analysis Ref. 50.

The necessary condition for a to be a relative point of extremum of the function $f\colon E \to R$ is that its partial derivatives $\dfrac{\partial f}{\partial x_i}$ be all null, at the point a or that its differential at the point a be identical to null that is

$$\sum_{i \in I} \frac{\partial f}{\partial x_i}(a)\, \mathrm{d}x_i = 0.$$

3.3.2 Fixed point theorems

We shall now give some results which we shall use for the study of the implicit functions.

Definition. Let E be a metric space [Vol. 1, Section 3.3.1] *and $f\colon E \to E$. We say that f is a contraction if there is a positive constant $k < 1$ so that for every $x, y \in E$*

$$d(f(x), f(y)) \leqslant k\, d(x, y).$$

Definition. We say that a is a fixed point of a mapping f if $f(a) = a$.

Theorem 1. *Every contraction of a complete metric space in itself admits a unique fixed point.*

Let us first prove the existence of a fixed point. We shall use the familiar method of the succesive approximations.

We take $x_0 \in E$ and we put

$$x_1 = f(x_0),\ x_2 = f(x_1),\ \ldots,\ x_n = f(x_{n-1}),\ \ldots$$

Let us show that $\{x_n\}$ is a Cauchy sequence. Since f is a contraction it is easily seen that it is a continuous function. Further, we obtain the following sequence of inequalities

$$d(x_2, x_1) \leqslant kd(x_1, x_0),$$
$$d(x_3, x_2) \leqslant kd(x_2, x_1) \leqslant k^2 d(x_1, x_0),$$
$$\cdots\cdots\cdots\cdots\cdots\cdots\cdots\cdots$$
$$d(x_{n+1}, x_n) \leqslant k^n d(x_1, x_0).$$

So

$$d(x_{n+p}, x_n) \leqslant d(x_{n+p}, x_{n+p-1}) + d(x_{n+p-1}, x_{n+p-2}) + \dots$$
$$\dots + d(x_{n+1}, x_n) \leqslant (k^{n-1} + k^{n-2} + \dots + k + 1)\, d(x_1, x_0) \leqslant$$
$$\leqslant \frac{k^n}{1-k}\, d(x_1, x_0).$$

Hence $d(x_{n+p}, x_n) \to 0$ as $n \to \infty$ i.e. $\{x_n\}$ is a Cauchy sequence. Since E is complete it follows that the sequence has a limit a. Since f is a continuous function, from $x_{n+1} = f(x_n)$ we have $a = f(a)$ as $n \to \infty$.

Let us now prove the unicity of the fixed point. Let us suppose a and b to be two fixed points. Then

$$d(a, b) \leqslant kd(a, b) < d(a, b) \text{ and } d(a, b) \neq 0.$$

Hence $d(a, b) = 0$ that is $a = b$.

Let us now suppose that the contraction f depends on a parameter λ. Then it results from the above theorem that for every value of λ, f has a fixed point a_λ. We shall examine in what way a_λ depends on the parameter λ.

Theorem 2. *Let E be a complete metric space, Λ a topological space and f: $E \times \Lambda \to E$. Let us suppose that the mapping $\lambda \to f(x, \lambda)$ of $\Lambda \to E$ is continuous for every point $x \in E$ and that the mapping $f_\lambda : x \to f(x, \lambda)$ of E in E is a contraction for every $\lambda \in \Lambda$, that is for every $x, y \in E$*

$$d(f_\lambda(x), f_\lambda(y)) \leqslant kd(x, y)$$

and k does not depend on λ. In this case the mapping $\lambda \to a_\lambda$ of Λ in E is continuous, where a_λ is the fixed point corresponding to the mapping f_λ.

Indeed, given $\lambda_0 \in \Lambda$ and $\varepsilon > 0$, we have the following sequence of inequalities

$$d(a_\lambda, a_{\lambda_0}) = d(f_\lambda(a_\lambda), f_{\lambda_0}(a_{\lambda_0})) \leqslant d(f_\lambda(a_\lambda), f_\lambda(a_{\lambda_0})) + d(f_\lambda(a_{\lambda_0}), f_{\lambda_0}(a_{\lambda_0})) \leqslant$$
$$\leqslant kd(a_\lambda, a_{\lambda_0}) + d(f(a_{\lambda_0}, \lambda),\ f(a_{\lambda_0}, \lambda_0)) \tag{3.17}$$

hence we deduce

$$(1-k)\, d(a_\lambda, a_{\lambda_0}) \leqslant d(f(a_{\lambda_0}, \lambda),\ f(a_{\lambda_0}, \lambda_0)).$$

From the continuity of f with respect to λ it follows that there is a neighbourhood $\mathscr{V} \ni \lambda_0$ so that, as soon as $\lambda \in \mathscr{V}$,

$$d(f(a_{\lambda_0}, \lambda),\ f(a_{\lambda_0}, \lambda_0)) \leqslant \varepsilon\,(1-k).$$

Taking this inequality into account in formula (3.17) we obtain

$$d(a_\lambda, a_{\lambda_0}) \leqslant \varepsilon$$

which proves the continuity of the mapping $\lambda \to a_\lambda$ at the point λ_0.

3.3.3 Implicit functions

Let E, F be two sets and $f: E \to F$. The set of the points $x \in E$ so that $f(x) = b$ is the reciprocal image $f^{-1}(\{b\})$. To determine this set means to solve the equation $f(x) = b$. If f: $E \times F \to G$ and $c \in G$ let us consider the equation

$$f(x, y) = c. \tag{3.18}$$

For every given x, the equation in y, $f(x, y) = c$ may have one and only one solution.

Definition. We say that equation (3.18) *defines* y *as a function of* x, *and we write* $y = g(x)$. *This function is called the implicit function defined by equation* (3.18) *and characterized by the following property*

$$f(x, g(x)) = c. \tag{3.18'}$$

We shall study the following particular case.

Theorem 1. *Let E be a topological space F and G normed affine spaces $\Omega \subset E \times F$ an open set and $a, b \in \Omega$. Let f: $\Omega \to G$ and $f(a, b) = c$. Let us suppose that for every fixed x, f admits a partial derivative $\frac{\partial f}{\partial y}(x, y) \in \mathfrak{L}(\overline{F}, \overline{G})$ and that the function $\frac{\partial f}{\partial y}: \Omega \to \mathfrak{L}(\overline{F}, \overline{G})$ is continuous. Let us also suppose that $Q = \frac{\partial f}{\partial y}(a, b)$ is an invertible mapping of $\overline{F}$ on $\overline{G}$, that is a one to one mapping, whose inverse Q^{-1} is linear and continuous. In this case if F is complete, there exist the open sets $A \subset E$ and $B \subset F$ with $a \in A$ and $b \in B$, so that for every $x \in A$ the equation* (3.18) *in y, should have one, and only one, solution in B. This solution is a function of x and the function $y = g(x)$, thus defined, is a continuous function from A in B.*

Proof. Let us denote by $\overline{\Omega}_1 = \{(x, \overline{y-b}),$ with $(x, y) \in \Omega\}$ and let $\overline{f}_1 : \overline{\Omega}_1 \to \overline{G}$ be defined by

$$\overline{f_1(x, \overline{Y})} = \overline{f(x, b + \overline{Y})} - \overline{c}. \tag{3.19}$$

Equation (3.18) is equivalent with

$$\overline{f_1(x, \overline{Y})} = \overline{0} \tag{3.20}$$

in the neighbourhood of the particular solution $(a, \overline{0})$.

Let us consider a mapping $f_2 : \overline{\Omega}_1 \to F$ defined thus

$$\overline{Q^{-1}} \circ \overline{f}_1 : \quad (x, \overline{Y}) \to Q^{-1}(f_1(x, \overline{Y})). \tag{3.21}$$

From proposition 6, Section 3.2.6 it follows at once that the partial derivative of the function f_2 with respect to the second variable is given by the formula:

$$\frac{\partial f_2}{\partial \overline{Y}} (a, \overline{0}) = Q^{-1} \circ Q = I \in \mathfrak{L}\ (F, \ F), \tag{3.22}$$

where I is the identical mapping.

Taking the image of (3.20) through Q^{-1} we obtain

$$f_2(x, \overline{Y}) = \overline{0} \tag{3.23}$$

and taking the image of equation (3.23) through Q we obtain equation (3.20). Consequently equation (3.23) is equivalent to equation (3.18).

We now define the mapping $\Phi : \overline{\Omega}_1 \to \overline{F}$ by

$$\overline{\Phi}(\lambda, \overline{Y}) = \overline{Y} - f_2(\lambda, \overline{Y}) \tag{3.24}$$

and we notice that its partial derivative with respect to the second variable at the origin is equal to $I - I = 0$. But equation (3.18) is equivalent to the equation

$$\overline{\Phi}(\lambda, \overline{Y}) = \overline{Y}, \quad \lambda = x, \quad \overline{Y} = \overline{y - b}. \tag{3.25}$$

Thus we reduced the problem to the solving of equation (3.25) in the neighbourhood of $\lambda = a$, $\overline{Y} = \overline{0}$.

Let A_1 and $\overline{B}_1$ be the neighbourhoods of a and $\overline{0}$ in E and $\overline{F}$ respectively so that $A_1 \times \overline{B}_1 \in \overline{\Omega}_1$. The spaces E and Λ from theorem 2, Section 3.3.2

are in our case replaced by $\overline{B}_1$ and A_1. The function $\overline{\Phi}$ is continuous on $A_1 \times \overline{B}$ and hence it is separately continuous. The mapping $\overline{\Phi}$ must also fulfil the following conditions to enable us to apply theorem 2, Section 3.2.7:

(a) It should map $\overline{A}_1 \times \overline{B}_1$ in $\overline{B}_1$. But the mapping Φ maps $A_1 \times \overline{B}_1$ in $\overline{F}$ and not necessarily in $\overline{B}_1$.

Consequently we shall have to consider some neighbourhoods $A_2 \ni e$ and $\overline{B}_2 \ni \overline{0}$ so that $A_2 \subset A_1$ and $\overline{B}_2 \subset \overline{B}_1$ with the property that tha restriction of Φ to $A_2 \times \overline{B}_2$ should map $A_2 \times \overline{B}_2$ in $\overline{B}_2$.

(b) It is necessary that for λ fixe din A_2, the mapping $\overline{\Phi}_\lambda : \overline{Y} \to \Phi(\lambda, \overline{Y})$ of $\overline{B}_2 \to B_2$ be a contraction and $k < 1$ be independent of $\lambda \in A_2$.

(c) The metric space B_2 must be complete.

Let us first consider condition (b). We saw that $\dfrac{\partial \overline{\Phi}}{\partial \overline{Y}}$ is continuous and null at $(a, \overline{0})$. Hence we can find a neighbourhood $a \in A_2'$ with $A_2' \subset A_1$ and a closed ball $\overline{B}_2$ with the centre in the origin and the radius $\beta > 0$ in B_1 so that if $\lambda \subset A_2'$ and $\overline{Y} \in \overline{B}_2$ then

$$\left\| \frac{\partial \overline{\Phi}}{\partial \overline{Y}}(\lambda, \overline{Y}) \right\| \leqslant \frac{1}{2}\cdot \tag{3.26}$$

Since the closed ball $\overline{B}_2$ is convex and hence if $\overline{Y}', \overline{Y}'' \in \overline{B}_2$ then the segment which joins $\overline{Y}'$ and $\overline{Y}''$ belongs to $\overline{B}_2$ and we can apply the mean-value theorem (Section 3.2.7) to the function $\overline{\Phi}$ and

$$\| \overline{\Phi}(\lambda, \overline{Y}') - \overline{\Phi}(\lambda, \overline{Y}'') \| \leqslant \frac{1}{2} \| \overline{Y}' - \overline{Y}'' \| \cdot \tag{3.27}$$

Consequently $\overline{\Phi}$ is a contraction and $k = \dfrac{1}{2}\cdot$

Let us pass now to condition (a). As the neighbourhoods A' and $\overline{B}_2$ are determined and taking into consideration that $\overline{\Phi}$ is continuous and $\overline{\Phi}(a, \overline{0}) = \overline{0}$ we can find a neighbourhood $A_2 \subset A_2'$ of a so that

$$\| \overline{\Phi}(\lambda, \overline{0}) \| \leqslant \frac{\beta}{2} \quad \text{for } \lambda \in A_2. \tag{3.28}$$

Hence for $\lambda \in A_2$ and $\|\overline{Y}\| \leqslant \beta$ we have

$$\|\overline{\Phi}(\lambda, \overline{Y})\| \leqslant \|\overline{\Phi}(\lambda, \overline{0})\| + \|\overline{\Phi}(\lambda, \overline{Y}) - \overline{\Phi}(\lambda, \overline{0}\| \leqslant \frac{\beta}{2} + \frac{1}{2}\|\overline{Y}\| \leqslant \beta \quad (3.29)$$

hence we have $\overline{\Phi}\colon A_2 \times \overline{B}_2 \to \overline{B}_2$.

Finally condition (c) is satisfied since $\overline{B}_2$ is a closed subset of a complete metric space, hence it is a complete metric space (Ref. 50). Consequently we can apply theorem 2, Section 3.3.2. Thus there is one and only one solution in $\overline{Y}$ of equation (3.25) so that $\overline{Y} = \varphi(\lambda)$ is a continuous function of λ and $\|\overline{Y}\| \leqslant \beta$. Consequently equation (3.18) admits one and only one solution y in the ball $\overline{B}_2$ for every $x \in A_2$. Let us denote by $g\colon A_2 \to \overline{B}_2$ the mapping $y = g(x)$ then $g(x) = b + \overline{\varphi}(x)$. Consequently g is a continuous mapping from A_2 in $\overline{B}_2$ and hence writing $B = \overset{\circ}{B}_2$ there is an open set $A \ni a$, from E so that $x \in A$ should imply $g(x) \in B$ and the theorem is completely proved.

We shall now suppose that the function f is differentiable and try to determine the conditions in which the function g defined by f is differentiable.

Theorem 2. *Let E, F and G be normed affine spaces, Ω an open set from $E \times F$ and $f\colon \Omega \to G$. Let A and B be open sets from E and F respectively so that $A \times B \subset \Omega$ and $g\colon A \to B$ so that $f(x, g(x)) = c$. Let $a \in A$ and $b = g(a)$. Then if the following conditions are fulfilled:*

(1) *The mapping f is differentiable at the point (a, b);*

(2) *Its partial derivatives P and Q at the point (a, b) are continuous linear mappings from $\overline{E}$ and $\overline{F}$ respectively in $\overline{G}$;*

(3) *Q is invertible;*

(4) *g is continuous at a.*

g is differentiable at a and its derivative is given by the formula

$$g'(a) = Q^{-1} \circ P = -\left(\frac{\partial f}{\partial y}(a, b)\right)^{-1} \circ \left(\frac{\partial f}{\partial x}(a, b)\right). \qquad (3.30)$$

Let us first show that g is differentiable at the point a.

Let $\overline{dx} = \overline{x - a}$, $\overline{\Delta y} = \overline{g(x) - g(a)}$. If we consider that g is an implicit function defined by equation (3.18) and that f satisfies condition (1), we have

$$\overline{0} = \overline{\Delta f} = -P\,\overline{dx} + Q\overline{\Delta y} + \overline{\alpha}\,(\|\overline{dx}\| + \|\overline{\Delta y}\|), \qquad (3.31)$$

where $\bar{\alpha} \to \bar{0}$ as $\overline{dx}$ and $\overline{\Delta y} \to \bar{0}$.* We have

$$Q\overline{\Delta y} = -P\,\overline{dx} - \bar{\alpha}(\|\overline{dx}\| + \|\overline{\Delta y}\|)$$

or taking condition (3) into account

$$\overline{\Delta y} = -(Q^{-1} \circ P)\,\overline{dx} - (Q^{-1}\alpha)(\|\overline{dx}\| + \|\overline{\Delta y}\|). \tag{3.32}$$

We deduce that

$$\|\overline{\Delta y}\| \leqslant \|Q^{-1}\|\,\|P\|\,\|\overline{dx}\| + \|Q^{-1}\|\,\bar{\alpha}\,\|\overline{dx}\| + \|Q^{-1}\|\,\|\bar{\alpha}\|\,\|\overline{\Delta y}\|. \tag{3.33}$$

Since $\bar{\alpha} \to \bar{0}$ when $\overline{dx} \to \bar{0}$, it follows that we can choose $\overline{dx}$ small enough such that

$$\|Q^{-1}\|\,\|\bar{\alpha}\| < \frac{1}{2}.$$

Using this majoration in formula (3.33) and putting in the left-hand side the term which contains $\overline{\Delta y}$ we have

$$\frac{1}{2}\|\overline{\Delta y}\| \leqslant \left(\|Q^{-1}\|\quad\|P\| + \frac{1}{2}\right)\|\overline{dx}\|$$

or

$$\|\overline{\Delta y}\| \leqslant (2\|Q^{-1}\|\,\|P\| + 1)\,\|\overline{dx}\|$$

hence

$$(\|\overline{dx}\| + \|\overline{\Delta y}\|) \leqslant (2\|Q^{-1}\|\,\|P\| + 2)\,\|\overline{dx}\| = k\,\|\overline{dx}\| \tag{3.34}$$

where k is a fixed constant.

Taking formula (3.34) into account we obtain from formula (3.32)

$$\overline{\Delta y} = -(Q^{-1} \circ P)\,\overline{dx} + \bar{h}\,\|\overline{dx}\| \tag{3.35}$$

with $\|\bar{h}\| \leqslant k\,\|Q^{-1}\|\,\|\bar{\alpha}\|$, hence $\bar{h} \to \bar{0}$ as $\overline{dx} \to \bar{0}$ and thus we proved

*) Let us notice that if $\overline{dx} \to \bar{0}$ then $\overline{\Delta y} \to 0$ from condition (4).

the differentiability of the function g at the point a. Let us now apply proposition 6, Section 3.2.6 concerning the differentiation of the composed functions to expression (3.24). We obtain

$$\frac{\partial f}{\partial x}(a,\ b) + \frac{\partial f}{\partial y}(a,\ b) \circ g'(a) = 0 \in \mathfrak{L}(\overline{E},\ \overline{G}) \tag{3.36}$$

or

$$P + Q \circ g'(a) = 0. \tag{3.37}$$

Composing to the left with Q^{-1} relationship (3.36) or (3.37) we obtain relationship (3.30) and thus the theorem is completely proved.

Definition. *We shall call an element* $v \in \mathfrak{L}(\overline{G},\ \overline{F})$ *with the property* $u \circ v = I$ *an inverse to the right of* $u \in \mathfrak{L}(\overline{F},\ \overline{G})$.

Definition. *An element* $v \in \mathfrak{L}(\overline{G},\ \overline{F})$ *with the property* $v \circ u = I$ *is called inverse to the left of an element* $u \in \mathfrak{L}(\overline{F},\ \overline{G})$.

Remark. Such elements do not exist always, for instance, if $u = 0$.

Remark. If u is invertible then $v = u^{-1}$. This follows from $uv = I$ (or $vu = I$ respectively), composing to the left (or right respectively) with u^{-1}.

Theorem 3. *Let* $\overline{F}, \overline{G}$ *be Banach spaces* (Vol. 1, Section 3.3.3) and U (*and* U^{-1} *respectively*) *the set of invertible elements of* $\mathfrak{L}(\overline{F},\ \overline{G})$ (*and* $\mathfrak{L}(\overline{G},\ \overline{F})$ *respectively*). In that case:

(a) U, U^{-1} are open sets.
(b) the one-to-one mapping $u \to u^{-1}$ of U on U^{-1} is a homeomorphism.
(c) The mapping $u \to u^{-1}$ and the inverse mapping are differentiable.
(d) The derivative of the mapping $u \to u^{-1}$ is the mapping:

$$du \to -u^{-1} \circ du \circ u^{-1}$$

of $\mathfrak{L}(\overline{F},\ \overline{G})$ in $\mathfrak{L}(\overline{G},\ \overline{F})$.

Since $\overline{F}$ and $\overline{G}$ are complete it follows that $\mathfrak{L}(\overline{F},\ \overline{G})$ and $\mathfrak{L}(\overline{G},\ \overline{F})$ are complete (Vol. 1, Chapter 3, theorem 3).

Let $u_0 \in U$. To look for an inverse to the right, of u, means to solve the equation $uv = I$ in v.

Let us first notice that from proposition 7, Section 3.2.5 it results that the mapping $(u, v) \to u \circ v$ is continuously differentiable as a continuous bilinear mapping. At the point (u_0, v_0), $v_0 = u_0^{-1}$, the partial derivative in v is the continuous linear mapping

$$V \to W = u_0 V \text{ of } \mathfrak{L}(\overline{G}, \overline{F}) \text{ in } \mathfrak{L}(\overline{G}, \overline{G}).$$

This mapping is invertible and its inverse mapping is the continuous mapping:

$$W \to V = u_0^{-1} W \text{ of } \mathfrak{L}(\overline{G}, \overline{G}) \text{ in } \mathfrak{L}(\overline{G}, \overline{F}).$$

From theorem 1 it then follows that there is an open neighbourhood $(U_0)_r \ni u_0$ in $\mathfrak{L}(\overline{F}, \overline{G})$ and an open neighbourhood $(U_0)_r^{-1} \ni u_0^{-1}$ in $\mathfrak{L}(\overline{G}, \overline{F})$ with the property that every $u \in (U_0)_r$ has one, and only one, inverse to the right u_r^{-1} and $u_r^{-1} \ni (U_0^{-1})_r$. Applying theorem 2 it follows that the mapping $u \to u_r^{-1}$ of $(U_0)_r$ in $(U_0^{-1})_r$ is continuous and differentiable at the point u_0. But we can follow the same reasoning in the case of the inverse to the left. We determine the open sets $(U_0)_l$ and $(U_0^{-1})_l$ and the mapping $u \to u_l^{-1}$ of $(U_0)_l$ in $(U_0)_l^{-1}$. Let us then consider the set $U_0 = (U_0)_l \circ (U_0)_l$ which is obviously a neighbourhood of u_0. If $u \in U_0$ then there is $u_l^{-1} \in (U_0)_l^{-1}$ and $u_r^{-1} \in (U_0)_r^{-1}$.

But if an element admits an inverse to the left and an inverse to the right, they coincide and hence the element is invertible. Indeed $uu_r^{-1} = I$ and $u_l^{-1} u = I$ implies that $u_l^{-1} uu_r^{-1} = (u_l^{-1} u) u_r^{-1} = u_r^{-1}$ and $u_l^{-1} (uu_r^{-1}) = u_l^{-1}$. Consequently every element in U_0 is invertible and U is a neighbourhood of each of its points hence U is open. If we exchange $\overline{F}$ and $\overline{G}$ it follows that U^{-1} is open that is we have condition (*a*). The mapping $u \to u^{-1}$ coinciding in U_0 with the mapping $u \to u_r^{-1}$ is continuous at the point u_0. It follows that $u \to u^{-1}$ is continuous in U. If we exchange $\overline{F}$ and $\overline{G}$ it follows that its reciprocal mapping is continuous in U^{-1}. Consequently we also proved (b) and by repeating the reasoning we obtain (c). By differentiating the relation $uv = I$ we have $\mathrm{d}u \circ v + u \circ \mathrm{d}v = 0$, and hence we obtain (d) by composing to the left with u^{-1}.

Corollary. Let E be an affine space. $\overline{F}$, $\overline{G}$ Banach spaces, Ω an open set in E and $x \to u(x)$ a mapping of E in $\mathfrak{L}(\overline{F}, \overline{G})$. If for every $x \in \Omega$, $u(x)$

is invertible then $x \to u^{-1}(x)$ in a differentiable mapping of the set Ω in $\mathfrak{L}(\overline{G}, \overline{F})$ and its derivative is

$$(u^{-1})'(x)\overline{X} = -u^{-1}(x) \circ (u'(x)\overline{X}) \circ u^{-1}(x) \in \mathfrak{L}(\overline{G}, \overline{F}), \tag{3.38}$$

where $\overline{X} \in \overline{E}$.

Particular case. If $\overline{F} = \overline{G} = K$ then we come again across the familiar formula

$$\left(\frac{1}{f}\right)' = -\frac{f'}{f^2}, \quad f \neq 0.$$

Let E, F, G be normed affine spaces, Ω an open set in $E \times F$ and $f\colon \Omega \to G$ a mapping *of* class C^1. We shall prove the following theorems.

Theorem 4. *Let $A \subset E$ and $B \subset F$ be open spaces with the property that $A \times B \subset \Omega$ and let $g\colon A \to B$ be a continuous mapping which verifies equation* (3.18′). *If for every $x \in A$, $\dfrac{\partial f}{\partial y}(x, g(x))$ is an invertible element in $\mathfrak{L}(\overline{F}, \overline{G})$, then g is a function of class C^1.*

Indeed from theorem 2 it follows that g is differentiable at every point in A and that for $x \in A$ we have

$$g'(x) = -\left(\frac{\partial f}{\partial y}(x, g(x))\right)^{-1} \circ \frac{\partial f}{\partial x}(x, g(x)).$$

Since $x \to g(x)$ is continuous on A and $\dfrac{\partial f}{\partial x}, \dfrac{\partial f}{\partial y}$ are continuous on Ω it follows that $x \to \dfrac{\partial f}{\partial x}(x, g(x))$ and $x \to \dfrac{\partial f}{\partial y}(x, g(x))$ are continuous mappings from A in $\mathfrak{L}(\overline{E}, \overline{G})$ and $\mathfrak{L}(\overline{F}, \overline{G})$ respectively. But from theorem 3 the mapping $u \to u^{-1}$ is continuous from U in U^{-1}, consequently the function $x \to \left(\dfrac{\partial f}{\partial y}(x, g(x))\right)^{-1}$ defined on A with values in $\mathfrak{L}(\overline{G}, \overline{F})$ is continuous. Then it follows that the composed function $(u, v) \to v \circ u$ defined in $\mathfrak{L}(\overline{E}, \overline{G}) \times \mathfrak{L}(\overline{G}, \overline{F})$ with values in $\mathfrak{L}(\overline{E}, \overline{F})$ is continuous (the proof is obvious) consequently g' is continuous and hence g is of class C^1.

Theorem 5. *Let $(a, b) \in \Omega$ and $f(a, b) = c$ then, if the following two conditions are fulfilled:*

(1) $\dfrac{\partial f}{\partial y}(a, b)$ *is an invertible element in $\mathfrak{L}(\overline{F}, \overline{G})$,*

(2) *F and G are complete,*
there are two open sets A, B with $a \in A$, $b \in B$ and $A \times B \subset \Omega$, so that for every $x \in A$ the equation (3.18) *in y should have one, and only one, solution in B and the function $y = g(x)$ thus defined should be of class C^1.*

From theorem 3 it follows that the set U of the invertible elements from $\mathfrak{L}(\bar{F}, \bar{G})$ is open. Since the mapping $(x, y) \to \left(\frac{\partial f}{\partial y}(x, y)\right)$ from Ω in $\mathfrak{L}(\bar{F}, \bar{G})$ is continuous, it follows that the reciprocal image of u through this mapping is an open set $\Omega_1 \subset \Omega$ which contains (a, b).

If we apply theorem 1 to the restriction of f on Ω_1 then it follows that we can determine the open sets $A \ni a$, $B \ni b$ with $A \times B \subset \Omega_1$ so that for every $x \in A$ there should be one and only one element $y \in B$ solution of equation (3.18). The function $y = g(x)$ thus defined is continuous. But $(x, g(x)) \in \in A \times B \subset \Omega_1$ for every $x \in A$, hence applying theorem 4 it follows that g is of class C^1.

Corollary 1. *In the conditions of theorem* 5, *the mapping $u \to u^{-1}$ of the open set U on the open set U^{-1} and its inverse mapping are of class C^1.*

We apply theorem 4 to equation $uv = I$.

Corollary 2. *In the conditions of theorem* 5, *the derivative of g is given by the formula*

$$g'(a) = -Q^{-1} \circ P = -\left(\frac{\partial_f}{\partial y}(a, b)\right)^{-1} \circ \left(\frac{\partial f}{\partial x}(a, b)\right).$$

It follows immediately by applying theorem 2.

3.3.4 Maxima and minima with constraints

Let E be a normed affine space on the field R of the real numbers and Ω an open set of E. Let us consider $m + 1$ real functions of class C^1 defined on Ω let them be $f, g_1, \ldots, g_m$. Problem: determine, if there are, the relative maxima and minima of the function f on the subset $A \subset \Omega$ defined by the equations

$$g_1(x) = 0, \quad g_2(x) = 0, \ldots, \quad g_m(x) = 0.$$

Remark. Consequently the relative maxima and minima of f on one of its closed subsets A, and not on Ω, are required.

Definition. *The maxima and the minima defined above are called maxima and minima with constraints.*

Theorem. *Let f, $g_1, \ldots, g_m$ be real functions of class C^1 defined on an open set $\Omega \subset E$ where E is a normed affine space and let $a \in \Omega$ be a point which verifies the equations $g_1(a) = 0, \ldots, g_m(a) = 0$. Let us suppose that $g_1'(a)$, $g_2'(a), \ldots, g_m'(a) \in \mathfrak{L}(E, R)$ are linearly independent. Then a necessary condition for a to be a relative maximum or minimum of f on the subset A defined above is that there should exist m real constants $\lambda_1, \ldots, \lambda_m$ which satisfy the relationship*

$$f'(a) = \lambda_1 g_1'(a) + \lambda_2 g_2'(a) + \ldots + \lambda_m g_m'(a).$$

We shall prove this theorem in two steps.

(1) We first suppose that E is of finite dimension and we consider in E a frame of reference $\bar{0}, \bar{e}_1, \ldots, \bar{e}_m$.

In this case the hypothesis that at the point a the differentials dg_i are linearly independent is reduced to the existence of m coordinates $x_1, \ldots, x_m$ so that the Jacobean $\frac{D(g_1, \ldots, g_m)}{D(x_1, \ldots, x_m)}(a) \neq 0$. Hence $n \geqslant m$. From theorem 5 it follows that one can calculate $x_1, \ldots, x_m$ with respect to $g_1 \ldots, g_m$, $x_{m+1}, \ldots, x_n$ in the neighbourhood of the point a. The values of $x_1, \ldots, x_m$ thus obtained are introduced in f and we obtain a function of class C^1 let it be

$$F(g_1, \ldots, g_m, \quad x_{m+1}, \ldots, x_m).$$

Let us write that a is an extremum of the function F, when the coordinates $x_{m+1}, x_{m+2}, \ldots, x_n$ vary in the neighbourhood of a and $g_1, \ldots, g_n$ are equal to zero.

From the theorem in Section 3.3.1 it follows that a necessary condition for this is that the derivatives of F with respect to $x_{m+1}, \ldots, x_n$ at the point a be null. Hence the differential dF at the point a has the following form

$$dF = \lambda_1 \, dg_1 + \lambda_2 \, dg_2 + \ldots + \lambda_m \, dg_m$$

where $\lambda_1, \ldots, \lambda_m$ are real constants. But it is known that the differential df is obtained from the above relationship by replacing the differentials dg_i with their expressions in terms of dx_i, $(i = 1, \ldots, n)$. Hence

$$df = \lambda_1 \, dg_1 + \ldots + \lambda_m \, dg_m$$

and the theorem is proved in this case.

(2) Let us now suppose that E is of infinite dimension. Since $g'_i(a)$ are linearly independent we shall assign them given values. Consequently for every i we can find a vector $\overline{X}_i \in \overline{E}$ so that

$$g'_i(a) \cdot \overline{X}_i = \delta_{ij} = \begin{cases} 0 \text{ for } i \neq j, \\ 1 \text{ for } i = j. \end{cases}$$

The vectors $\overline{X}_1, \overline{X}_2, \ldots, \overline{X}_m$ thus formed are obviously linearly independent because from the relation $\sum_{i=1}^{m} \lambda_i \overline{X}_i = 0$ we deduce $g'_j(a) \sum_{i=1}^{m} \lambda_i \overline{X}_i = \lambda_j = 0$. Let us take $\lambda_i = f'(a)\ (\overline{X}_i)$. Then the linear form (Vol. 1, Section 3.1.7) $f'(a) - \sum_{i=1}^{m} \lambda_i g'_i(a)$ is null on the vectors $\overline{X}_1, \overline{X}_2, \ldots, \overline{X}_m$ hence on the vector subspace generated by them.

Let $\overline{X} \in \overline{E}$ be arbitrary and let $\overline{E}_0$ be the vector subspace generated by $\overline{X}_1, \overline{X}_2, \ldots, \overline{X}_m, \overline{X}$. On the affine subspace E_0, f must have an extremum in a on the set $A_0 = A \cap E_0$ defined by the equations $g_i = 0$ $(i = 1, \ldots, m)$. But since E_0 is of finite dimension (m or $m + 1$) it follows from the first part of the theorem that there are m constants $\mu_1, \ldots, \mu_m$ so that $f'(a) - \sum_{i=1}^{m} \mu_i g'_i(a) = 0$ on $\overline{E}_0$. If we write that this relationship is in particular zero on the vectors $\overline{X}_1, \ldots, \overline{X}_m$ we obtain $\mu_i = \lambda_i$.

Consequently

$$f'(a) - \sum_{i=1}^{m} \lambda_i g'_i(a)$$

is zero on $\overline{X}$ and since $\overline{X}$ is an arbitrary vector in $\overline{E}$, it follows that it is zero on $\overline{E}$ and the theorem is completely proved.

Definition. The constants λ_i *determined above are called Lagrange multipliers.*

Remark. From the proof of the theorem it results that Lagrange's multipliers are uniquely determined. But this was obvious from the beginning since $g'_i(a)$ are linearly independent.

3.4 EXERCISES

(1) If f is a mapping of $\Omega \subset E$ on F having a derivative $f'(a)$ at $a \in \Omega$ and g is a continuous affine mapping of F on G, the composed mapping $g \circ f$ of Ω on G has a derivative at a given by

$$(g \circ f)'(a) = g \circ f'(a).$$

(2) If E, F and G are affine spaces of finite dimension, then the derivative matrix of the mapping $h = g \circ f$ at the point a is the product of the derivative matrix of the mapping at the point $b = f(a)$ and of the derivative matrix of the mapping f at the point a.

4

Some Elements of the Calculus of Variations

4.1 SOME ELEMENTS OF THE CALCULUS OF VARIATIONS

4.1.1 Setting the problem

Our starting point will be the following problem of mechanics:

Let A, B be two points in the three dimensional space and let $x = f(\xi)$, where $f \in C^1$, be the equation of a curve whose extremities are A and B. Let us consider on this curve a free material point frictionless. Under the influence of gravity, the point starts from A with an initial speed equal to zero and arrives in B after a time t. The points A and B being given, the time t will depend on the considered curve.

Problem. How should the curve be chosen for the time t to be a minimum. A curve with this property is called a brachistocronic curve. Let us now assume that the curve which realizes the minimum is placed in the vertical plane of the segment AB. Let us take the horizontal axis Ox in this plane and the vertical axis Oz directed downwards. Let $z = a$, $x = \alpha$ be the coordinates of A and $z = b$, $x = \beta$ the coordinates of B. It is known that the speed v of the material point is given by the formula

$$v^2 = 2g(z - a).$$

We deduce the differential $\mathrm{d}t$ as function of $\mathrm{d}s$ (the differential of the curvilinear abscissa)

$$\mathrm{d}t = \frac{\mathrm{d}s}{\sqrt{2g(z-a)}}.$$

It follows that the necessary time for the mobile to go from A to B is given by the integral

$$\int_a^b \frac{\sqrt{1 + x'^2}}{\sqrt{2g(z-a)}}\,\mathrm{d}z = \int_a^b \frac{\sqrt{1 + f'^2(z)}}{\sqrt{2g(z-a)}}\,\mathrm{d}z. \tag{4.1}$$

Consequently the problem is reduced to finding a real function f, of real variable z, so that the integral (4.1) be a minimum. We notice that a, b, g are given and $f(a) = \alpha$, $f(b) = \beta$.

We pass now to the generalization of this problem.

Let $[a, b] \subset R$, F be a normed affine space on the field of the real numbers R and $U \subset F \times \overline{F}$ an open set. Let $L\colon (x, (y, \overline{Y})) \to L(x, y, \overline{Y})$ be a real function defined on $[a, b] \times U$ and let $f\colon [a, b] \to F$ be a function of class C^1 with the property that the pair $(f(x), f'(x)) \in U$ for every $x \in [a, b]$. It then follows that the mapping $x \to L(x, f(x), f'(x))$ is a real function continuous on the interval $(a, b]$.

Let us consider the integral of this function

$$J(f) = \int_a^b L(x, f(x), \overline{f}'(x))\,\mathrm{d}x. \tag{4.2}$$

$J(f)$ is a real number which depends on f and hence defines a function $J\colon f \to J(f)$. We have the following problem.

Problem. Supposing that the extremities a, b of the interval $[a, b]$ are fixed and L is a given function find the function f which takes given values α, β at a and b respectively, ($f(a) = \alpha, f(b) = \beta$) for which the integral $J(f)$ is maximum or minimum.

Particular case. The space F is of finite dimension m (hence we can identify it with R^m). The function f: $[a, b] \to F$ is in this case a system of m real functions f_i, $i = 1, 2, \ldots, m$, defined on $[a, b]$ and its derivative is the system of m derivative functions f_i'. The mapping L is then a real function of $2\,m + 1$ independent variables x, y_i, Y_i, $i = 1, 2, \ldots, m$. Hence the above problem is reduced to finding a system f_i, $i = 1, 2, \ldots, m$ of m continuously differentiable functions for which the integral

$$J(f_1, f_2, \ldots, f_m) = \int_a^b L(x, f_1(x), f_2(x), \ldots, f_m(x), f_1'(x), \ldots, f_m'(x))\, dx$$

is minimum or maximum. Let the normed affine space be $E = (F^{[a,\, b]})_{cd}$ (proposition 5, Section 3.2.6) and let Ω be the subset of E formed by the functions f which have the property that the pair $(f, \bar{f}') \in U$ for every $x \in [a, b]$, where U is an open set from $F \times \bar{F}$. We shall prove the following proposition.

Proposition. *The set Ω defined above is an open set of the affine space $E = (F^{[a,\, b]})_{cd}$.*

Indeed let $f_0 \in \Omega$ and let K be the image through $(f_0, \bar{f_0'})$ of the interval $[a, b]$. Since $[a, b]$ is a compact interval it follows that K is a compact set since the image of a compact through a continuous function is a compact (Ref. 50). But then it can be easily shown that the distance from K to $\complement U$ is a number $\delta > 0$.

Let us consider the ball of E, with the centre f_0 and the radius δ, that is $\{f\colon |||f - f_0||| < \delta\}$. For an element f of this ball $(f, \bar{f}) \in U$. Hence the ball is contained in Ω and hence Ω is an open set.

Let us denote by E_0 the affine subspace of the space E consisting of the functions $f \in E$ which verify the relationships $f(a) = \alpha$, $f(b) = \beta$ and put $\Omega_0 = \Omega \cap E_0$, then from the above proposition we have the following corollary.

Corollary. The set Ω_0 is an open set in E_0. With the above notations it follows that $J\colon f \to J(f)$ is a real function defined on the set Ω_0.

4.1.2 The differentiability of the function J

Let us denote by $\overline{\delta f}$ a given increase of f. The mapping $x \to \overline{\delta f}(x)$ is a mapping of $[a, b]$ with values in $\overline{F}$.

Theorem. *The real function J, defined on Ω is of class C^1 and its derivative $J'(f_0) \in \mathfrak{L}(\overline{E}, R) = \overline{E}'$ at the point f_0 is given by*

$$\overline{\delta f} \to \delta J = \overline{J}'(f_0)(\overline{\delta f}) = \int_a^b L'(x, f_0(x), f_0'(x))(0, \overline{\delta f}(x), \overline{\delta f}'(x))\, dx =$$

$$= \int_a^b \frac{\overline{\partial}}{\partial y} L(x, f_0(x), \overline{f_0'}(x)(\overline{\delta f}(x)) + \frac{\overline{\partial}}{\partial \overline{Y}} L(x, f_0(x), f_0'(x), (\overline{\delta f}'(x))\ dx, \tag{4.3}$$

*where L' is the total derivative and $\frac{\overline{\partial L}}{\partial y}$, $\frac{\overline{\partial L}}{\partial \overline{Y}}$ are the partial derivatives of the function L with respect to the second and the third variable *.*

Indeed, the variation ΔJ of the integral (4.2) corresponding to the variation $\overline{\delta f}$ of f is given by the formula:

$$\Delta J = \int_a^b [L(x, f_0(x) + \overline{\delta f}(x), \overline{f_0'}(x) + \overline{\delta f}'(x)) - L(x, f_0(x), \overline{f_0'}(x))]\ dx. \tag{4.4}$$

* For every

$$x \in [a, b], \quad \frac{\overline{\partial}}{\partial y} L(x, f_0(x), \overline{f_0'}(x)) \in L(\overline{F}, R) \text{ and } \overline{\delta f}(x) \in \overline{F},$$

hence

$$\frac{\overline{\partial}}{\partial \overline{Y}} L(x, f_0(x), \overline{f_0'}(x))(\overline{\delta f}(x)) \text{ is a real number.}$$

Similarly

$$\frac{\partial}{\partial \overline{Y}} L(x, f_0(x), f_0'(x))(\overline{\delta f}(x)) \text{ is a real number.}$$

From the corollary of the mean-value theorem (Section 3.2.7) we have

$$L(x, f_0(x) + \overline{\delta f}(x), \overline{f'_0}(x) + \overline{\delta f'}(x)) - L(x, f_0(x), \overline{f'_0}(x)) =$$

$$= L'(x, f_0(x), f'_0(x))(0, \overline{\delta f}(x), \overline{\delta f'}(x)) + R(x)) = \tag{4.5}$$

$$= \frac{\overline{\partial L}}{\partial y}(x, f_0(x), \overline{f'_0}(x))\,\overline{\delta f}(x) + \frac{\partial L}{\partial \overline{Y}}(x, f_0(x), \overline{f'_0}(x))\,\overline{\delta f'_0}(x) + R(x)$$

where

$$|R(x)| \leqslant \sup \| L'(x, f_0(x) + \overline{\xi}, \overline{f'_0}(x) + \overline{\xi'}) -$$

$$- L(x, f_0(x)), \overline{f'_0}(x) \| \, \| 0, \overline{\delta f}(x); \overline{\delta f'}(x) \|. \tag{4.6}$$

Let us use the theorem of the uniform continuity taking into account the fact that L' is continuous on $[a, b] \times U$ and that the set of the points $(x, f_0(x), \overline{f'_0}(x))$, $x \in [a, b]$ is contained in the compact $[a, b] \times K \subset [a, b] \times U$. Consequently $\varepsilon > 0$ given, there is $\eta > 0$, $0 < \eta < \delta$ so that if $x \in [a, b]$, $|\overline{\xi}\| \leqslant \eta$, $\|\overline{\xi'}\| < \eta$ then *

$$\| L'(x, f_0(x) + \overline{\xi}, \overline{f'_0}(x) + \overline{\xi'}) - L'(x, f_0(x), \overline{f'_0}(x)) \| \leqslant \frac{\varepsilon}{b-a}. \tag{4.7}$$

Hence $|||\overline{\delta f}|||_1 < \eta$ implies

$$|R(x)| \leqslant \frac{\varepsilon}{b-a} |||\overline{\delta f}|||_1. \tag{4.7'}$$

Consequently

$$\Delta J = \int_a^b \left(\frac{\overline{\partial}}{\partial y} L(\overline{\delta f}) + \frac{\overline{\partial}}{\partial \overline{Y}} L(\overline{\delta f}) \right) \mathrm{d}x + \int_a^b R(x)\, \mathrm{d}x. \tag{4.8}$$

Let us first notice that the mappings

$$x \to \frac{\partial}{\partial y} L(x, f_0(x), \overline{f'_0}(x))(\overline{\delta f}(x)) \text{ and } x \to \frac{\partial L}{\partial \overline{Y}}(x, f_0(x), \overline{f'_0}(x))(\overline{\delta f}(x))$$

are continuous. Indeed $\dfrac{\partial L}{\partial y}$ is a continuous function defined in $[a, b] \times U$

* In this theorem we kept the notations from Section 4.1.1.

with values in $\mathfrak{L}(\overline{F}, R) = \overline{F}'$ since L is of class C^1, f_0, $\overline{f_0'}$, $\overline{\delta f}$ are continuous functions defined in $[a, b]$ with values in $\overline{F}$, F and $\overline{F}'$ respectively and hence

$$x \to \frac{\overline{\partial}}{\partial y} L(x, f_0(x), \overline{f_0'}(x)) (\overline{\delta f}(x))$$

is a continuous mapping of $[a, b]$ in $\overline{F}'$.

Hence $x \to \frac{\overline{\partial}}{\partial y} L(x, f_0(x), \overline{f_0'}(x)) (\overline{\delta f}(x))$ is a continuous function from $[a, b]$ in R.

It follows from equations (4.5) that the function $x \to R(x)$ is als continuous.

The second integral from equation (4.8) is dominated owing to inequality (4.7′) by $\varepsilon \, ||| \overline{\delta f} |||_1$ if $||| \overline{\delta f} |||_1 \leqslant \eta$.
Let us show that the first integral from equation (4.8) defines a continuous linear form on $\overline{E}$. Indeed

$$\left| \int_a^b \left(\frac{\overline{\partial L}}{\partial y} (\overline{\delta f}) + \frac{\overline{\partial L}}{\partial \overline{Y}} (\overline{\delta f'}) \right) dx \right| \leqslant$$

$$\leqslant \int_a^b | L'(x, f_0(x), f_0'(x)) (0, \overline{\delta f}(x), \overline{\delta f'}(x)) \, dx \, | \leqslant (b - a) \; M \, ||| \overline{\delta f} |||_1 \qquad (4.9)$$

where

$$M = \sup_{a \leqslant x \leqslant b} \| L'(x, f_0(x), \overline{f_0'}(x)) \|.$$

Consequently J is differentiable with respect to $f_0 \in \Omega$ and its derivative is given by formula (4.3).

We have to show that J is of class C^1 on Ω. Let f_0, $f_1 \in \Omega$ and le

$$\overline{J'(x)} \; (\overline{\delta f}) - \overline{J'(f_0)} \; (\overline{\delta f}) =$$

$$= \int_a^b L'(x, f(x), \overline{f'}(x)) - L'(x, f_0(x), \overline{f_0'}(x)) (0, \overline{\delta f}(x), \; \overline{\delta f'}(x)) \, dx.$$

From formula (4.7) it follows that $|||\bar{f}-\bar{f}_0|||_1 \leqslant \eta$ implies

$$|\overline{(J'(f)} - \overline{J'(f_0))}\ (\overline{\delta f})| \leqslant \varepsilon\, |||\overline{\delta f}|||_1 \tag{4.10}$$

and hence

$$\|\overline{J'(f)} - \overline{J'(f_0)}\| = \sup_{|||\delta f|||_1 \leqslant 1} |J'(f) - J'(f_0)(\overline{\delta f})| \leqslant \varepsilon. \tag{4.11}$$

Consequently $\overline{J'}(f) - \overline{J'}(f_0)$ converges to $\bar{0} \in \bar{E}'$ as $f \to f_0$ in E hence J is of class C^1 and thus the theorem is completely proved.

Particular case 1. $F = R$, $U = [a, b] \times R \times R$. The function L is of class C^1 of three real variables x, y, y', $x \in [a, b]$, $y, y' \in R$ and J is a function defined on $(R^{[a,\, b]})_{cb,\, 1}$ thus

$$J(f) = \int_a^b L(x, f(x), f'(x))\, dx.$$

If δf is an increase of f, then δf is a function of class C^1 defined on $[a, b]$. Let us consider the expression

$$\Delta J = \int_a^b (L(x, f_0 + \delta f, f_0' + \delta f') - L(x, f_0, f_0'))\, dx \tag{4.12}$$

whose principal part is the differential

$$\delta J = \int_a^b \left(\frac{\partial L}{\partial y}(x, f_0(x), f_0'(x))(\delta f(x)) + \frac{\partial L}{\partial y'}(x, f_0(x), f'(x)(\delta f'(x)) \right) dx. \tag{4.13}$$

Applying the mean-value theorem to the function L we obtain

$$|\delta J - \Delta J| \leqslant \varepsilon\, |||\delta f|||_1 \quad \text{for } |||\delta f|||_1 \leqslant \eta.$$

For the proposition which follows we need a formula of integration by parts which we shall now give without proof.

Lemma. *(Formula of integration by parts). Let $\bar{u}$, $\bar{v}$ be differentiable mappings of an open set Ω of a normed affine space E in the normed vector* *spaces $\overline{F}_1$ and $\overline{F}_2$ and let B be a bilinear continuous mapping of $\overline{F}_1 \times \overline{F}_2$ in $\overline{G}$ where $\overline{G}$ is a normed vector space. Then*

$$\int_a^b B(\overline{u}(x), \overline{v}'(x))\, \mathrm{d}x = [B(\overline{u}(x), \overline{v}(x))]_{x=a}^{x=b} - \int_a^b B(\overline{u}'(x), \overline{v}'(x))\, \mathrm{d}x.$$

The formula above follows immediately from proposition 7, Section 3.2.6.

Let us see what the above formula becomes if we make a restrictive hypothesis over L and f_0.

Proposition. *In the condition of the theorem, if L is a function of class C^2 on $[a, b] \times U$ and f_0 is of class C^2 on $[a, b]$, then*

$$\delta J = \overline{J}'(f)\ (\overline{\delta f}) = \left[\frac{\partial}{\partial \overline{Y}} L(x, f_0(x), (f_0'(x))\ \ (\overline{\delta f(x)})\right]_{x=a}^{x=b} +$$

$$+ \int_a^b \left[\frac{\partial}{\partial y} L(x, f_0(x), f_0'(x)) - \frac{\mathrm{d}}{\mathrm{d}x}\frac{\partial}{\partial \overline{Y}} L(x, f_0(x), \overline{f}_0'(x))\right] (\overline{\delta f}(x))\, \mathrm{d}x. \quad \textbf{(4.14)}$$

In order to prove formula (4.14) we apply the formula of integration by parts to relationship (4.3). In formula (4.14) $\dfrac{\mathrm{d}}{\mathrm{d}x}$ is the derivative of the mapping $x \to \dfrac{\partial L}{\partial \overline{Y}}(x, f(x), f'(x))$ of $[a, b]$ in $\mathfrak{L}(\overline{F}, R)$.

This is a total derivative which is obtained by taking into account the fact that f_0 and f_0' are functions of x. The mapping B from the lemma is replaced by the continuous bilinear mapping $(\overline{u}, \overline{v}) \to \overline{u}(\overline{v})$ of $\mathfrak{L}(F, R) \times F$ in R. This integration by parts cannot be done unless u and $\overline{u}$ are of class C^1. This is why we have to assume that f_0 and L are of class C^2 because then $\overline{u}$: $x \to \dfrac{\partial \overline{L}}{\partial \overline{Y}}(\overline{x}, f_0(x), f_0'(x))$ is of class C^1 (f_0' being of class C^1).

Particular case 2. $F = R$, $U = R \times R$ then formula (4.14) becomes

$$\delta J = \left[\frac{\partial L}{\partial y'}(x, f_0(x), f_0'(x))\ \ (\delta f(x))\right]_{x=a}^{x=b} +$$

$$+ \int_a^b \left[\frac{\partial L}{\partial y}(x, f_0(x), f_0'(x)) - \frac{\mathrm{d}}{\mathrm{d}x}\left(\frac{\partial L}{\partial y'}(x, f_0(x), f_0'(x))\right)\right] (\delta f(x))\, \mathrm{d}x. \quad (4.15)$$

4.1.3 The necessary condition for an extremum

Lemma. *Let $\overline{F}$ be a normed vector space and $\overline{\Phi}$: $[a, b] \rightarrow \mathfrak{L}(\overline{F}, R)$ a continuous function. Then if the relationship*

$$\int_a^b \langle \overline{\Phi}(x), \overline{\eta}(x) \rangle \, dx = 0 \tag{4.16}$$

takes place for every function $\overline{\eta}$: $[a, b] \rightarrow \overline{F}$ of class C^1 which vanishes at the extremities of the interval $[a, b]$ it follows that $\overline{\Phi}$ is identical to zero on $[a, b]$.

To prove the lemma we shall suppose that $\overline{\Phi}$ is not identical to zero and thus obtaining a contradiction. Let $x_0 \in [a, b]$ with the property that $\overline{\Phi}(x_0) \neq 0$. We can suppose that $x_0 \neq a, b$ since if $\overline{\Phi}$ is identical to zero in (a, b), it follows from the continuity of the function $\overline{\Phi}$ that it is identical to zero on the closed interval $[a, b]$. But $\overline{\Phi}(x_0) \in \mathfrak{L}(\overline{F}, R)$, hence $\overline{\Phi}(x_0)$ is a linear form continuous on $\overline{F}$ different from 0. Consequently there is an element $\bar{e} \in \overline{F}$ so that the value of $\overline{\Phi}(x_0)$ at $\bar{e}$ is different from zero, and obviously we can choose $\bar{e}$ so that the value of $\overline{\Phi}(x_0)$ at $\bar{e}$ should be > 0 (in the contrary case $\bar{e}$ is replaced with $-\bar{e}$).

From the continuity of $\overline{\Phi}$ it follows that there is a neighbourhood of x_0 i.e. a number α small enough so that $[x_0 - \alpha, x_0 + \alpha] \subset [a, b]$ and so that the continuous function $x \rightarrow \langle \overline{\Phi}(x), \bar{e} \rangle$ be > 0 for every x of this neighbourhood. Let φ be a scalar function, of class C^1, strictly positive in the interval $(x_0 - \alpha, x_0 + \alpha)$ and vanishing outside this interval. Then taking $\overline{\eta}(x) = \bar{e}\ \varphi(x)$ it follows that $\overline{\eta}(x)$ is of class C^1 in $[a, b]$ and vanishes at the extremities.

The function $x \rightarrow \langle \overline{\Phi}(x), \overline{\eta}(x) \rangle$ is continuous, strictly positive in the interval $(x_0 - \alpha, x_0 + \alpha)$ and vanishes outside this interval. Then we have

$$\int_a^b \langle \overline{\varphi}(x), \overline{\eta}(x) \rangle \, dx > 0$$

which contradicts the hypothesis.

We are now able to give an extremum criterion for $J(f)$. Keeping the notation used in Section 4.1.1 we have the following theorem.

Theorem. *In order that $f_0 \in \Omega_0$ of class C^2 be an extremum for the integral $J(f)$ it is necessary that f_0 be a solution of the differential equation of second order*

$$\frac{\overline{\partial L}}{\partial y} - \frac{\mathrm{d}}{\mathrm{d}x}\left(\frac{\overline{\partial L}}{\partial y'}\right) = 0. \tag{4.17}$$

Indeed if $\overline{\delta f} \in \overline{E}_0$, then $\overline{\delta f}(x) = \overline{\delta f}(b) = \overline{0}$ and hence

$$\frac{\overline{\partial L}}{\partial y'}(x, f_0(x), \overline{f_0'}(x), \overline{\delta f}(x))\Big|_{x=a}^{x=b} = 0.$$

But the necessary condition for the extremum is $\delta J = 0$ and consequently from equation (4.14) we have

$$\int_a^b \left[\frac{\overline{\partial}}{\partial y} L(x, f_0(x), \overline{f_0'}(x)) - \frac{\mathrm{d}}{\mathrm{d}x}\left(\frac{\overline{\partial}}{\partial y'} L(x, f_0(x), \overline{f_0'}(x)\right)\right](\overline{\delta f}(x))\, \mathrm{d}x = 0$$

for every $\overline{\delta f} \in \overline{E}_0$. To deduce that f_0 satisfies equation (4.17) we apply the above lemma to the function

$$x \to \overline{\Phi}(x) = \frac{\overline{\partial}}{\partial y} L(x, f_0(x), \overline{f_0'}(x)) - \frac{\mathrm{d}}{\mathrm{d}x}\left(\frac{\overline{\partial}}{\partial y'} L(x, f_0(x), \overline{f_0'}(x))\right).$$

Definition. Equation (4.17) *is called Euler's equation and a solution of this equation is called as extremal of the integral J.*

Particular case. In the conditions of the particular case 1 for $f_0 \in R^{[a,b]}$ of class C^1 and such that $f_0(a) = \alpha$, $f_0(b) = \beta$ to be an extremum for the integral J, it is necessary and sufficient that f_0 should be of class C^2 and should verify the relation

$$\int_a^b \left[\frac{\partial L}{\partial y}(x, f_0(x), f_0'(x)) - \frac{\mathrm{d}}{\mathrm{d}x}\left(\frac{\partial L}{\partial y'}(x, f_0(x), f_0'(x))\right)\right](\delta f(x))\, \mathrm{d}x = 0 \tag{4.18}$$

for every $\delta f \in R^{[a,b]}$ of class C^1 and such that $\delta f(a) = \delta f(b) = 0$. From the above lemma it results that for this purpose it is necessary and sufficient that f_0 should verify Euler's equation

$$\frac{\partial L}{\partial y} - \frac{\mathrm{d}}{\mathrm{d}x}\left(\frac{\partial L}{\partial y'}\right) = 0. \tag{4.19}$$

Corollary. *Let f_0 be a function which represents an extremum for J in the conditions of the previous theorem. If $a_1 \geqslant a$, $b_1 \leqslant b$, then f_0 represents an extremum also for the integral*

$$J_1(f) = \int_{a_1}^{b_1} L(x, f(x), \overline{f'}(x))\, dx,$$

on the space of all the functions f of class C^1 defined on $[a_1, b_1]$ with values in F, so that $(f, \overline{f'})$ maps $[a_1, b_1]$ in U and $f(a_1) = \alpha_1 = f_0(a_1)$, $f(b_1) = \beta_1 = f_0(b_1)$.

Indeed, the extremities of the interval do not appear in Euler's equation.

Remark. It can be proved that a function of class C^1 which represents an extremum for J is necessarily of a class C^2.

Particular case. Coming back to the particular case from Section 4.1.1 we notice that $\overline{\delta f}$ is the system $(\delta f_1, \ldots, \delta f_m)$, the partial derivative $\dfrac{\overline{\partial f}}{\partial y}(x, y, \overline{Y})$, as vector from $\overline{F'} = R^m$ is the system of m derivatives

$$\frac{\partial f}{\partial y_i}(x, y_1, y_2, \ldots, y_m, Y_1, \ldots, Y_m), \quad i = 1, 2, \ldots, m,$$

and $\dfrac{\overline{\partial f}}{\partial \overline{y'}}(x, y, \overline{Y})$ is the system

$$\frac{\partial f}{\partial Y_i}(x, y_1, \ldots, y_m, Y_1, \ldots, Y_m), \qquad i = 1, 2, \ldots, m.$$

Finally the scalar product $\langle \overline{\alpha}, \overline{x} \rangle$ where

$$\overline{\alpha} = (\alpha_1, \ldots, \alpha_m),\ \overline{x} = (x_1, \ldots, x_m) \text{ is } \alpha_1 x_1 + \ldots + \alpha_m x_m.$$

Euler's equation with respect to the function

$f = (f_1, \ldots, f_m)$ is identical to the system of m equations

$$\frac{\partial L}{\partial y_i} - \frac{d}{dx}\left(\frac{\partial L}{\partial y_i'}\right) = 0, \quad i = 1, 2, \ldots, m \tag{4.20}$$

or

$$\frac{\partial L}{\partial y_i} - \frac{\partial^2 L}{\partial y'_i\, \partial x} - \sum_{j=1}^{m} \frac{\partial^2 L}{\partial y'_i\, \partial y_j} y'_j - \sum_{j=1}^{m} \frac{\partial^2 L}{\partial y'_i \partial y'_j} y''_j = 0. \tag{4.21}$$

The solution of this system which depends on $2m$ arbitrary constants must satisfy the $2m$ equations $f_i(a) = \alpha_i$, $f_i(b) = \beta_i$, $i = 1, 2, \ldots, m$.

Remark. In the above case

$$\delta J = \int_a^b \sum_{i=1}^{m} \left(\frac{\partial L}{\partial y_i} - \frac{\mathrm{d}}{\mathrm{d}x} \left(\frac{\partial L}{\partial y'_i} \right) \right) \delta f_i \, \mathrm{d}x. \tag{4.22}$$

4.1.4 Simple cases of elementary integrability of Euler's equations

(1) Let us suppose that L does not depend on y. Then $\dfrac{\partial \overline{L}}{\partial y} = 0$. Euler's equation is reduced in this case to a differential equation of the first order

$$\frac{\partial \overline{L}}{\partial y'} (x, y') = \overline{c} \text{ (constant)} \in \mathfrak{L}(\overline{F}, R). \tag{4.23}$$

Particular case. If $\overline{F} = R$, $\dfrac{\partial L}{\partial y'}(x, y') = c$ (real constant).

(2) Let us suppose that L does not depend on x. Then Euler's equations from the particular case of equation (4.21) are written

$$\frac{\partial L}{\partial y_i} - \sum_{j=1}^{m} \frac{\partial^2 L}{\partial y_j\, \partial y'_i} y'_i - \sum_{j=1}^{m} \frac{\partial^2 L}{\partial y'_i\, \partial y'_j} y''_j = 0, \ i = 1, 2, \ldots, m. \tag{4.23$'$}$$

Multiplying the ith equation from equations (4.23′) by y'_i and summin from 1 to m, we obtain

$$\sum_{i=1}^{m} \frac{\partial L}{\partial y_i} y'_i - \sum_{i,\, j=1}^{m} \frac{\partial^2 L}{\partial y_j\, \partial y'_i} y'_i y'_j - \sum_{i,\, j=1}^{m} \frac{\partial^2 L}{\partial y'_i\, \partial y'_j} y'_i y''_j = 0, \tag{4.24}$$

or

$$\frac{\mathrm{d}}{\mathrm{d}x} \left(L - \sum_{j=1}^{m} \frac{\partial L}{\partial y'_j} y'_j \right) = 0. \tag{4.25}$$

From equation (4.25) we immediately obtain the prime integral

$$L - \sum_{j=1}^{m} \frac{\partial L}{\partial y'_j} \bar{y}'_j = C \tag{4.26}$$

or written differently

$$L - \overline{\frac{\partial L}{\partial y'}} \cdot \bar{y}' = C. \tag{4.27}$$

Particular case. If $m = 1$, equation (4.26) becomes

$$L - \frac{\partial L}{\partial y'} y' = C. \tag{4.28}$$

4.1.5 The brachistocronic curve

Let us consider again the example from Section 4.1.1.

We have to determine the minimum of the integral (4.1). Euler's equation (formula (4.23)) is written in this case

$$\frac{\partial L}{\partial x'} = \frac{x'}{\sqrt{1 + x'^2}} \frac{1}{\sqrt{2g(z - a)}} = \text{constant} = \frac{\pm 1}{\sqrt{4gc}}, \tag{4.29}$$

where c is a strictly positive arbitrary constant. In this way we leave the case of the null constant aside, which cannot be written as $\pm \frac{1}{\sqrt{4gc}}$; in this case we have $x = \text{constant} = \alpha$ which means that B is on the vertical of A, when the solution of the problem is evidently given by the vertical segment AB. Equation (4.29) is written as

$$x'^2 = \frac{z - a}{2c} (1 + x'^2) \tag{4.30}$$

or

$$x'^2 \left(1 - \frac{z - a}{2c}\right) = \frac{z - a}{2c}. \tag{4.31}$$

We can make a change of variable

$$z = a + c(1 - \cos u) \tag{4.32}$$

since necessarily

$$0 \leqslant \frac{z-a}{c} \leqslant 2$$

and then equation (4.30) becomes

$$\frac{dx}{dz} = \pm \sqrt{\frac{z-a}{2c-(z-a)}} = \pm \sqrt{\frac{1-\cos u}{1+\cos u}}, \quad \frac{dx}{du} = \frac{dx}{dz}\frac{dz}{du} =$$

$$= \pm c \sin u \sqrt{\frac{1-\cos u}{1+\cos u}} = \pm 2\, c \sin^2 \frac{u}{2} = \pm c(1-\cos u). \tag{4.33}$$

The solution of equation (4.33) is

$$z = a + c(1-\cos u), \quad x = \pm c(u - \sin u) + \text{constant}. \tag{4.34}$$

The constant c and the constant from equation (4.34) must be determined so, that the curve should pass through A and B. In order to make it more simple let us suppose that A is in the origin of the coordinates. We have $z = a = 0$ for $u = 2k\pi$. If we change u into $u - 2k\pi$ we take $u = 0$. In this case, if the constant from equation (4.34) is null we have $x = 0$. We notice that we can always take the sign $+$, the sign $-$ is obtained by changing u into $-u$. Consequently the brachistocronic curve is a cycloid Γ having a turning point in A with a vertical tangent. When c varies we obtain a family of cycloids Γ_c which are homothetic to Γ_1 and which correspond to $c = 1$. Consequently c can be determined geometrically in the following way; the line AB cuts the curve Γ_1 at a finite number of points $C_1, C_2, \ldots$ (except for the case when B is on the horizontal of A and when we have an infinity of solutions). If C_i is such a point, then the homothetic of Γ_1 of centre A and of ratio $\frac{AB}{AC_i}$ passes through B and answers the problem.

Almost all the difficulties of the variational calculus are included in this example. Indeed we have remark 1.

Remark 1. From the beginning we admitted that the curve is in the vertical plane of AB. This hypothesis can now be avoided. Let us take three coordinates and the axis z downwards in R^3. Let us determine the brachistocronic curve as an arc of a parametric curve of class C^2, $x(w)$, $y(w)$, $z(w)$. We suppose that A and B correspond to $w = w_1$ and $w = w_2$

respectively which are fixed. Then the necessary time to cover the distance from A to B is

$$t = \int_{w_1}^{w_2} \frac{ds}{\sqrt{2g(z-a)}} = \int_{w_1}^{w_2} \frac{\sqrt{x'^2 + y'^2 + z'^2}}{\sqrt{2g(z-a)}}\, dw. \tag{4.35}$$

The functions x, y, z take given values for $w = w_1$, and $w = w_2$. As we have three unknown functions, we shall have three equations of Euler's type.

From equation (4.23) it follows that the first two are

$$\begin{aligned} \frac{x'}{\sqrt{x'^2 + y'^2 + z'^2}} \cdot \frac{1}{\sqrt{2g(z-a)}} &= \text{constant}, \\ \frac{y'}{\sqrt{x'^2 + y'^2 + z'^2}} \cdot \frac{1}{\sqrt{2g(z-a)}} &= \text{constant}, \end{aligned} \tag{4.36}$$

hence we deduce a relationship of the form

$$\lambda x' + \mu y' = 0,$$

with the coefficients λ, μ constant, hence we have

$$\lambda x + \mu y + \nu = 0.$$

Consequently the curve is in a vertical plane, the plane of AB.

Remark 2. We admitted from the beginning that the brachistocronic curve can be represented by expressing x as function of z. This hypothesis is not justified. Indeed after obtaining Euler's equation, we transformed it by effecting a change of variable given by formula (4.32). The solutions of the transformed equation are cycloids, where x is not necessarily expressed as function of z. In the same way as above, if AB cuts Γ_1 at a single point situated on the first branch of the cycloid, then there is a unique solution corresponding to the points A and B and for this solution c is not expressed as function of z. In this case we shall proceed as above and we shall look for Γ in the vertical plane AB, as a parametric curve of class C^2. We shall thus obtain the two equations of Euler in x and z

$$x = -\frac{d}{dw}\left(\frac{x}{\sqrt{x'^2 + y'^2}}\, \frac{1}{\sqrt{2g(z-a)}}\right) = 0$$

or

$$\frac{x}{\sqrt{x'^2 + z'^2}} \frac{1}{\sqrt{2g(z-a)}} = \text{constant} \tag{4.37}$$

and

$$z = -\frac{1}{2} \cdot \frac{\sqrt{x'^2 + z'^2}}{\sqrt{2g}(z-a)^{\frac{3}{2}}} - \frac{d}{dw}\left(\frac{z'}{\sqrt{x'^2 + z'^2}} \frac{1}{\sqrt{2g(z-a)}}\right) = 0.$$

But in this case there are an infinity of solutions for given A and B since the two equations (4.37) are not independent.

Remark 3. We return to the problem of the minimization of integral (4.1) through a function $x = f(z)$ of class C^1. The only possible solutions are the cycloids we found. If we are in the case discussed in remark 2, when no cycloid for which x can be expressed as function of z passes through AB, then the minimum we are looking for does not exist. Integral (4.1) will have a lower bound > 0 in the set $\Omega_0 \subset E_0$, but not a minimum.

Remark 4. In the case of the brachistocronic curve we are not in the conditions of the theorem in Section 4.1.3 because the function $L = \frac{\sqrt{1 + x'^2}}{\sqrt{2g(z-a)}}$ is singular for $z = a$, that is exactly for one of the limits of the integral.

Remark 5. We defined above an extremal as being a curve which verifies Euler's equation between arbitrary limits.

This seems to contradict the result above, namely that the curve necessarily possesses a vertical tangent at the initial point which is immediately seen (for $z = a$, we have $x' = 0$). The explanation is that the curve is extremal for the integral

$$\int_{a_1}^{b_1} \frac{\sqrt{1 + x'^2}}{\sqrt{2g(z-a)}} dz$$

for every two points a_1, b_1 but with the same integrand that is, with the same constant a in $\sqrt{2g(z-a)}$. Consequently we no longer solve the problem of mechanics we considered, in which we supposed the initial speed at the initial point to be null. As A_1 is no longer a turning point with vertical tangent, the same cycloid limited at the extremities A_1 and B_1, minimizes integral (4.35) which represents the necessary time for the point to fall along the curve when the initial speed in A_1 is equal to $\sqrt{2g(a_1 - a)}$, $a_1 > a$ speed which corresponds to a speed which is null at A.

From the above remarks it follows that by writing Euler's equations we only solved a small part of the set problems of maximum or minimum.

In order to solve the problems of variational calculus powerful instruments like the theory of Hilbert spaces, algebraic topology and the theory of differential equations must be used.

4.1.6 Problems of extremum with constraints

Keeping the notations from Section 4.1.1 let M_i $(i = 1, 2, \ldots, m)$ be m real functions of class C^2 defined on $[a, b] \times U$ and let k_i, $(i = 1, \ldots, m)$ be m real given numbers. Let K_i $(i = 1, \ldots, m)$ be m functions on Ω_0 with values in R, defined thus

$$f \to K_i(f) = \int_a^b M_i(x, f(x), \overline{f(x)})\,\mathrm{d}x. \tag{4.38}$$

We have the following theorem.

Theorem. *If for $f_0 \in \Omega_0$, the derivatives $K_i'(f_0)$ are independent and $K_i(f_0) = k_i$, $(i = 1, \ldots, m)$, then, the necessary condition that among all the functions $f \in \Omega_0$ for which $K_i(f) = k_i$, $(i = 1, \ldots, m)$, f_0 should represent a maximum or a minimum for J is that the real numbers λ_i $(i = 1, 2, \ldots, m)$ exist so that*

$$\frac{\overline{\partial L}}{\partial y} - \frac{\mathrm{d}}{\mathrm{d}x}\left(\frac{\overline{\partial L}}{\partial y'}\right) = \sum_{i=1}^{m} \lambda_i \left(\frac{\overline{\partial M_i}}{\partial y} - \frac{\mathrm{d}}{\mathrm{d}x}\left(\frac{\overline{\partial M_i}}{\partial y'}\right)\right). \tag{4.39}$$

To prove this theorem it is sufficient to apply the results obtained in Section 3.3.4 (the theory of Lagrange multipliers).

Remark. If F is of finite dimension, relation (4.39) represents a system of n differential equations of the second order, whose solutions would depend on $2n$ constants, if $\lambda_i (i = 1, \ldots, m)$ were known. But since λ_i are not known the general solution of this system depends on $2n + m$ arbitrary constants.

These constants will thus be determined as to satisfy the $2n$ conditions to the limit and the m conditions $K_i = k_i$ $(i = 1, 2, \ldots, m)$.

Application. Among all the arcs of curve of class C^1 of given length l, which enjoin A and B, determine, that arc which together with the segment AB encloses a maximum area.

Let us look for this arc by expressing y as function of x and take AB parallel to the axis Ox. In this case the length is given by the integral

$$l = \int_{x(A)}^{x(B)} \sqrt{1 + y'^2}\, dx$$

and the area is given by

$$S = \int_{x(A)}^{x(B)} y\, dx.$$

We notice that there must exist a Lagrange multiplier λ so that we have Euler's equation

$$-\lambda \frac{d}{dx}\left(\frac{y'}{\sqrt{1 + y'^2}}\right) = 1 \quad \text{or} \quad -\frac{\lambda y''}{(1 + y'^2)^{\frac{3}{2}}} = 1.$$

As the curvature is constant it follows that the curve is an arc of the circle. Its centre and radius are unknown. If we write that the arc of the circle passes through A and B, that it is above the axis $Ox (y \geqslant 0)$ and that it has the given length l, then we shall determine it in a unique way. If $l > \frac{\pi}{2} |AB|$, we can see that the arc of the circle is greater then the semicircle and that the circle cannot be represented by expressing y in terms of x. If we suppose that A and B coincide then we obtain the following proposition.

Proposition. *Among all the closed curves of class C^1 and length l the curve which encloses a maximum area is a circle.*

Remark. The radius of this circle is $\frac{l}{2\pi}$, hence its area is $S = \frac{l^2}{4\pi}$. This enables us to say that every closed curve of class C^1 and length l encloses an area

$$S \leqslant \frac{l^2}{4\pi},$$

where the inequality is strict except when the curve is a circle.

Conversely, if a closed curve of class C^1 encloses an area S, its length satisfies the inequality

$$l \geqslant 2\sqrt{\pi S}$$

where the inequality is strict except when the curve is a circle.

4.1.7 The change of variable

Let F_1 be a normed affine space, and let h: $[a, b] \times F_1 \to F$ be a function of class C^1. If f_1: $[a, b] \to F$ is of class C^1 then the composed function

$$x \to f(x) = h(x, f_1(x))$$

is a function of class C^1 defined on $[a, b]$ with values in F. Consequently h defines a mapping $f_1 \to f$ of $E_1 = (F_1^{[a, b]})_{cd}$ in $E = (F^{[a, b]})_{cd}$. As it will follow from the following theorem this mapping is differentiable *if h* is of class C^2.

Theorem. *The mapping $f_1 \to f$ defined above is differentiable if h is of class C^2 and its derivative at the point $f_1 \in (F_1^{[a, b]})_{cd}$ is given by*

$$\delta f_1 \to \delta f$$

where

$$\overline{\delta f} = \partial_2 h(x, f_1(x))\, \overline{\delta f_1}(x). \tag{4.40}$$

Let us first show that formula (4.40) defines, for f_1 fixed, a linear continuous mapping $\overline{\delta f_1} \to \overline{\delta f}$ of $\overline{E}_1$ in $\overline{E}$. Indeed since h is of class C^2, it follows that $\partial_2 h$ is of class C^1, and since from proposition 6, Section 3.2.6 it follows that the composed mapping of two mappings of class C^2 is of class C^2, we deduce that if $\overline{\delta f_1}$ is in $\overline{E}_1$, $\overline{\delta f} \in \overline{E}$ (proposition 7, Section 3.2.6). The mapping $\overline{\delta f_1} \to \overline{\delta f}$ is obviously linear. We have

$$|||\, \delta f \,|||_0 \leqslant \sup_{a \leqslant x \leqslant b} \|\, \partial_2 h(x, f_1(x)) \,\|\, |||\, \delta f_1 \,|||_0 ,$$

$$|||\, \overline{\delta f'} \,|||_0 \leqslant \sup_{a \leqslant x \leqslant b} \|\, \partial_1\partial_2 h(x, f_1(x)) + \partial_2^2 h(x, f_1(x), f_1'(x)) \,\|\, |||\, \overline{\delta f_1} \,|||_0 + $$
$$+ \sup_{a \leqslant x \leqslant b} \|\, \partial_2 h(x, f_1(x)) \,\|\, |||\, \delta f_1' \,|||_0$$

hence the continuity of the mapping $\overline{\partial f_1} \to \overline{\delta f}$.

Let us now show that $\overline{\delta f}$, given by formula (4.40) is the differential we are looking for. We give the increase $\overline{\delta f_1}$ to f_1. Then

$$\overline{\Delta f}(x) = h(x, f_1(x) + \overline{\delta f_1}(x)) - h(x, f_1(x)).$$

Let us calculate $\overline{\Delta f} - \overline{\delta f}$ and $\overline{\Delta f'} - \overline{\delta f'}$ where $\overline{\delta f}$ is given by formula (4.40). But using the mean-value theorem for x fixed (Section 3.2.7) and the theorem of the uniform continuity it follows that for given $\varepsilon > 0$ we can determine $\eta > 0$ so that $|||\, \delta f_1 \,||| < \eta$ implies

$$|||\, \overline{\Delta f} - \overline{\delta f} \,|||_1 \leqslant \varepsilon \,|||\, \delta f_1 \,|||,$$

which proves that $\overline{\delta f}$ is indeed the differential we are looking for.

4.1.8 The canonical equations of Hamilton

If F is the space R^m, then the function f is equivalent to the system of m functions $f_1, f_2, \ldots, f_m$ of variable x. If we take $z_i = y_i'$ and consider L as a function of $2m + 1$ variables x, y_i, z_i it follows that the equations of Euler are reduced to

$$y_i' = z_i, \; i = 1, \ldots, m.$$
$$\frac{\partial L}{\partial y_i} - \frac{\mathrm{d}}{\mathrm{d}x}\left(\frac{\partial L}{\partial z_i}\right) = 0, \; i = 1, \ldots, m. \tag{4.41}$$

We take

$$y_i = q_i$$

and

$$p_i = \frac{\partial L}{\partial z_i}. \tag{4.42}$$

Let us suppose that in equation (4.42) we can calculate z_i in terms of x, y_i and p_i and let us write

$$H(x, q_1, \ldots, q_m, p_1, \ldots, p_m) = \sum_{i=1}^{m} p_i z_i - L. \tag{4.43}$$

The function H defined in formula (4.43) is called the *Hamiltonian*. Its differential is

$$dH = \sum_{i=1}^{m}(p_i dz_i + z_i\, dp_i) - \frac{\partial L}{\partial x}\, dx - \sum_{i=1}^{m}\frac{\partial L}{\partial y_i}\, dy_i - \sum_{i=1}^{m}\frac{\partial L}{\partial z_i}\, dz_i =$$

$$= \sum_{i=1}^{m} z_i\, dp_i - \sum_{i=1}^{m}\frac{\partial L}{\partial y_i}\, dq_i - \frac{\partial L}{\partial x}\, dx, \tag{4.44}$$

hence the partial derivatives of H with respect to x, q_i, p_i are

$$\frac{\partial H}{\partial x} = -\frac{\partial L}{\partial x}, \quad \frac{\partial H}{\partial q_i} = -\frac{\partial L}{\partial y_i}, \quad \frac{\partial H}{\partial p_i} = z_i. \tag{4.45}$$

Taking into account formulae (4.42), (4.43), (4.44) and (4.45) it follows that Euler's equations are written in the form

$$q_i' = \frac{\partial H}{\partial p_i}, \quad p_i' = -\frac{\partial H}{\partial q_i}. \tag{4.46}$$

Definition. The equations (4.46) *are called Hamilton's equations with respect to the function H.*

4.1.9 Applications to mechanics

Hamilton's equations have many important applications in mechanics and theoretical physics.

Let us consider, for example, a problem of mechanics with fixed constraints, frictionless, characterized by a field of forces, independent of time, deriving from a potential. The position of a system can be then represented by means of a finite number of parameters $q_1, \ldots, q_m$. The potential energy U will be a known function of these parameters and the kinetic energy $T = \Sigma \frac{1}{2} mv^2$ is a quadratic form with respect to the derivatives $q_1', q_2', \ldots q_m'$, $\left(q_i' = \frac{dq_i}{dt}\right)$. Solving the problem of mechanics means finding the trajectories, a trajectory being defined by the functions of t, $t \to q_i(t)$.

We prove that the trajectory of the problem of mechanics is a solution of a problem of extremum. If t_1 and t_2 are two determined moments we can consider on the real trajectory or on every fictitious trajectory the integral

$$\int_{t_1}^{t_2} L(q_i, q_i') \, \mathrm{d}t = \int_{t_1}^{t_2} (T(q_i, q_i') - U(q_i)) \, \mathrm{d}t. \tag{4.47}$$

We prove that the real trajectory is, among all the fictitious trajectories which at the considered moments t_1 and t_2 pass through the same points $q_i(t_1)$, $q_i(t_2)$, the one for which integral (4.47) is stationary. In other words every trajectory is an extremal and the equation which gives the trajectories of the considered problem of mechanics is represented by the system of Euler's equations

$$\frac{\partial L}{\partial q_i} - \frac{\mathrm{d}}{\mathrm{d}t}\left(\frac{\partial L}{\partial q_i'}\right) = 0,$$

$$i = 1, 2, \ldots, m \tag{4.48}$$

$$\frac{\partial T}{\partial q_i} - \frac{\mathrm{d}}{\mathrm{d}t}\left(\frac{\partial T}{\partial q_i'}\right) = \frac{\partial U}{\partial q_i}.$$

Definition. Equations (4.48) *are called Lagrange's equations of the problem of mechanics and* $L = T - U$ *is called the Lagrangian.*

In this case

$$p_i = \frac{\partial L}{\partial q_i'} = \frac{\partial T}{\partial q_i'} \tag{4.49}$$

and the Hamiltonian H becomes

$$H(q_i, p_i) = \sum_{i=1}^{m} q_i' \frac{\partial T}{\partial q_i'} - L. \tag{4.50}$$

If we consider that T is a quadratic form in q_i' then, from Euler's indentity of the homogeneous functions, we have

$$\sum_{i=1}^{m} q_i' \frac{\partial T}{\partial q_i'} = 2T. \tag{4.51}$$

If in equation (4.50) we take equation (4.51) into account we obtain

$$H = 2T - (T - U) = T + U \tag{4.52}$$

and hence H is exactly the energy of the system, that is the sum of the potential energy and of the kinetic energy expressed in terms of q_i and p_i. The Hamiltonian H does not depend on time and consequently represents a prime integral of system (4.46). In other words it follows that along a trajectory of the system, the energy H, the sum of the kinetic and potential energies, remains constant, a property very well known in elementary mechanics.

4.1.10 The calculus of variations with respect to multiple integrals

To be able to put the problem in a general way we must quickly explain the notion of a manifold with a border. Let us first give the definition of a manifold of class C^m.

Let E be an affine space on the field K of the real or complex numbers, of dimension N, and let V be a set in E.

Definition. *We say V is a manifold of dimension n, of class C^m, if for every point $a \in V$ there is frame of reference $0, e_1, e_2, \ldots, e_N$, an open set B of the subspace of the first n axes of coordinates, a system of $N-n$ functions G_k, $k = 1, 2, \ldots, N-n$ of class O^m, defined on B with scalar values and an open set $\mathcal{V}$ of E which contains a, whose projection on the subspace of the first n axes of coordinates is B, so that the intersection $V \cap \mathcal{V}$ is the set of the points $x = (x_1, \ldots, x_N)$ of E which verify the equations*

$$x_{n+k} = G_k(x_1, \ldots, x_n), \qquad k = 1, \ldots, N-n.$$

Consequently in $\mathcal{V}$, the last $N-n$ coordinates $x_{n+1}, x_{n+2}, \ldots, x_N$ on the manifold V are expressed as functions of class C^m of the first n coordinates $x_1, x_2, \ldots, x_n$ which can be arbitrarily chosen in B.

Remark. The special importance of the first n coordinates could seem odd. But it is normal, since for the point $a \in V$ we chose a frame of reference of E. Consequently if n of the coordinates are particularly important, we can return to the case of the first n coordinates by changing the order of the vectors of the frame of reference.

We shall now introduce the notion of a manifold with a border.

Definition. *A part V of a manifold $\tilde{V}$ of class C^m and of dimension n, closed, identical to the adherence of its interior and whose boundary Σ is a* *hypersurface of $\widetilde{V}$, submanifold of class C^m and of dimension $n - 1$, is called the manifold with a border of class C^m and of dimension n. The boundary Σ is called the border of V.*

Example. For the closed ball of an affine Euclidian space, the border is the corresponding sphere.

We can now pass to the setting of the problem. Let there be a compact curve $\mathcal{C}$ of class C^1 in an affine Euclidean space, with three dimensions, in the field of the real numbers.

Problem. Among all the surfaces of class C^1, bordered by this curve, find the surface which has the minimum area. This problem is called the problem of the minimum surface.

Remark. $\mathcal{S}$ and $\mathcal{C}$ are manifolds in the above sense of dimensions 2 and 1 respectively, they are without a common point and $\overline{\mathcal{S}} = \mathcal{S} \cup \mathcal{C}$.

Let us now consider the coordinates (x, y, z) in space R^3 and let $z = z(x, y)$ be the equation of a surface. The area of the surface is then expressed thus

$$S = \iint_{\Sigma} \sqrt{1 + p^2 + q^2}\, \mathrm{d}x\, \mathrm{d}y, \quad p = \frac{\partial z}{\partial x}, \quad q = \frac{\partial z}{\partial y}, \tag{4.53}$$

where the integral is taken on the surface Σ, the projection of the surface $\mathcal{S}$ on the plane (x, y).

The surface Σ is bordered by the curve Γ, the projection of the curve $\mathcal{C}$.

We must look for a function $z = f(x, y)$ of class C^1 which takes given values on the contour Γ of R^2, so that integral (4.53) is a minimum.

Let us now consider, in the space R^n, the open set θ bounded by a compact hypersurface Γ of class C^1.

Let F be a normed affine space and V an open set from $F \times \overline{F}^n$. Let L be a real function on $\overline{\theta} \times U$, of class C^2, which we denote by

$$L(x_1, x_2, \ldots, x_n, z, \overline{p}_1, \overline{p}_2, \ldots, \overline{p}_n).$$

Then if f is a mapping of class C^1 of $\overline{\Sigma}$ in F, it admits partial derivatives $\overline{p}_i = \dfrac{\partial f}{\partial x_i}$ which are functions defined on $\overline{\Sigma}$ with values in $\overline{F}$. If the image of the set Σ through $(f, \overline{p_1}, \overline{p_2}, \ldots, \overline{p_n})$ is in U, then we can consider the integral

$$Y(f) = \int \underset{\theta}{\cdots} \int L(x_1, \ldots, x_n; \quad f(x_1, \ldots, x_n);$$

$$\overline{p}_1(x_1, \ldots, x_n), \ldots, \overline{p}_n(x_1 \ldots, x_n)) \, \mathrm{d}x_1 \ldots \mathrm{d}x_n =$$

$$= \int \underset{\theta}{\cdots} \int L(x_i, f, p_i) \, \mathrm{d}x_1 \ldots \mathrm{d}x_n. \tag{4.54}$$

Problem. Determine among the functions f which take given values on Σ the function for which the integral $Y(f)$ is a minimum or a maximum.

To solve this problem we use a reasoning analogous to the one followed in the case of the simple integrals.

Theorem 1. *The function $Y: f \to Yf$ is of class C^1 and its differential is*

$$\delta Y = \int \underset{\theta}{\cdots} \int \left(\overline{\frac{\partial L}{\partial z}} \, \overline{\delta f} + \sum_{i=1}^{n} \overline{\frac{\partial L}{\partial p_i}} \, \overline{\delta p_i} \right) \mathrm{d}x_1 \ldots \mathrm{d}x_n \tag{4.55}$$

where $\delta p_i = \dfrac{\partial}{\partial x_i} \delta f$.

If L and f_0 are of class C^2 and if we restrict ourselves to the subspace of the functions f which take given values in the contour Γ, formula (4.55) becomes

$$\delta Y = \int \underset{\theta}{\cdots} \int \left(\overline{\frac{\partial L}{\partial z}} - \sum_{i=1}^{n} \frac{\mathrm{d}}{\mathrm{d}x_i} \overline{\frac{\partial L}{\partial p_i}} \right) \overline{\delta f} \, \mathrm{d}x_1 \ldots \mathrm{d}x_n, \tag{4.56}$$

where $\dfrac{\mathrm{d}}{\mathrm{d}x_i}\left(\overline{\dfrac{\partial L}{\partial p_i}}\right)$ *is the partial derivative with respect to x_i of the composed function*

$$(x_1, \ldots, x_n) \to$$

$$\to \overline{\frac{\partial L}{\partial p_i}} \left(x_1, \ldots, x_n, f_0(x_1, \cdots, x_n); \; \frac{\partial f}{\partial x_1}(x_1, \ldots, x_n), \ldots, \frac{\partial f}{\partial x_n}(x_1, \ldots, x_n) \right). \tag{4.57}$$

Theorem 2. *In order that a function f_0 of class C^2 be a maximum or a minimum for Y, among all the functions f of class C^1 which take given values* *on the contour* Γ, *it is necessary that* f_0 *should satisfy Euler's equations with partial derivatives of the second-order*

$$\overline{\frac{\partial L}{\partial z}} - \sum_{i=1}^{n} \frac{d}{dx_i}\left(\overline{\frac{\partial L}{\partial p_i}}\right) = 0. \tag{4.58}$$

The first part of the theorem is proved in the same way as the theorem in Section 4.1.2. To prove formula (4.56) we write the relationship (4.55) thus

$$\delta Y = \int \underset{\theta}{\cdots} \int \left(\overline{\frac{\partial L}{\partial z}}\, \overline{\delta f} + \sum_{i=1}^{m} \frac{d}{dx_i}\left(\overline{\frac{\partial L}{\partial p_i}} \cdot \overline{\delta f}\right) - \right.$$
$$\left. - \sum_{i=1}^{n}{}' \frac{d}{dx_i}\left(\overline{\frac{\partial L}{\partial p_i}} \cdot \overline{\delta f}\right)\right) dx_1 \ldots dx_n. \tag{4.59}$$

If we show that

$$\int \underset{\theta}{\cdots} \int \sum_{i=1}^{n}{}' \frac{d}{dx_i}\left(\overline{\frac{\partial L}{\partial p_i}} \cdot \overline{\delta f}\right) dx_1 \ldots dx_n = 0, \tag{4.60}$$

then from equation (4.59) we have relationship (4.56) and theorem 1 is completely proved. But to prove formula (4.60) we use a formula of the Ostrogradski type, namely the integral of volume (4.60) is replaced by an integral of surface

$$\int \underset{\theta}{\cdots} \int \left(\sum_{i=1}^{n}{}' \overline{\frac{\partial L}{\partial p_i}}\, \overline{\delta f} \cos \alpha_i \right) dS. \tag{4.61}$$

We will not give the proof of Ostrogradski's formula here, it can be found in Ref. 50.

But since from the hypothesis we restrict ourselves to the subspace of the functions f which take given values on Γ, it follows that $\overline{\delta f}$ is zero on Γ and hence we obtain equation (4.60).

A lemma of the type of the lemma in Section 4.3 is necessary to prove theorem 2. But such a lemma is proved exactly in the same way as the lemma in 4.3.

Example 1. Let us return to the problem of finding minimum surfaces. Such a surface in R^3, for which z is expressed in terms of x and y, satisfies the equation with partial derivatives

$$\frac{d}{dx}\left(\frac{p}{\sqrt{1 + p^2 + q^2}}\right) + \frac{d}{dy}\left(\frac{q}{\sqrt{1 + p^2 + q^2}}\right) = 0 \tag{4.62}$$

which can also be written thus

$$\frac{r(1+p^2+q^2)-p(pr+qs)}{(1+p^2+q^2)^{\frac{3}{2}}}+\frac{t(1+p^2+q^2)-q(ps+qt)}{(1+p^2+q^2)^{\frac{3}{2}}}=0 \qquad (4.63)$$

or

$$(r+t)(1+p^2+q^2)-(rp^2+2spq+tq^2)=0. \qquad (4.64)$$

Remark. In the rigorous solution of this problem we meet greater difficulties than in the case of the simple integrals. Let us first notice that the areas of the surfaces $\mathcal{S}$ bordered by $\mathcal{C}$ have a strictly positive lower bound but we do not know yet if it is attained by a surface for which z can be expressed as a function of x and y of class C^1. But if all these conditions are fulfilled and if the function z satisfies equation (4.64) we have to determine the solution of equation (4.64) for which the surface $\mathcal{S}$ passes through the contour $\mathcal{C}$. In other words the function z must take given values on the projection Γ of $\mathcal{C}$ in the plane (x, y).

The problem of finding such a minimum suface, bordered by a given curve, is called Plateau's problem.

Example 2. Let θ be an open set in R^n bounded by a hypersurface Γ of class C^1. Find the function f of class C^1 on $\overline{\theta}$ which takes given values on Γ and which realizes a minimum for the integral

$$\int \cdots_{\theta} \int \sum_{i=1}^{n} \left| \frac{df}{dx_i} \right|^2 dx_1 \ldots dx_n = \int \cdots_{\theta} \int \left(\sum_{i=1}^{n} p_i^2 \right) dx_1 \ldots dx_n .$$

For this, f must be a solution of Euler's equation

$$\sum_{i=1}^{n} \frac{\partial p_i}{\partial x_i} = 0$$

that is, f must be a solution of Laplace's equation

$$\Delta f = \sum_{i=1}^{n} \frac{\partial^2 f}{\partial x_i^2} = 0.$$

In other words f must be a harmonic function and the above problem is reduced to finding a function f of class C^2 in $\overline{\theta}$, which is harmonic, and which takes given values on the contour Γ. Such a problem is called the Dirichlet problem.

4.1.11 Exercises

1. Find the balance (equilibrium) position of a heavy flexible and inextensible chain whose extremities slide on two given curves φ and ψ.

2. Determine an arc $y = y(x)$, $z = z(x)$ of minimum length, that is the so called geodesic arc which joins the points $A(x_0, y_0, z_0)$ and $B(x_1, y_1, z_1)$ on the surface $\varphi(x, y, z) = 0$.

5

Second-order Partial Differential Equations

5.1 CLASSIFICATION OF THE EQUATIONS

5.1.1 Preliminary definitions

A relationship of the form

$$F(x, y, \ldots, u, u_x, u_y, u_{xx}, \ldots) = 0 \tag{5.1}$$

where F is a given function of variables $x, y, \ldots, u, u_x, u_y, u_{xx}, \ldots$ is called a partial diferential equation.

The function $u(x, y, \ldots)$ is called the solution of the equation (5.1) if, in a region from the space of its independent variables, it satisfies, together with its derivatives, equation (5.1).

Equation (5.1) is called nth order equation if the highest order derivative in the equation is of nth order. If F is a linear function of u and of its derivatives then equation (5.1) is called linear. If F is a linear function in the highest order derivatives of u then equation (5.1) is quasilinear.

5.1.2 Cauchy's problem. Characteristics

Let us consider the quasilinear equation of mth order

$$\sum A_{i_1, \ldots, i_n} \frac{\partial^m u}{\partial x_1^{i_1} \ldots \partial x_n^{i_n}} = B, \quad \sum_{j=1}^{n} i_j = m, \tag{5.2}$$

where the summation is made for all the mth order derivatives with respect to x and where the coefficients $A_{i_1, \ldots, i_n}$ and B depend on x_i, u and the derivatives of u of order less than m.

For such an equation we can give some initial conditions. For example we consider a $n - 1$ dimensional manifold and we suppose that on it u and its $(m - 1)$th order derivatives have given values.

Let us consider, for example, as initial manifold the hyperplane $x_n = 0$. We suppose that for $x_n = 0$, u and its derivatives with respect to x_n up to the $n - 1$ order have given values. This means that for $x_n = 0$

$$
\begin{aligned}
u &= \Phi_0(x_1, \ldots, x_{n-1}), \\
\frac{\partial u}{\partial x_n} &= \Phi_1(x_1, \ldots, x_{n-1}), \\
&\ldots\ldots\ldots\ldots\ldots\ldots \\
\frac{\partial^{m-1} u}{\partial x_n^{m-1}} &= \Phi_{m-1}(x_1, \ldots, x_{n-1}).
\end{aligned}
\tag{5.3}
$$

Cauchy's problem consists of finding a solution of equation (5.2) which satisfies the initial conditions shown in equation (5.3).

We suppose that the functions Φ_i have differentials of a high enough order. In this case the initial conditions (5.3) will give initial data for all the other derivatives with respect to x_n, of order $< m$. Then if we write equation (5.2) in the form

$$
a \frac{\partial^m u}{\partial x_n^m} = -\sum b_{i_1, \ldots, i_n} \frac{\partial^m u}{\partial x_1^{i_1} \ldots \partial x_n^{i_n}} + B, \tag{5.4}
$$

it follows that the derivatives from the right-hand side of the above equation are given on the manifold $x_n = 0$, and the values of $\dfrac{\partial^m u}{\partial x_n^m}$ are not known. We can have two cases at each point of the initial manifold $x_n = 0$.

Case 1. $a \neq 0$. In this case we say that the initial conditions are not characteristic.

Case 2. $a = 0$. In this case the initial conditions are characteristic. Let us examine both cases:

Case 1. If $a \neq 0$ at a point P of the manifold $x_n = 0$, relationship (5.4) can be written

$$
\frac{\partial^m u}{\partial x_n^m} = -\sum \frac{b_{i_1 \ldots i_n}}{a} \frac{\partial^m u}{\partial x_1^{i_1} \ldots \partial x_n^{i_n}} + \frac{B}{a} \tag{5.5}
$$

and in this way we obtain all the derivatives of u at the point P of order $\leqslant m$.

Case 2. If $a = 0$ we can not carry out the operations from case 1 and the initial conditions can not determine u and its derivatives.

Example. Consider the equation

$$u_{xy} = 0$$

with the initial conditions

$$u(x, 0) = \Phi(x), \ u_y(x, 0) = \psi(x).$$

We are in case 2, $y = 0$ is a characteristic. We have

$$u_{xy} = u_{yx} = \psi'(x) = 0 \text{ for } y = 0,$$

which implies $\psi(x) = \text{constant} = c_1$ and

$$u_{xyy} = u_{yyx} = 0, \text{ that is } u_{yy} = \text{constant} = c_2.$$

The expansion in series of the function u has the form

$$u(x, y) = \Phi(x) + c_1 y + c_2 \frac{y^2}{2!} + \ldots = \Phi(x) + f(y).$$

This is a solution of the partial differential equation for every constants c_i and thus $f(y)$ can not be determined by means of the initial conditions.

5.1.3 Derivatives along a direction *

Let us consider in the space R^n a surface S of equation

$$f(x_1, \ldots, x_n) = 0, \tag{5.6}$$

where we suppose that f has continuous first-order derivatives which are not all equal to zero, that is

$$\sum_{i=1}^{n} \left(\frac{\partial f}{\partial x_i}\right)^2 \neq 0.$$

The directional parameters of the normal S are given by $\dfrac{\partial f}{\partial x_i}$, $i = 1, \ldots, n$, and the directional cosines are

$$\xi_i = \frac{\partial f}{\partial x_i} \bigg/ \sqrt{\sum_{j=1}^{n} \left(\frac{\partial f}{\partial x_j}\right)^2}, \ i = 1, 2, \ldots, n.$$

---

* or directional derivatives

Let us suppose that the function $u(x_1, \ldots, x_n)$ is defined in a neighbourhood of a point $P \in S$ and has first-order continuous derivatives at P.

Definition. *An expression of the form*

$$\sum_{i=1}^{n} a_i \frac{\partial u}{\partial x_i}$$

where a_i are real numbers, defines the derivative, along a direction, of the function u at the point P, or the directional derivative of u at the point P.

The derivative is along the direction of the tangent or is tangential if

$$\sum_{i=1}^{n} a_i \xi_i = 0. \tag{5.7}$$

Proposition 1. *The tangential derivatives of u on S are determined by the values of u on the surface S.*

Indeed let us suppose that $\dfrac{\partial f}{\partial x_n} \neq 0$. Then in equation (5.6) we can solve with respect to x_n and we obtain

$$x_n = g(x_1, \ldots, x_{n-1})$$

where

$$f(x_1, \ldots, x_{n-1}, \; g(x_1, \ldots, x_{n-1})) = 0.$$

Differentiating the above expression we obtain

$$\frac{\partial f}{\partial x_k} + \frac{\partial f}{\partial x_n} \frac{\partial g}{\partial x_k} = 0, \quad k = 1, \ldots, n-1.$$

or

$$\frac{\partial g}{\partial x_k} = -\frac{\dfrac{\partial f}{\partial x_k}}{\dfrac{\partial f}{\partial x_n}} = -\frac{\xi_k}{\xi_n}, \quad k = 1, \ldots, n-1.$$

As the function u is given on S it follows that the function

$$v(x_1, \ldots, x_{n-1}) = u(x_1, \ldots, x_{n-1}, \; g(x_1, \ldots, x_{n-1}))$$

and its derivatives are given

$$v_{x_k} = \frac{\partial u}{\partial x_k} + \frac{\partial u}{\partial x_n}\frac{\partial g}{\partial x_k} = \frac{\partial u}{\partial x_k} - \frac{\xi_k}{\xi_n}\frac{\partial u}{\partial x_n}. \tag{5.8}$$

Taking equation (5.8) into account we obtain

$$\sum_{k=1}^{n} a_k \frac{\partial u}{\partial x_k} = \sum_{k=1}^{n} a_k \left(v_{x_k} + \frac{\xi_k}{\xi_n}\cdot\frac{\partial u}{\partial x_n}\right) + a_n \frac{\partial u}{\partial x_n} = \sum_{k=1}^{n-1} a_k v_{x_k} +$$

$$+ \frac{1}{\xi_n}\left(\frac{\partial u}{\partial x_n}\right)\sum_{k=1}^{n} a_k\cdot\xi_k. \tag{5.9}$$

But if the derivative along a direction from the left-hand side of relationship (5.9) is tangential then relationship (5.7) is satisfied and we obtain

$$\sum_{k=1}^{n} a_k \frac{\partial u}{\partial x_k} = \sum_{k=1}^{n-1} a_k v_{x_k}$$

where the right-hand side is known and in this way the theorem is completely proved.

Let

$$u' = \sum_{k=1}^{n} \xi_k \frac{\partial u}{\partial x_k}$$

be the derivative of u along the normal which shall be called the norma derivative. We shall prove the following proposition.

Proposition 2. *An arbitrary directional derivative on S is the sum of a multiple of the normal derivative and the tangential derivative.*

In order to prove the proposition it is sufficient to show that the expression

$$\sum_{k=1}^{n} a_k \frac{\partial u}{\partial x_k} - \left(\sum_{k=1}^{n} a_k \xi_k\right) u' \tag{5.10}$$

is a tangential derivative. We have

$$\sum_{k} a_k \frac{\partial u}{\partial x_k} - \left(\sum_{k=1}^{n} a_k \xi_k\right) u' = \sum_{l} a_l \frac{\partial u}{\partial x_l} - \sum_{r,l} a_r \xi_r \xi_l \frac{\partial u}{\partial x_l} =$$

$$= \sum_{l}\left[a_l - \sum_{r} a_r \xi_r \xi_l\right]\frac{\partial u}{\partial x_l}.$$

In order to show that expression (5.10) is a tangential derivative it is necessary to prove that

$$\sum_l \Big[a_l - \sum_r a_r \xi_r \xi_l \Big] \xi_l = 0.$$

We have

$$\sum_l \Big[a_l - \sum_r a_r \xi_r \xi_l \Big] \xi_l = \sum_l a_l \xi_l - \sum_{r,l} a_r \xi_r \xi_l^2 =$$

$$= \sum_l a_l \xi_l - \sum_l a_l \xi_l = 0.$$

Corollary. If u and u' are known on S then all the first-order derivatives of u on S are known.

Definition. The expression

$$\sum_{i,k} a_{ik} \frac{\partial^2 u}{\partial x_i \, \partial x_k} \tag{5.11}$$

is called the second derivative along a direction (or second-order directional derivative).

Generally the derivative of kth order along a direction is

$$\sum_{i_1 + i_2 + \dots + i_n = k} a_{i_1, \dots, i_n} \frac{\partial^k u}{\partial x_1^{i_1} \dots \partial x_n^{i_n}} \tag{5.12}$$

where the summation is made with respect to all kth order derivatives of u.

Definition. We say that the derivative (5.12) *is tangential if:*

$$\sum_{i_1 + i_2 + \dots + i_n = k} a_{i_1, \dots, i_n} \xi_1^{i_1} \dots \xi_n^{i_n} = 0.$$

Definition. The second-order normal derivative is given by the formula

$$u'' = \sum_{i,k} \xi_i \xi_k \frac{\partial^2 u}{\partial x_i \, \partial x_k} .$$

The kth order normal derivative is

$$u^{(k)} = \sum_{i_1, \ldots, i_n} \xi_1^{i_1} \ldots \xi_n^{i_n} \frac{\partial^k u}{\partial x_1^{i_1} \ldots \partial x_n^{i_n}}, \quad \sum_{j=1}^{n} i_j = k.$$

Proposition 3. *If u and its normal derivatives up to the $(m-1)$th order are zero on S then the mth order derivative along a direction, has on S the expression*

$$\sum a_{i_1, \ldots, i_n} \frac{\partial^m u}{\partial x_1^{i_1} \ldots \partial x_n^{i_n}} = \left[\sum a_{i_1, \ldots, i_n} \xi_1^{i_1} \ldots \xi_n^{i_n}\right] u^{(m)}. \tag{5.13}$$

We have to show that

$$\frac{\partial^m u}{\partial x_1^{i_1} \ldots \partial x_n^{i_n}} = \xi_1^{i_1} \ldots \xi_n^{i_n} u^{(m)}. \tag{5.14}$$

We shall use the method of induction. The relationship (5.14) is true for $m = 1$, that is

$$\frac{\partial u}{\partial x_k} = \xi_k u'. \tag{5.15}$$

Let us suppose that it is true for $k \leqslant m-1$. In this case it follows that all the derivatives of u on S of order $\leqslant m-1$ are equal, up to a numerical factor, with the normal derivatives of order $\leqslant m-1$ and thus are zero on S. Also it follows that for every derivative of u on S, let it be $\frac{\partial u}{\partial x_i}$, we have

$$\left(\frac{\partial u}{\partial x_i}\right)' = \left(\frac{\partial u}{\partial x_i}\right)'' = \ldots = \left(\frac{\partial u}{\partial x_i}\right)^{(m-2)} = 0.$$

For $k = m$, using the hypothesis of induction for $\frac{\partial u}{\partial x_k}$, we obtain

$$\frac{\partial^m u}{\partial x_1^{i_1} \ldots \partial x_n^{i_n}} = \frac{\partial^{m-1}}{\partial x_1^{i_1} \ldots \partial x_k^{i_k - 1} \ldots \partial x_n^{i_n}} \left(\frac{\partial u}{\partial x_k}\right) =$$

$$= (\xi_1^{i_1} \ldots \xi_k^{i_k-1} \ldots \xi_n^{i_n}) \left(\frac{\partial u}{\partial x_k}\right)^{(m-1)}. \tag{5.16}$$

But using the hypothesis and relationship (5.15) we obtain

$$(\xi_1^{i_1} \dots \xi_k^{i_k-1} \dots \xi_n^{i_n}) \left(\frac{\partial u}{\partial x_k}\right)^{(m-1)} = (\xi_1^{i_1} \dots \xi_k^{i_k-1} \dots \xi_n^{i_n}) \times$$

$$\times \sum_{j_1+\dots+j_n=m-1} \xi_1^{j_1} \dots \xi_n^{j_n} \frac{\partial^{m-1}}{\partial x_1^{j_1} \dots \partial x_n^{j_n}} \left(\frac{\partial u}{\partial x_k}\right) = (\xi_1^{i_1} \dots \xi_k^{i_k-1} \dots \xi_n^{i_n}) \times$$

$$\times \sum \xi_1^{j_1} \dots \xi_n^{j_n} \frac{\partial}{\partial x_k} \left(\frac{\partial^{m-1} u}{\partial x_1^{j_1} \dots \partial x_n^{j_n}}\right) = (\xi_1^{i_1} \dots \xi_k^{i_k-1} \dots \xi_n^{i_n}) \times$$

$$\times \sum \xi_1^{j_1} \dots \xi_n^{j_n} \xi_k \left(\frac{\partial^{m-1} u}{\partial x_1^{j_1} \dots \partial x_n^{j_n}}\right). \tag{5.17}$$

But

$$(\xi_1^{i_1} \dots \xi_k^{i_k-1} \dots \xi_n^{i_n}) \xi_i \sum \xi_1^{j_1} \dots \xi_n^{i_n} \left(\frac{\partial^{m-1} u}{\partial x_1^{j_1} \dots \partial x_n^{j_n}}\right)' =$$

$$= (\xi_1^{i_1} \dots \xi_n^{i_n}) \sum \xi_1^{j_1} \dots \xi_n^{j_n} \sum_{l=1}^{n} \xi_l \frac{\partial}{\partial x_l} \left(\frac{\partial^{m-1} u}{\partial x_1^{j_1} \dots \partial x_n^{j_n}}\right) =$$

$$= (\xi_1^{i_1} \dots \xi_n^{i_n}) \sum_l (\xi_1^{j_1} \dots \xi_l^{j_l+1} \dots \xi_n^{j_n}) \qquad \frac{\partial^m u}{\partial x_1^{j_1} \dots \partial x_l^{j_l+1} \dots \partial x_n^{j_n}} =$$

$$= \xi_1^{i_1} \dots \xi_n^{i_n} u^{(m)},$$

where

$$j_1 + j_2 + \dots + j_m = m - 1.$$

From formulae (5.16), (5.17) and (5.18) it follows that relationship (5.14) is true for $k = m$ hence the proposition is proved completely.

Corollary 1. If u and its normal derivatives up to the $(m-1)$th *order are given on S, then the tangential derivatives of u of order* $\leqslant m$ *are known on S.*

Indeed if u, u', ..., $u^{(m-1)}$ are zero on S then from formula (5.14) it follows that the tangential derivatives are also zero.

Corollary 2. If u and its normal derivatives up to the mth *order are known, then all the derivatives of order* $\leqslant m$ *are known on S.*

Corollary 3. The derivative along a direction on S can be written in the following form

$$\sum a_{i_1}, \ldots, {}_{i_n} \frac{\partial^m u}{\partial x_1^{i_1} \ldots \partial x_n^{i_n}} = \left[\sum a_{i_1}, \ldots, {}_{i_n} \xi_1^{i_1} \ldots \xi_n^{i_n}\right] u^{(m)} +$$

$$+ \Phi(u, u', \ldots, u^{(m-1)}).$$

5.1.4 The generalized Cauchy problem

Let

$$\sum A_{i_1}, \ldots, {}_{i_n} \frac{\partial^m u}{\partial x_1^{i_1} \ldots \partial x_n^{i_n}} = B, \tag{5.18}$$

be a quasilinear equation of the mth order.

To establish initial conditions for a surface

$$S : f(x_1, \ldots, x_n) = 0,$$

means to suppose that on S the function u and its normal derivatives up to $(m - 1)$th order are known.

From the results we obtained in Section 5.1.3 we deduce that all the derivatives of u of order $\leqslant m - 1$ and all the tangential derivatives of order $\leqslant m$ will be known on S.

If the left-hand side of equation (5.18) is considered as a directional derivative along a given direction then the equation can be written as follows

$$\left[\sum a_{i_1}, \ldots, {}_{i_n} \xi_1^{i_1} \ldots \xi_n^{i_n}\right] u^{(m)} = B + R \tag{5.19}$$

where R is known.

If we suppose that

$$\sum a_{i_1}, \ldots, {}_{i_n} \xi_1^{i_1} \ldots \xi_n^{i_n} \neq 0,$$

then we can obtain directly $u^{(m)}$ from equation (5.19) and thus all the derivatives of u on S of order $\leqslant m$. In this case we say that the initial data are not characteristic. We say that the initial conditions are characteristic if

$$\sum a_{i_1}, \ldots, {}_{i_n} \xi_1^{i_1} \ldots \xi_n^{i_n} = 0.$$

Theorem. *Let there be equation* (5.18) *and the initial non-characteristic conditions. Let us suppose that the coefficients of this equation have continuous derivatives. Then, if a solution of the equation has continuous derivatives of order* $\leqslant s$, *these derivatives are uniquely determined on* S.

Let us suppose that the equation has two solutions and let u and v be these solutions and

$$w = u - v, \text{ or } u = v + w.$$

On S, $w = w' = \ldots = w^{(m-1)} = 0$, let us prove that $w^{(k)} = 0$ for all $k \leqslant s$. Let $w^{(r)}$ be the first normal derivative which is not null on S; we have $m \leqslant r \leqslant s$. Applying Taylor's theorem to the function $w(x_1 + t\xi_1, \ldots$ $\ldots, x_n + t\xi_n)$ we obtain

$$w\,(x_1 + t\;\xi_1, \ldots, x_n + t\xi_n) = \frac{t^r}{r!}\,w^{(r)}\,(0) + O(t^r).$$

The derivatives of w are given by

$$\frac{\partial w}{\partial x_k} = \frac{t^{r-1}}{(r-1)!}\,\xi_k\,w^{(r)}\,(0) + O(t^{r-1}),\ k = 1, \ldots, n$$

and

$$\frac{\partial^m w}{\partial x_1^{i_1} \ldots \partial x_n^{i_n}} = \frac{t^{r-m}}{(r-m)!}\,\xi_1^{i_1} \ldots \xi_n^{i_n}\,w^{(r)}\,(0) + O(t^{r-m}). \tag{5.20}$$

Let us write that u and v are solutions of equation (5.18)

$$\sum A\,(u)\,\frac{\partial^m u}{\partial x_1^{i_1} \ldots \partial x_n^{i_n}} = B\,(u)$$

and

$$\sum A\,(v)\,\frac{\partial^m v}{\partial x_1^{i_1} \ldots \partial x_n^{i_n}} = B\,(v).$$

Substracting these relationships side by side we obtain

$$A\,(u)\,\frac{\partial^m\,(u-v)}{\partial x_1^{i_1} \ldots \partial x_n^{i_n}} + \left(\sum A\,(u) - \sum A\,(v)\right)\,\frac{\partial^m v}{\partial x_1^{i} \ldots \partial x_n^{i_n}} = B\,(u) - B\,(v)$$

or

$$\sum A\,(u)\,\frac{\partial^m w}{\partial x_1^{i_1} \ldots \partial x_n^{i_n}} = -\left(\sum A\,(v+w) - \sum A\,(v)\right)\,\frac{\partial^m v}{\partial x_1^{i_1} \ldots \partial x_n^{i_n}} +$$

$$+ B\,(v+w) - B\,(v),$$

whence taking into account formula (5.20) we obtain

$$\left[\sum A(u)\ \xi_1^{i_1}\dots\xi_n^{i_n}\right]\frac{t^{r-m}}{(r-m)!}\,w^{(r)}(0)=-\sum A(v+w)-$$

$$-\sum A(v)\frac{\partial^m v}{\partial x_1^{i_1}\dots\partial x_n^{i_n}}+B(v+w)-B(v)+K, \tag{5.21}$$

where we denoted the terms of $O(t^{r-m})$ order by K.

Let us denote by

$$\Delta B=B(v+w)-B(v)$$

and let us consider that B can depend on the derivatives of u of order $\leqslant m-1$ as well. Applying the mean value theorem for functions of several variables we obtain

$$\Delta B=\left(\frac{\partial B}{\partial v}\right)w+\frac{\partial B}{\partial\left(\frac{\partial v}{\partial\dots}\right)}\frac{\partial w}{\partial\dots}+\dots+\frac{\partial B}{\partial\left(\frac{\partial^{m-1}v}{\partial\dots}\right)}\frac{\partial^{m-1}w}{\partial\dots}$$

whence it follows immediately that

$$\Delta B=O(t^{r-m}). \tag{5.22}$$

By analogy we obtain

$$\Delta A=O(t^{r-m}). \tag{5.23}$$

Substituting equations (5.22) and (5.23) into equation (5.21) we obtain

$$\left(\sum A\,\xi_1^{i_1}\dots\xi_n^{i_n}\right)\frac{t^{r-m}}{(r-m)!}\,w^{(r)}(0)=R, \tag{5.24}$$

where we denoted the terms of $O(t^{r-m})$ order by R. Dividing equation (5.24) by t^{r-m} and considering that t tends to zero we obtain

$$\left(\sum A\xi_1^{i_1}\dots\xi_n^{i_n}\right)\frac{1}{(r-m)!}\,w^{(r)}(0)=0.$$

But since the initial conditions are non-characteristic i.e. $\sum A\xi_1^{i_1}\dots\xi_n^{i_n}\neq 0$ it follows that

$$w^{(r)}(0)=0$$

which is contrary to the hypothesis. Consequently, all the normal derivatives of w of order $\leqslant s$ are null on S, whence it follows that the derivatives of u on S of order $\leqslant s$ are unique.

Corollary. If equation (5.18) *with initial non-characteristic data has an analytic solution, this solution is uniquely determined.*

Let us consider the equation

$$Q = \sum A_{i_1, \ldots, i_n} \xi_1^{i_1} \ldots \xi_n^{i_n} = 0,$$

where Q is a homogeneous polynomial of degree m in the directional cosines ξ_i, $i = 1, \ldots, n$ of the normal on the surface $f(x_1, \ldots, x_n) = 0$. Q is the characteristic form of equation (5.18).

Considering $\xi_i = \dfrac{\partial f}{\partial x_i}$, $i = 1, \ldots, n$ we obtain a first-order partial differential equation with respect to f

$$\sum A_{i_1, \ldots, i_n} \left(\frac{\partial f}{\partial x_1}\right)^{i_1} \ldots \left(\frac{\partial f}{\partial x_n}\right)^{i_n} = 0. \tag{5.25}$$

Equation (5.25) is called the characteristic equation of equation (5.18).

Example. Let us write the equation

$$a(x, y)\, u_{xx} + 2\, b(x, y)\, u_{xy} + c(x, y)\, u_{yy} = 0.$$

The characteristic form of this equation is

$$Q = a\xi_1^2 + 2b\, \xi_1\, \xi_2 + c\xi_2^2$$

and the characteristic equation is

$$a\left(\frac{\partial f}{\partial x}\right)^2 + 2b\left(\frac{\partial f}{\partial x}\right)\left(\frac{\partial f}{\partial y}\right) + c\left(\frac{\partial f}{\partial y}\right)^2 = 0. \tag{5.26}$$

Since $f(x, y) = 0$ and $y = y(x)$ it follows that $\dfrac{dy}{dx} = -\dfrac{\dfrac{\partial f}{\partial x}}{\dfrac{\partial f}{\partial y}}$ whence if we take into acount equation (5.26) we obtain

$$a\left(\frac{dy}{dx}\right)^2 - 2b\frac{dy}{dx} + c = 0. \tag{5.27}$$

Equation (5.27) can be written as two ordinary differential equations

$$\frac{dy}{dx} = \frac{b \pm \sqrt{b^2 - ac}}{a}.$$

If $b^2 - ac > 0$ we have two families of solutions for equation (5.27) representing the characteristic curves $y = y(x)$. If $b^2 - ac < 0$ we have no real solutions.

5.1.5 The Cauchy-Kowalewski theorem

Let there be the second-order partial differential equation

$$F(x_1, \ldots, x_n, u, p_1, \ldots, p_n, p_{11}, p_{12}, \ldots, p_{nn}) = 0,$$

where F is an analytic function of x_i, u, $p_i = \dfrac{\partial u}{\partial x_i}$, $p_{ik} = \dfrac{\partial^2 u}{\partial x_i \, \partial x_k}$, $i, k = 1, \ldots, n$.

Let $f(x_1, \ldots, x_n) = 0$ be the initial surface S with $\sum_{i=1}^{n} \left(\dfrac{\partial f}{\partial x_i}\right)^2 \neq 0$ and f an analytic function of x_i, $i = 1, \ldots, n$. Let us suppose that u and u' are given analytic functions on S and that at a point $P \in S$ the characteristic form $\sum_{i,k=1}^{n} \dfrac{\partial F}{\partial p_{ik}} \xi_i \xi_k$ is not zero. In this case one can prove that there is an analytic solution $u(x_1, \ldots, x_n)$ in a neighbourhood of the point P which has on S the given initial values u and u'. We shall not prove this theorem, the proof can be found in Ref. 25.

5.1.6 Second-order equations with constant coefficients

Let us consider a linear second-order partial differential equation with constant coefficients

$$au_{xx} + 2bu_{xy} + cu_{yy} + 2du_x + 2eu_y + fu = h(x, y). \tag{5.28}$$

Its characteristic equation is

$$a \, dy^2 - 2b \, dx \, dy + c \, dx^2 = 0. \tag{5.29}$$

If we denote by λ_1, λ_2 the roots of the equation $a\lambda^2 - 2b\lambda + c = 0$, then we can write equation (5.29) in the following form

$$(\mathrm{d}y - \lambda_1 \mathrm{d}x)\,(\mathrm{d}y - \lambda_2 \mathrm{d}x) = 0. \tag{5.30}$$

The solutions of this equation are

$$\begin{aligned} y - \lambda_1 x &= \text{constant}, \\ y - \lambda_2 x &= \text{constant}, \end{aligned} \tag{5.31}$$

and they are called the characteristics of equation (5.28). The roots λ_1, λ_2 can be real and distinct, real and coinciding or complex conjugated.

Let us consider the change of variables

$$\begin{aligned} \xi &= \alpha x + \beta y, \\ \eta &= \gamma x + \delta y. \end{aligned} \tag{5.32}$$

One can immediately verify that the transformed partial differential equation is of the same type as equation (5.28) i.e.

$$Au_{\xi\xi} + 2Bu_{\xi\eta} + Cu_{\eta\eta} + 2Du_{\xi} + 2Eu_{\eta} + Fu = H(\xi, \eta) \tag{5.33}$$

whose characteristic equation is

$$A\ \mathrm{d}\eta^2 - 2\ B\ \mathrm{d}\eta\ \mathrm{d}\xi + C\ \mathrm{d}\xi^2 = 0. \tag{5.34}$$

The characteristics (given by equations (5.31)) of equation (5.28) become by the transformation (5.32) the characteristics of equation (5.33), that is solutions of equation (5.34).

This suggest us to consider the transformation

$$\begin{aligned} y - \lambda_1 x &= \xi, \\ y - \lambda_2 x &= \eta, \end{aligned} \tag{5.35}$$

since in this case the transformed equation has the characteristics

$$\begin{aligned} \xi &= \text{constant}, \\ \eta &= \text{constant}. \end{aligned} \tag{5.36}$$

Let us consider the following cases:

(1) *The hyperbolic case when* $b^2 - ac > 0$. Then λ_1, λ_2 are real and distinct. But by the transformation (5.35) it follows that equation (5.34)

has the solutions shown in formulae (5.36), and we deduce that $A = C = 0$, whence dividing equation (5.33) by 2 B we obtain

$$u_{\xi\eta} + \frac{D}{B} u_{\xi} + \frac{E}{B} u_{\eta} + \frac{F}{2B} u = \frac{H}{2B}. \tag{5.37}$$

In the hyperbolic case we can use the transformation

$$\begin{aligned} y - \lambda_1 x &= \xi + \eta, \\ y - \lambda_2 x &= \xi - \eta, \end{aligned} \tag{5.38}$$

which reduces the partial differential equation to the canonical form

$$u_{\xi\xi} - u_{\eta\eta} + \ldots = H. \tag{5.39}$$

If we set

$$u = e^{-D\xi - E\eta}\, v \tag{5.40}$$

equation (5.39) becomes

$$v_{\xi\xi} - v_{\eta\eta} + kv = f(\xi, \eta). \tag{5.41}$$

(2) *The elliptic case when* $b^2 - ac < 0$. We shall use the transformation

$$\begin{aligned} y - \lambda_1 x &= \xi + i\eta, \\ y - \lambda_2 x &= \xi - i\eta, \end{aligned} \tag{5.42}$$

which, one can notice immediately, is real. Indeed since λ_1 and λ_2 are complex conjugated, formulae (5.42) can be written as

$$\begin{aligned} y - (\text{Re}\,\lambda_1)x &= \xi, \\ y - (\text{Im}\,\lambda_1)x &= \eta. \end{aligned} \tag{5.43}$$

The transformed partial differential equation has the characteristics

$$\xi \pm i\eta = \text{constant}$$

that is

$$d\xi \pm i\, d\eta = 0,$$

has to satisfy the characteristic equation (5.34). Whence we obtain

$$\begin{aligned} -A + 2B + C &= 0, \\ -A - 2B + C &= 0, \end{aligned}$$

whence we have $A = C$ and $B = 0$. Consequently the transformed equation will have the following form

$$u_{\xi\xi} + u_{\eta\eta} + \ldots = H. \tag{5.44}$$

If we use transformation (5.40) equation (5.44) can be written also as:

$$v_{\xi\xi} + v_{\eta\eta} + kv = f(\xi, \eta). \tag{5.45}$$

(3) *The parabolic case.* When $b^2 - ac = 0$, then $\lambda_1 = \lambda_2 = \lambda$. In this case we consider the transformation

$$y - \lambda x = \eta,$$

$$\alpha y + \beta x = \xi,$$

where α, β are arbitrary real numbers satisfying the condition $\beta + a\lambda \neq 0$. In this case the characteristic equation (5.34) is satisfied only by $d\eta = 0$. Hence $C = B = 0$ and the transformed equation will be of the following form

$$u_{\xi\xi} + \ldots = H. \tag{5.46}$$

By means of transformation (5.40) equation (5.46) becomes:

$$v_{\xi\xi} - v_{\eta} = f(\xi, \eta). \tag{5.47}$$

5.2 EQUATIONS OF THE HYPERBOLIC TYPE

5.2.1 Cauchy's problem for the one-dimensional wave equation

Let there be the equation of hyperbolic type:

$$\left(\frac{\partial^2}{\partial t^2} - c^2 \frac{\partial^2}{\partial x^2}\right) u(x, t) = 0, \tag{5.48}$$

where c is a given constant. Equation (5.48) is called the wave equation which can be considered as the equation of the vibrating string. The variable x represents the position of a point of the string and t the time. The operator $\dfrac{\partial^2}{\partial t^2} - c^2 \dfrac{\partial^2}{\partial x^2}$ can be decomposed thus

$$\frac{\partial^2}{\partial t^2} - c^2 \frac{\partial^2}{\partial x^2} = \left(\frac{\partial}{\partial t} - c\frac{\partial}{\partial x}\right)\left(\frac{\partial}{\partial t} + c\frac{\partial}{\partial x}\right).$$

In this case equation (5.48) can be written as:

$$\left(\frac{\partial}{\partial t} - c\frac{\partial}{\partial x}\right) u = v, \quad \left(\frac{\partial}{\partial t} + c\frac{\partial}{\partial x}\right) v = 0. \tag{5.49}$$

Let us suppose that the functions u and v satisfy equation (5.49) in a convex region R of the plane (x, t). Then R can be covered by a set of parallel segments S_γ of equations:

$$x + ct = \gamma$$

Along such a segment S_γ,

$$dv = \frac{\partial v}{\partial x} dx + \frac{\partial v}{\partial t} dt = \frac{\partial v}{\partial x} dx + c\frac{\partial v}{\partial x} dt = \frac{\partial v}{\partial x} d(x + ct) = 0.$$

Thus v is constant along the segment S_γ, let us denote this constant by $\Phi(\gamma)$. Thus

$$v(x, t) = \Phi(\gamma) = \Phi(x + ct).$$

Consequently

$$\left(\frac{\partial}{\partial t} + c\frac{\partial}{\partial x}\right) u = \Phi(x + ct).$$

Let us consider a function $\alpha(\gamma)$ so that

$$\alpha'(\gamma) = \frac{1}{2c}\Phi(\gamma)$$

then

$$\left(\frac{\partial}{\partial t} + c\frac{\partial}{\partial x}\right) u = \left(\frac{\partial}{\partial t} + c\frac{\partial}{\partial x}\right) \alpha(x + ct).$$

By a similar reasoning to the above one it follows that $u(x, t) - \alpha(x + ct)$ is constant along the lines $x - ct =$ constant and hence it is of the form $\beta(x - ct)$. Thus the general solution of equation (5.48) is of the form

$$u(x, t) = \alpha(x + ct) + \beta(x - ct), \tag{5.50}$$

in a convex region R, where α and β are functions of a single argument $x + ct$ and $x - ct$ respectively. For a point from the axis x which moves with the speed $\frac{dx}{dt} = c$ to the right, the value of $x - ct$ and hence of the function $\beta(x - ct)$ does not change. The graph of the function $\beta(x - ct)$ as function of x defines a translation with the speed c along the axis x. We say that $\beta(x - ct)$ represents a plane wave which moves with the speed c.

We can consider formula (5.50) as being obtained by a superposition of two plane waves which move along the axis Ox with the speeds c and $-c$ respectively.

Let us now consider for equation (5.48) the following initial conditions:

$$u = f(x), \quad \frac{\partial u}{\partial t} = g(x) \text{ for } t = 0. \tag{5.51}$$

In this case

$$\alpha(x) + \beta(x) = f(x), \quad c(\alpha'(x) - \beta'(x)) = g(x). \tag{5.51'}$$

From the obvious equality

$$\alpha(x+ct) + \beta(x-ct) = \frac{1}{2}[\alpha(x+ct) + \beta(x+ct) + \alpha(x-ct) + \beta(x-ct)] + \frac{1}{2}\int_{x-ct}^{x+ct} (\alpha'(\xi) - \beta'(\xi))\, d\xi$$

the following formula follows

$$u(x, t) = \frac{f(x+ct) + f(x-ct)}{2} + \frac{1}{2c}\int_{x-ct}^{x+ct} g(\xi)\, d\xi. \tag{5.52}$$

Every solution u of equation (5.48) with the initial conditions given by formulae (5.51) of class C^2 in the triangle (Figure 5.1) of summits $(x+ct, 0)$,

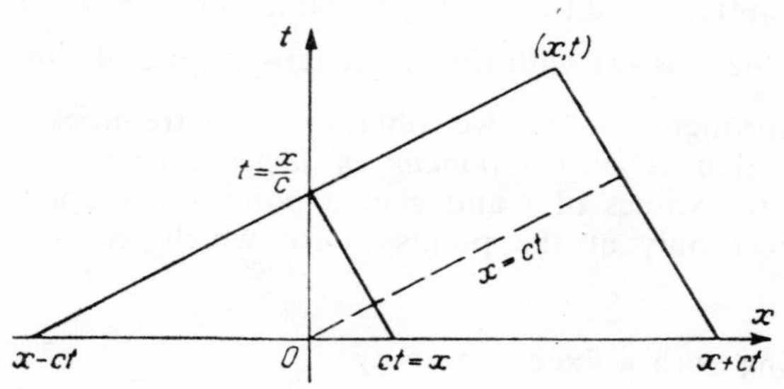

Figure 5.1.

$(x-ct, 0)$, (x, t) is of form shown in formula (5.52) and is uniquely determined by the initial values f, g. Conversely if f is of class C^2 and g is of class C^1 then the function u given by formula (5.52) is a solution of equation (5.48) of class C^2 with the conditions given by formulae (5.51).

Remark 1. Problem (5.48) with the conditions given by formulae (5.51) is *well set*. Indeed let us notice that u depends continuously on the initial conditions given by formulae (5.51) that is, if f and g become $f + \varepsilon$, $g + \varepsilon$, then u varies with $(1 + t)\ \varepsilon$.

Remark 2. Let $f \in C^2$, $g \in C^1$, then for $t = t_1 > 0$ from formula (5.52) we have

$$u = f_1(x), \quad \frac{\partial u}{\partial t} = g_1(x),$$

where $f_1 \in C^2$ and $g_1 \in C^1$. As the partial diferential equation does not change by substituting t by $t - t_1$, the value of $u(x, t)$ for every $t > t_1$ is given by

$$u(x, t) = \frac{f_1(x + c\ (t - t_1)) + f_2(x - c(t - t_1))}{2} + \frac{1}{2c} \int\limits_{x-c(t-t_1)}^{x+c(t-t_1)} g_1(\xi)\ d\xi. \quad (5.53)$$

Formula (5.53) is reduced to formula (5.52) if we substitute f_1 and g_1 by their expressions which are functions of f and g.

Remark 3. Formula (5.52) shows that $u(x, t)$ depends on the initial values $f(y)$, $g(y)$ (where we put $y = x + ct$ or $y = x - ct$) for y in the interval $| y - x | \leqslant ct$.

The interval $x - ct \leqslant y \leqslant x + ct$ is called the domain of dependence of u on the initial conditions at the point (x, t). If we consider the intersection of the axis Ox with the characteristic lines of slope $\frac{1}{c}$ and $-\frac{1}{c}$ which pass through (x, t) we obtain the extremities of the interval $| y - x | \leqslant ct$, that is of *the domain of dependence.*

Conversely the values of f and g at a point y act upon the function u at the moment t only at the points x, for which $| x - y | \leqslant ct$.

5.2.2 The string with a fixed extremity

In applications one seldom comes across initial conditions in which x varies from $-\infty$ to ∞. More often x varies in an interval $[0, L]$. In this case the initial conditions make sense only for $0 \leqslant x \leqslant L$. In order to determine the solution of the equation in this case, we have to have boundary conditions for $x = 0$, $x = L$, when $t \geqslant 0$.

We shall now suppose that $0 \leqslant x \leqslant \infty$. Let $u(x, t)$ be a solution of equation (5.48) of class C^2 for $x \geqslant 0$, $t \geqslant 0$. We consider the initial conditions and the boundary conditions:

$$\begin{aligned} u &= f(x),\ u_t = g(x) \text{ for } x \geqslant 0,\ t = 0, \\ u &= h(t), \text{ for } x = 0,\ t \geqslant 0. \end{aligned} \tag{5.54}$$

For the existence of u it is necessary to have $f \in C^2$, $g \in C^1$ for $x \geqslant 0$ and $h \in C^2$ for $t \geqslant 0$.

In order that $u \in C^2$ for $x = t = 0$, from the conditions given by formulae (5.54) we obtain

$$h(0) = f(0),\ h'(0) = g(0),\ h''(0) = c^2 f''(0). \tag{5.55}$$

Theorem. *Equation* (5.48) *with the conditions given by formulae* (5.54) *and* (5.55) *has a unique solution.*

Indeed, if u exists, then it is of the form (5.50). From conditions (5.54) we obtain:

$$\begin{aligned} \alpha(x) + \beta(x) = f(x),\ c\alpha'(x) - c\beta'(x) = g(x) \text{ for } x \geqslant 0, \\ \alpha(ct) + \beta(-ct) = h(t) \text{ for } t \geqslant 0. \end{aligned} \tag{5.56}$$

If $\alpha(0) = \beta(0) = \frac{1}{2} f(0)$ then from the relationships (5.51′) we determine α and β uniquely that is

$$\alpha(x) = \frac{1}{2} f(x) + \frac{1}{2c}\int_0^x g(\xi)\, d\xi,\quad \beta(x) = \frac{1}{2} f(x) - \frac{1}{2c}\int_0^x g(\xi)\, d\xi \text{ for } x \geqslant 0.$$

From relationship (5.56) it follows that

$$\beta(t) = -\alpha(-t) + h(-t/c) = h(-t/c) - \frac{1}{2} f(-t) + $$

$$+ \frac{1}{2c}\int_{-t}^{0} g(\xi)\, d\xi \text{ for } t \leqslant 0.$$

Substituting this in formula (5.50) we obtain

$$u(x, t) = \frac{f(x + ct) + f(x - ct)}{2} + \frac{1}{2c}\int_{x-ct}^{x+ct} g(\xi)\, d\xi \text{ for } x \geqslant ct \tag{5.57}$$

and

$$u(x, t) = h\left(t - \frac{x}{c}\right) + \frac{f(x+ct) - f(ct - x)}{2} + \frac{1}{2c}\int_{ct-x}^{ct+x} g(\xi)\, d\xi \quad \text{for } x \leqslant ct. \tag{5.57'}$$

One can easily verify that u is a solution of equation (5.48).

In particular the conditions given by formulae (5.55) are necessary in order that u and its first two derivatives should be continuous along the characteristics $x - ct = 0$.

Particular case. $h(t) = 0$ for $t \geqslant 0$. Then if $f \in C^2$, $g \in C^1$ and

$$f(0) = g'(0) = f''(0) = 0,$$

we shall have a solution $u(x, t)$ for $x \geqslant 0$, $t \geqslant 0$.

If we extend f and g for $x < 0$ as even functions of x

$$-f(x) = f(-x), \quad g(-x) = -g(x)$$

the two formulae (5.57) and (5.57′) become identical and the solution $u(x, t)$, defined for every x and $t \geqslant 0$, is an even function of x.

5.2.3 The string with both extremities fixed. D'Alembert's method

We have to obtain a solution of equation (5.48) for $0 \leqslant x < L$, $t \geqslant 0$ which must satisfy the initial conditions

$$u = f(x), \quad u_t = g(x) \quad \text{for} \quad 0 \leqslant x \leqslant L, \quad t = 0 \tag{5.58}$$

and the boundary conditions

$$\begin{aligned} u &= 0, \quad \text{for} \quad x = 0, \quad t \geqslant 0, \\ u &= 0, \quad \text{for} \quad x = L, \quad t \geqslant 0. \end{aligned} \tag{5.58'}$$

From formulae (5.50) and (5.58) we obtain

$$\alpha(ct) + \beta(-ct) = 0,$$

$$\alpha(L + ct) + \beta(L - ct) = 0.$$

From the first relationship we obtain β as function of α, hence

$$u(x, t) = \alpha(x + ct) - \alpha(-x + ct),$$

where the function α verifies the condition

$$\alpha(L + S) - \alpha(-L + S) = 0.$$

That means that α is a periodic function of period $2L$. Consequently every function u which is a solution of the equation of the vibrating string defined for $0 \leqslant x \leqslant L$, $-\infty < t < \infty$ and verifying the conditions $u(0, t) = u(L, t) = 0$ can be extended uniquely as solution $\bar{u}$ defined for every x and t and satisfying the relationships

$$\bar{u}(-x, t) = -\bar{u}(x, t), \tag{5.59}$$

$$\bar{u}(x + 2L, t) = \bar{u}(x, t) \quad \text{and} \quad \bar{u}\left(x, t + \frac{2L}{c}\right) = \bar{u}(x, t).$$

Conversely relationships (5.58) imply

$$u(0, t) = u(L, t) = 0.$$

From the second relationship (5.59) it follows that u is periodic with respect to x of period $2L$. This is natural because u is an even function of x that is antisymmetrical with respect to $x = 0$, but u has to be antisymmetric with respect to $x = L$ as well, hence it has to be a periodic function of period $2L$. This result is interesting only from the theoretical point of view because the length of the string is L. The last relationship (5.59) shows that u is periodic of period $\frac{2L}{c}$ with respect to t.

In order to obtain the solution u which also satisfies the initial conditions (5.58) we have to extend f and g so that the equations

$$f(x) + f(-x) = 0, \quad g(x) + g(-x) = 0$$

$$f(x) + f(2L - x) = 0, \quad g(x) + g(2L - x) = 0 \tag{5.59'}$$

should be satisfied.

This extension can be easily made if f and g are represented in the interval $[0, L]$ by the corresponding Fourier series

$$f(x) = \sum_{n=1}^{\infty} a_n \sin \frac{n\pi x}{L}, \quad g(x) = \sum_{n=1}^{\infty} b_n \sin \frac{n\pi x}{L}.$$

From formula (5.52) it follows that the solution of the equation of the vibrating string with the initial conditions, given by formulae (5.58), and the boundary conditions given by formulae (5.58′), is

$$u(x, t) = \frac{1}{2}\sum_{n=1}^{\infty} a_n \left(\sin\frac{\pi n(x+ct)}{L} + \sin\frac{\pi n(x-ct)}{L}\right) -$$

$$-\frac{L}{2\pi c}\sum_{n=1}^{\infty}\frac{1}{n} b_n\left(\cos\frac{n\pi(x+ct)}{L} - \cos\frac{n\pi(x-ct)}{L}\right).$$

5.2.4 The string with both extremities fixed. Fourier's method

Definition. If $u(x, t)$ is the product of a function of x and a function of t we say that the string has a stationary movement.

Then let there be

$$u(x, t) = U(x)\ V(t). \tag{5.60}$$

Let us determine the solutions of equation (5.48) of the form shown in formula (5.60) which have to satisfy the initial conditions (5.58) and the boundary conditions given by formulae (5.58′). Substituting formula (5.60) in equation (5.48) we obtain

$$\frac{1}{c^2}UV'' = U''V$$

or

$$\frac{1}{c^2}\frac{V''}{V} = \frac{U''}{U} = -\lambda \tag{5.61}$$

where λ is a constant. If we suppose that $V(t)$ is not identically zero then from the boundary conditions given by formulae (5.58′) we obtain

$$U(0) = U(L) = 0. \tag{5.62}$$

From relationships (5.61) we have

$$V'' + \lambda c^2 V = 0,$$

$$U'' + \lambda U = 0. \tag{5.63}$$

this means two differential equations with constant coefficients. From the second equation (5.63) we obtain

$$U(x) = A \cos \sqrt{\lambda}\, x + B \sin \sqrt{\lambda}\, x.$$

Using the conditions given by formulae (5.62) we obtain

$$U(0) = A = 0 \quad \text{and} \quad U(L) = B \sin \sqrt{\lambda}\, L = 0. \tag{5.64}$$

From the last relationship we have

$$\sin \sqrt{\lambda}\, L = 0 \tag{5.65}$$

Let us prove that $\lambda > 0$. Indeed if $\lambda < 0$, $\sqrt{\lambda}$ would be complex and $\cos \sqrt{\lambda}\, L$ and $\sin \sqrt{\lambda}\, L$ would become $\operatorname{ch} \sqrt{\lambda}\, L$ and $\operatorname{sh} \sqrt{\lambda}\, L$ respectively. But as $\operatorname{sh} \sqrt{\lambda}\, L$ cannot be null for any real value of the argument it follows that $\lambda > 0$. Consequently from equation (5.65) we obtain

$$\sqrt{\lambda}\, L = k\pi, \ k \text{ integral}$$

or

$$\sqrt{\lambda} = \frac{k\pi}{L}.$$

Consequently the solution of the second equation (5.63) with the conditions shown in formulae (5.64) is

$$U(x) = C \sin k\pi \frac{x}{L}$$

in which we can consider the arbitrary constant $C = 1$ because every constant factor of U can be introduced in V. Hence we can consider $k > 0$ because if we substitute k by $-k$ it follows that we multiply U by -1.

The general solution of the first equation (5.63) is

$$V(t) = A \cos c\sqrt{\lambda}\, t + B \sin c\sqrt{\lambda}\, t = A \cos k\pi \frac{ct}{L} + B \sin k\pi \frac{ct}{L} \cdot \tag{5.66}$$

Consequently the general solution of equation (5.48) in the case of stationary motion is

$$u(x, t) = \sin k\pi \frac{x}{L} \left(A \cos k\pi \frac{ct}{L} + B \sin k\pi \frac{ct}{L} \right).$$

We can notice that $u(x, t)$ is a periodic function with respect to t of period

$$T = \frac{2L}{kc}.$$

We call the frequency the number

$$N = \frac{1}{T} = k\frac{c}{2L}$$

and the number

$$\Lambda = cT = \frac{2L}{k}, \tag{5.67}$$

represents the wave length. We notice that this wave length is exactly the period of the extended solution $\bar{u}$ of u.

From formula (5.67) it follows that L must be a multiple of half of the wave length. We notice that U was extended to an even periodic function of period $2L$.

Let us now consider the initial conditions given by formulae (5.58) supposing that $u(x, t)$ is of the form

$$u(x, t) = \sum_{k=1}^{\infty} \sin k\pi\frac{x}{L}\left(A_k \cos k\pi\frac{ct}{L} + B_k \sin k\pi\frac{ct}{L}\right). \tag{5.68}$$

If such a solution can be found from equations (5.7) and (5.9), it follows that it is unique.

If we write that the function $u(x, t)$ given by formula (5.68) satisfies the conditions given by formulae (5.58), we obtain the following relationships:

$$f(x) = u(x, 0) = \sum_{k=1}^{\infty} A_k \sin k\pi\frac{x}{L},$$

$$g(x) = \frac{\partial u}{\partial t}(x, 0) = \sum_{k=1}^{\infty} k\pi\frac{c}{L} \sin k\pi\frac{x}{L}. \tag{5.69}$$

Remark. Every function of x, defined in an interval $(0, L)$ admits a (formal) unique expansion in a Fourier series of $\sin k\pi\frac{x}{L}$. Indeed if we extend such a function in $(-L, 0)$ to an even function and then we extend it on the whole axis Ox to a periodic function of period $2L$, then we can expand it in series of the form (5.69). But this extension was justified and made in formulae (5.59′). From formulae (5.69) we have

$$A_k = \frac{2}{L}\int_0^L f(\xi) \sin k\pi \frac{\xi}{L}\, d\xi,$$

$$B_k = \frac{2}{k\pi c}\int_0^L g(\xi) \sin k\pi \frac{\xi}{L}\, d\xi.$$

Consequently the explicit solution of equation (5.48) with the initial conditions (5.58) and the boundary conditions given by formulae (5.58′) is

$$u(x, t) = \sum_{k=1}^{\infty} \sin k\pi \frac{x}{L}\left[\left(\cos k\pi \frac{ct}{L}\right)\frac{2}{L}\int_0^L f(\xi) \sin k\pi \frac{\xi}{L}\, d\xi + \right.$$

$$\left. + \left(\sin k\pi \frac{ct}{L}\right)\frac{2}{k\pi c}\int_0^L g(\xi) \sin k\pi \frac{\xi}{L}\, d\xi\right]. \tag{5.70}$$

From the conditions imposed on the functions f and g in Section 5.2.2 it follows that the series (5.69) are uniformly convergent. But it does not follow that the series (5.70) is convergent, nor that u can be differentiated up to the second-order. Therefore we cannot state that u is a solution of equation (5.48). We do not discuss over these problems in detail.

5.2.5 The wave equation in the three-dimensional space. The method of spherical means

Let the wave equation in the space R^3 be

$$L(u) = \left(\frac{\partial^2}{\partial t^2} - c^2\Delta\right) u(x_1, x_2, x_3, t) = 0 \tag{5.71}$$

where $\Delta = \frac{\partial^2}{\partial x_1^2} + \frac{\partial^2}{\partial x_2^2} + \frac{\partial^2}{\partial x_3^2}$ is the Laplace operator. We shall write $x = (x_1, x_2, x_3)$ hence follows the notation $u = u(x, t)$.

We shall use a method which will reduce equation (5.71) to an equation of the form of equation (5.48); this is called the method of the spherical means.

Let us consider a function $h(x) = h(x_1, x_2, x_3)$ and let us consider the mean of h on a sphere with radius r and the centre at x, that is

$$I(x, r) = \frac{1}{4\pi} \int_{|y|=1} h(x + ry)\, \mathrm{d}S_y.$$

Remark 1. If h is of class C^s then I is of class C^s in x and r for $r \geqslant 0$.

Remark 2. If h is continuous then

$$I(x, 0) = h(x).$$

Remark 3. Let us write $\mathrm{d}z = \mathrm{d}z_1\, \mathrm{d}z_2\, \mathrm{d}z_3$ and let us consider

$$\iiint_{|z| \leqslant R} h(x + z)\, \mathrm{d}z = \int_0^R 4\pi r^2\, I(x, r)\, \mathrm{d}r. \tag{5.72}$$

If $h \in C^2$, we obtain equation (5.73) applying the operator Δ to relationship (5.72) and then using the formula of divergence

$$\Delta \int_0^R 4\pi r^2 I(x, r)\, \mathrm{d}r = \iiint_{|z| \leqslant R} \Delta\, h(x + z)\, \mathrm{d}z = \iint_{|z| = R} \sum_i h_{x_i}(x + z) \frac{z_i}{R}\, \mathrm{d}S_z =$$

$$= \iint_{|y|=1} \sum_i h_{x_i}(x + Ry)\, y_i R^2\, \mathrm{d}S_y = 4\pi R^2 \frac{\partial}{\partial R} I(x, R). \tag{5.73}$$

Differentiating relationship (5.73) with respect to R we obtain

$$\frac{\partial}{\partial R} \Delta \int_0^R r^2 I(x, r)\, \mathrm{d}r = R^2 \Delta\, I(x, R) = \frac{\partial}{\partial R} R^2 \frac{\partial}{\partial R} I(x, R),$$

hence $I(x, r)$ is a solution of the following partial differential equation

$$\Delta r\, I(x, r) = \frac{\partial^2}{\partial r^2} r\, I(x, r). \tag{5.74}$$

Let us denote by Ω_r the operator which applied to the function h transforms it into $rI(x, r)$. The operator Ω_r transforms functions of x into functions of x for each value of the parameter r. We have

$$\Omega_r h(x) = rI(x, r).$$

From equation (5.74) we have

$$\Delta\Omega_r h = \Omega_r \Delta h = \frac{\partial^2}{\partial r^2}\Omega_r h.$$

Let us apply the above remarks in order to prove the following theorem.

Theorem. *If $u(x, t)$ is a solution of class C^2 of equation* (5.71) *with the initial conditions*

$$u = f(x), \frac{\partial u}{\partial t} = g(x) \text{ for } t = 0 \tag{5.75}$$

then

$$u(x, t) = \left(\frac{\partial}{\partial r}\Omega_r f\right)_{r = ct} + \frac{1}{c}\Omega_{ct} g = \frac{1}{4\pi}\frac{\partial}{\partial t} t \iint\limits_{|y|=1} f(x + cty)\, \mathrm{d}S_y +$$

$$+ \frac{1}{4\pi} t \iint\limits_{|y|=1} g(x + cty)\, \mathrm{d}S_y. \tag{5.76}$$

Indeed from the hypothesis of the theorem it follows that for every $r \geqslant 0$, $t \geqslant 0$ the function $\Omega_r\, u(x, t)$ satisfies the equation

$$0 = \Omega_r\, L(u) = L(\Omega_r u) = \frac{\partial^2}{\partial t^2} - c^2 \frac{\partial^2}{\partial r^2}\Omega_r u$$

with the initial conditions

$$\Omega_r\, u = \Omega_r f, \quad \frac{\partial}{\partial t}\Omega_r u = \Omega_r\, g \qquad \text{for } r \geqslant 0,\ t = 0$$

and with the boundary conditions

$$\Omega_r u = 0 \qquad \text{for } r = 0,\ t \geqslant 0.$$

Consequently this is the same case we studied in Section 5.2.2. From (5.57′) we get

$$\Omega_r u(x, t) = \frac{1}{2}[\Omega_{ct+r} f(x) - \Omega_{ct-r} f(x)] + \frac{1}{2c}\int_{ct-r}^{ct+r} \Omega_\xi\, g(x)\, \mathrm{d}\xi$$

for $0 \leqslant r \leqslant ct$. But from remark 2 it follows that

$$u(x,t) = \lim_{r \to 0} \frac{1}{r} \Omega_r u(x, t),$$

whence formula (5.76) is deduced and thus the theorem is proved.

Remark. From formula (5.76) one can notice immediately that the solution u at a point $P(x_1, x_2, x_3)$ at the moment t, depends only on the values of f and g and of the first derivatives of f on the surface of the sphere with radius ct and with its centre in P. This surface is the domain of dependence of the solution u on the initial conditions. Conversely the initial perturbation, f and g in a neighbourhood of a point $Q(y_1, y_2, y_3)$ will modify the values of u at the moment t for x close to the surface of the sphere with radius ct and with the centre at the point Q. This means that for the wave equation $L(u) = 0$ the perturbations propagate with the velocity c. For the three-dimensional wave equation the following suplementary phenomenon occurs: the effect at a point x due to an initial perturbation at a point y disappears for $ct > | y - x |$; in other words, the effect at the point x disappears after the spherical wave corresponding to the point y has left behind the point x.
This property, which is called Huygen's principle, does not belong to other equations of the hyperbolic type. A finite time is necessary for a perturbation to arrive at a point but it persists, in general, after the wave has left behind the respective point.

5.2.6 The equation of the vibrating membrane. The method of the descent

We shall notice below that unlike the three-dimensional case, for the wave equation in two dimensions the perturbation continues indefinitely.

In order to obtain the solution of the wave equation in two dimensions with given initial conditions, we shall consider this equation as a particular case of the equation in three dimensions. This method, due to Hadamard, is called the method of the descent by which we pass to solutions for equations in $n - 1$ dimensions, from solutions of the equations in n dimensions.

The following equation is given

$$\frac{\partial^2 v}{\partial x_1^2} + \frac{\partial^2 v}{\partial x_2^2} = \frac{1}{c^2} \frac{\partial^2 v}{\partial t^2}$$

with the initial conditions

$$v(x_1, x_2, 0) = f(x_1, x_2), \quad \frac{\partial v}{\partial t}(x_1, x_2, 0) = g(x_1, x_2).$$

It is obvious that $u = v(x_1, x_2, t)$ is a solution of the wave equation in three dimensions in the particular case when u, f and g do not depend on x_3. The solution of the equation is given by formula (5.76) where

$$\Omega_r f = \frac{r}{4\pi} \iint_{|y|=1} f(x_1 + ry_1, x_2 + ry_2)\, \mathrm{d}S_y =$$

$$= \frac{r}{2\pi} \iint_{y_1^2 + y_2^2 \leqslant 1} \frac{f(x_1 + ry_1, x_2 + ry_2)}{\sqrt{1 - y_1^2 - y_2^2}}\, \mathrm{d}y_1\, \mathrm{d}y_2.$$

Remark. We notice that in this case the solution v at a point (x_1, x_2) at the moment t depends on the initial conditions in a disk with the radius ct and the centre at (x_1, x_2).

5.2.7 The non-homogeneous wave equation

Let the non-homogeneous wave equation be

$$L(u) = \left(\frac{\partial^2}{\partial t^2} - c^2\Delta\right) u(x_1, x_2, x_3) = w(x_1, x_2, x_3, t), \tag{5.77}$$

where w is a given function, with the initial conditions

$$u(x_1, x_2, x_3, 0) = 0, \quad \frac{\partial u}{\partial t}(x_1, x_2, x_3, 0) = 0. \tag{5.77'}$$

Let us look for a solution of the form

$$u(x_1, x_2, x_3, t) = \int_0^t v(x_1, x_2, x_3, t, s)\, \mathrm{d}s \tag{5.77''}$$

with

$$L(v) = 0 \tag{5.77'''}$$

for every s.

Let us suppose that

$$v(x_1, x_2, x_3, s, s) = 0.$$

Then we obtain for $\frac{\partial u}{\partial t}$ and $\frac{\partial^2 u}{\partial t^2}$ the following expressions

$$\frac{\partial u}{\partial t} = v(x_1, x_2, x_3, t, t) + \int_0^t \frac{\partial v}{\partial t} (x_1, x_2, x_3, t, s)\, ds =$$

$$= \int_0^t \frac{\partial v}{\partial t} (x_1, x_2, x_3, t, s)\, ds,$$

$$\frac{\partial^2 u}{\partial t^2} = \frac{\partial v}{\partial t} (x_1, x_2, x_3, t, t) + \int_0^t \frac{\partial^2 v}{\partial t^2} (x_1, x_2, x_3, t, s)\, ds$$

and

$$\Delta u = \int_0^t \Delta v\, ds.$$

Using the above expressions in equation (5.77) we obtain

$$L(u) = \frac{\partial v}{\partial t} (x_1, x_2, x_3, t, t) + \int_0^t L(v)\, ds = w(x_1, x_2, x_3, t). \quad (5.78)$$

From equation (5.77″) and (5.78) it follows that for $t = s$

$$\frac{\partial v}{\partial t} (x_1, x_2, x_3, t, s) = w(x_1, x_2, x_3, t).$$

Thus we obtained the following result. The function v is a solution of the equation $L(u) = 0$, that is

$$L(v) = 0,$$

and for $t = s$

$$v(x_1, x_2, x_3, s, t) = 0, \quad \frac{\partial v}{\partial t} (x_1, x_2, x_3, s, t) = w(x_1, x_2, x_3, t).$$

It is obvious that u, given by formula (5.77″), satisfies the initial conditions shown in formulae (5.77′).

We shall proceed as follows:

First we look for a solution $\tilde{v}\,(x_1, x_2, x_3, t, s)$ of the equation $L(u) = 0$ which satisfies the initial conditions

$$\tilde{v}(x_1, x_2, x_3, t, s) = 0, \qquad \frac{\partial \tilde{v}}{\partial t} = w \text{ for } t = 0.$$

From formula (5.76) in Section 5.2.5 it follows that

$$\tilde{v}\,(x_1, x_2, x_3, t, s) = \frac{1}{4\pi c^2 t} \iint\limits_{(\xi_1-x_1)^2+(\xi_2-x_2)^2+(\xi_3-x_3)^2=c^2t^2} w(\xi_1, \xi_2, \xi_3, s)\, \mathrm{d}S$$

Denoting by $r^2 = (\xi_1 - x_1)^2 + (\xi_2 - x_2)^2 + (\xi_3 - x_3)^2$ one can then easily see that

$$v(x_1, x_2, x_3, t, s) = \tilde{v}\,(x_1, x_2, x_3, t - s, s) =$$

$$= \frac{1}{4\pi c^2(t-s)} \iint\limits_{r=c(t-s)} w(x_1, x_2, x_3, s)\, \mathrm{d}S$$

from which it follows that

$$u(x_1, x_2, x_3, t) = \int_0^t v\, \mathrm{d}s = \frac{1}{4\pi c^2} \int_0^t \int\limits_{r=c(t-s)} \frac{w(x_1, x_2, x_3, s)}{t-s}\, \mathrm{d}S\, \mathrm{d}s. \quad (5.79)$$

From formula (5.79) we notice that $u(x, t)$ (where $x = (x_1, x_2, x_3)$) depends only on the values of $w(\xi, s)$, $\xi = (\xi_1, \xi_2, \xi_3)$ on the characteristic cone

$$r^2 = c^2(t - s)^2,\ 0 \leqslant s \leqslant t,$$

with the summit at (x_1, x_2, x_3, t).

Relationship (5.79) can be written in a simpler way if we transform it in a triple integral and write $\xi_3 = c(t - s)$. Then

$$u(x_1, x_2, x_3, t) = \frac{1}{4\pi c^2} \int_0^{ct} \int\limits_{r=\xi_3} \frac{w\left(\xi_1, \xi_2, \xi_3, t - \dfrac{r}{c}\right)}{\xi_3}\, \mathrm{d}S\, \mathrm{d}\xi_3$$

whence

$$u(x_1, x_2, x_3, t) = \frac{1}{4\pi c^2} \iiint\limits_{r \leqslant ct} w\left(\xi_1, \xi_2, \xi_3, t - \frac{r}{c}\right) d\xi_1\, d\xi_2\, d\xi_3. \quad (5.79')$$

If we suppose that $w \in C^2$, then the above relationship can be written as

$$u = L^{-1}(w)$$

where

$$L[L^{-1}(w)] = w$$

and $L^{-1}(w)$ satisfies the following conditions

$$L^{-1}(w) = 0, \quad \frac{\partial}{\partial t} L^{-1}(w) = 0 \text{ for } t = 0.$$

5.3 EQUATIONS OF THE PARABOLIC TYPE

5.3.1 The heat equation

Let us consider the homogeneous equation of parabolic type, with two independent variables, reduced to the canonical form

$$\frac{\partial u}{\partial t} = \frac{\partial^2 u}{\partial x^2}. \quad (5.80)$$

This equation governs the one-dimensional heat propagation. In order to determine a solution of the heat equation (5.80), we shall use the method of the separation of the variables. Thus we shall write

$$u(x, t) = f(x)\, g(t). \quad (5.81)$$

Substituting this expression in equation (5.80) we obtain

$$\frac{g'(t)}{g(t)} = \frac{f''(x)}{f(x)}$$

whence it follows that both terms are equal to a constant λ. Consequently f and g satisfy the differential equations

$$g' - \lambda g = 0,$$
$$f'' - \lambda f = 0.$$

The solutions of these equations are

$$g(t) = e^{\lambda t} \text{ and } f(x) = e^{\pm\sqrt{\lambda}x}$$

whence is follows that

$$u(x, t) = Ae^{\pm\sqrt{\lambda}x + \lambda t}.$$

Let us now suppose that

$$u(x, 0) = Ae^{iax}. \tag{5.82}$$

In this case we obtain $\lambda = -a^2$ and thus

$$u(x, t) = Ae^{iax - a^2t}. \tag{5.83}$$

Let us now consider the heat equation (5.80) with the initial condition

$$u(x, 0) = f(x) \tag{5.84}$$

where

$$f \in C^1 \text{ and } \int_{-\infty}^{\infty} |f(x)| \, dx < \infty.$$

Then it follows (Vol. 1, Section 6.2) that f can be represented by a Fourier integral

$$f(x) = \frac{1}{2\pi} \int_{-\infty}^{\infty} \left[\int_{-\infty}^{\infty} f(\xi) e^{-ia\xi} d\xi \right] e^{iax} dx. \tag{5.85}$$

Let us consider the solution (5.83) of the heat equation with the initial condition (5.82) and let us suppose that A is a function of a, that is:

$$u_a(x, t) = A(a) e^{iax - a^2 t}.$$

We shall look for a solution of the form

$$u(x, t) = \int_{-\infty}^{\infty} u_a(x, t) \, da = \int_{-\infty}^{\infty} A(a) e^{iax - a^2 t} \, da. \tag{5.86}$$

Taking into account the initial condition (5.82) and using Fourier's integral (5.85) we obtain

$$A(a) = \frac{1}{2\pi}\int_{-\infty}^{\infty} f(\xi)\, e^{-ia\xi}\, d\xi.$$

Using again formula (5.85) we can verify that $u(x, t)$ given by expression (5.86), satisfies the initial condition given by formula (5.82). Indeed

$$u(x, 0) = \frac{1}{2\pi}\int_{-\infty}^{\infty}\left[\int_{-\infty}^{\infty} f(\xi)\, e^{-ia\xi}\, d\xi\right] e^{iax}\, da = f(x).$$

The solution we obtained

$$u(x, t) = \frac{1}{2\pi}\int_{-\infty}^{\infty} f(\xi)\, d\xi \int_{-\infty}^{\infty} e^{ia(x-\xi)^2 - a^2 t}\, da$$

can be written more simply if we consider that

$$I = \int_{-\infty}^{\infty} e^{ia(x-\xi)-a^2 t}\, da = \int_{-\infty}^{\infty} e^{-t\left[a - \frac{i(x-\xi)}{2t}\right]^2 - \frac{(x-\xi)^2}{4t}}\, da$$

and make the following change of variable

$$a - \frac{i(x-\xi)}{2t} = \frac{\alpha}{\sqrt{t}}.$$

We then obtain

$$I = \frac{e^{-\frac{(x-\xi)^2}{4t}}}{\sqrt{t}}\int_{-\infty}^{\infty} e^{-\alpha^2}\cdot d\alpha = \frac{\sqrt{\pi}\, e^{-\frac{(x-\xi)^2}{4t}}}{\sqrt{t}}.^{*}$$

* We took into account that $\int_{-\infty}^{\infty} e^{-\alpha^2}\, d\alpha = \sqrt{\pi}$.

Consequently

$$u(x, t) = \int_{-\infty}^{\infty} \frac{e^{-\frac{(x-\xi)^2}{4t}}}{\sqrt{4\pi t}} f(\xi) \, d\xi. \tag{5.87}$$

Remark 1. If we write

$$K(x-\xi, t) = \frac{e^{-\frac{(x-\xi)^2}{4t}}}{\sqrt{4\pi t}}$$

we then see that

$$K(x-\xi, t) = K(\xi - x, t).$$

Remark 2. If we consider that the relationship (5.84) represents the temperature at the initial time we notice that the integral of the temperature at the time $t > 0$ is equal to the integral of the temperature at the initial time.

Indeed if we take into account that

$$\int_{-\infty}^{\infty} K(x-\xi, t) \, d\xi = 1 \text{ for every } t > 0 \tag{5.87'}$$

and we integrate relationship (5.87) after applying Fubini's theorem [Vol 1, Section 5.2], we obtain the following formulae

$$\int_{-\infty}^{\infty} u(x, t) \, dx = \int_{-\infty}^{\infty} f(\xi) \left[\int_{-\infty}^{\infty} K(x-\xi, t) \, dx \right] d\xi =$$

$$= \int_{-\infty}^{\infty} f(\xi) \left[\int_{-\infty}^{\infty} K(\xi - x, t) \, dx \right] d\xi = \int_{-\infty}^{\infty} f(\xi) \, d\xi. \tag{5.88}$$

Remark 3. $K(x-\xi, 0) = \delta_0 (x-\xi)$ where δ_0 is Dirac's function.

Let us verify that formula (5.87) represents a solution of the heat equation with the initial condition given by equation (5.85). We shall show below that this is true for a function f more general than the function we have considered.

Lemma 1. *The function*

$$K(x-\xi, t) = \frac{e^{-\frac{(x-\xi)^2}{4t}}}{\sqrt{4\pi t}}$$

is a solution of the heat equation. This function is called the elementary solution of equation (5.80).

One can check it immediately.

Lemma 2. *If $f(x)$ is piecewise continuous and if*

$$|f(x)| \leqslant Me^{Nx^2} \tag{5.89}$$

then the integral

$$u(x, t) = \int_{-\infty}^{\infty} f(\xi)\, K(x-\xi, t)\, d\xi \tag{5.90}$$

is convergent for $t \in \left(0, \frac{1}{4N}\right)$ and the integrals obtained by differentiating formula (5.90) *under the integral sign are convergent.*

Indeed we have

$$|f(\xi)\, K(x-\xi, t)| \leqslant \frac{Me^{N\xi^2}\, e^{-\frac{x^2+2x\xi-\xi^2}{4t}}}{\sqrt{4\pi t}} = \frac{Me^{-\frac{x^2}{4t}}}{\sqrt{4\pi t}}\, e^{\frac{2x}{t}\xi + \left(N-\frac{1}{4t}\right)\xi^2}$$

whence it follows that the integral will be convergent if the coefficient of ξ^2 is negative that is if $N - \frac{1}{4t} < 0$, or if $t < \frac{1}{4N}\cdot$ Moreover one can immediately verify that for $0 < t < \frac{1}{4N}$ the integrand has continuous derivatives of arbitrary order with respect to x and t and that the integrals of the derivatives are convergent for $t \in \left(0, \frac{1}{4N}\right)\cdot$

Theorem 1. *If $f(x)$ fulfils the conditions from lemma 2, then the function*

$$u(x, t) = \int_{-\infty}^{\infty} f(\xi)\, K(x-\xi, t)\, d\xi, \tag{5.91}$$

where $K(x-\xi, t)$, is given by formula (5.88), is a solution of equation (5.80) for $0 < t < \frac{1}{4N}$ and

$$\lim_{(x,\, t) \to (x, +0)} u(x, t) = f(x) \tag{5.92}$$

at the points where $f(x)$ is continuous. Moreover, u is bounded for $t \geqslant 0$ and x bounded.

Indeed from lemma 2 it follows that for $t \in \left(0, \frac{1}{4N}\right)$ we can differentiate in formula (5.91) under the integral and by taking into account lemma 1 we obtain

$$\frac{\partial u}{\partial t} - \frac{\partial^2 u}{\partial x^2} = \int_{-\infty}^{\infty} f(\xi) \left[\frac{\partial K}{\partial t} - \frac{\partial^2 K}{\partial x^2}\right] d\xi = 0.$$

In order to prove relationship (5.92) let us consider a point x_0 where the function $f(x)$ is continuous and let us notice first that we may consider $f(x_0) = 0$. Indeed otherwise, taking into account formula (5.87′), we can write

$$u(x, t) = \int_{-\infty}^{\infty} [f(\xi) - f(x_0)] K(x-\xi, t) \, d\xi + f(x_0) \int_{-\infty}^{\infty} K(x-\xi, t) \, d\xi =$$

$$= \int_{-\infty}^{\infty} [f(\xi) - f(x_0)] \, K(x-\xi, t) \, d\xi + f(x_0). \tag{5.93}$$

If we denote by $h(\xi) = f(\xi) - f(x_0)$, then h is continuous at the point x_0 and $h(x_0) = 0$. Supposing that relationship (5.92) holds also in this case, we obtain from formula (5.93) the following formula

$$\lim_{(x,t) \to (x_0, +0)} u(x, t) = h(x_0) + f(x_0) = f(x_0)$$

hence relationship (5.92) holds when $f(x_0) \neq 0$ if we suppose that it holds when $h(x_0) = 0$. Hence we may always suppose that $f(x_0) = 0$.

By an obvious substitution relationship (5.91) can be written as

$$u(x, t) = \int_{-\infty}^{\infty} f(\xi + x) \; K(\xi, t) \; d\xi \tag{5.94}$$

and in addition it can be decomposed as follows:

$$u(x, t) = \int\limits_{|\xi| < H} f(x + \xi)\, K(\xi, t)\, d\xi + \int\limits_{|\xi| > H} f(x + \xi)\, K(\xi, t)\, d\xi = I_1 + I_2 \quad (5.94')$$

where H has to be determined. We then have the following expression

$$|I_1| = \left| \int\limits_{|\xi| < H} f(x + \xi)\, K(\xi, t)\, d\xi \right| \leqslant \max |f(x + \xi)| \int\limits_{|\xi| < H} K\, d\xi =$$

$$= \max_{|\xi| < H} |f(x + \xi)| \int\limits_{-\infty}^{\infty} K\, d\xi = \max_{|\xi| < H} |f(x + \xi)|. \quad (5.95)$$

In the above relationship we took into account formula (5.87′) and the fact that $K \geqslant 0$. Since $f(x)$ is continuous at the point $x = x_0$ and $f(x_0) = 0$ it follows that for

$$|x - x_0| < \frac{\delta(\varepsilon)}{2} \text{ and } H < \frac{\delta(\varepsilon)}{2}$$

we have

$$|I_1| \leqslant \max_{|\xi| < H} |f(x + \xi)| < \varepsilon. \quad (5.95')$$

Consequently H has to be less than $\frac{\delta(\varepsilon)}{2}$. Using inequality (5.89) we obtain

$$|I_2| < \left| \int\limits_{|\xi| > H} f(x + \xi)\, K(\xi, t)\, d\xi \right| \leqslant M \int\limits_{|\xi| > H} \frac{e^{N(x + \xi)^2 - \frac{\xi^2}{4t}}}{\sqrt{4\pi t}}\, d\xi$$

hence it follows that

$$|I_2| \leqslant \frac{M e^{\frac{Nx^2}{1-4Nt}}}{\sqrt{4\pi t}} \int\limits_{|\xi| > H} e^{-\left[\frac{1-4Nt}{4t}\right]\left[\xi - \frac{4Ntx}{1-4Nt}\right]^2} d\xi.$$

Writing $\mu = \sqrt{\frac{1-4Nt}{4t}} \left(\xi - \frac{4Ntx}{1-4Nt}\right) = A(\xi - B)$, we obtain

$$|I_2| \leqslant \frac{M e^{\frac{Nx^2}{1-4Nt}}}{\sqrt{\pi}\sqrt{1-4Nt}} \int\limits_{|\mu + AB| > AH} e^{-\mu^2}\, d\mu. \quad (5.96)$$

If x is bounded we have,

$$\lim_{t\to 0} \frac{4\,Ntx}{1-4Nt} = 0, \quad \lim_{t\to 0} H\sqrt{\frac{1-4Nt}{4t}} = \infty.$$

Since $\int_{-\infty}^{\infty} e^{-\mu^2}\,d\mu$ is convergent it follows that

$$\int_{|\mu + AB| > AH} e^{-\mu^2}\,d\mu \to 0$$

or written differently for t small enough and $t < \frac{1}{4N} - \varepsilon_1$, where δ' depends on δ through H, we have

$$\left| \int_{|\mu + AB| > AH} e^{-\mu^2}\,d\mu \right| < \varepsilon_2$$

and thus

$$|I_2| < \frac{Me^{\frac{Nx^2}{\varepsilon_1}}}{\sqrt{\pi}\sqrt{\varepsilon_1}}\,\varepsilon_2 < \varepsilon.$$

Consequently $|x - x_0| < \frac{\delta}{2}$ and $t < \delta' < \frac{1}{4N} - \varepsilon_1$ implies

$$|u(x, t)| \leqslant |I_1| + |I_2| < 2\,\varepsilon$$

and thus relationship (5.92) is completely proved.

We still have to show that u is bounded for $0 \leqslant t \leqslant \frac{1}{4N} - \varepsilon$ and for x bounded. Indeed if we suppose that $|x| < C$ and we take into account formulae (5.94′), (5.95′) and (5.96) we obtain

$$|u(x, t)| \leqslant \max_{|\xi| < H} Me^{N(x+\xi)^2} + \frac{Me^{\frac{Nx^2}{1-4Nt}}}{\sqrt{\pi}\sqrt{1-4Nt}} \int_{|\mu + A| > B} e^{-\mu^2}\,d\mu \leqslant$$

$$\leqslant Me^{N(C+H)^2} + \frac{Me^{\frac{NC^2}{\varepsilon_1}}}{\sqrt{\pi}\sqrt{\varepsilon_1}} \int_{-\infty}^{\infty} e^{-\mu^2}\,d\mu = Me^{N(C+H)^2} + \frac{Me^{\frac{NC^2}{\varepsilon_1}}}{\sqrt{\varepsilon_1}}$$

and thus the theorem is completely proved.

5.3.2 The principle of the maximum value

Theorem. *If a function u defined and continuous in the closed domain* $0 \leqslant t \leqslant T$ and $0 \leqslant x \leqslant l$ *satisfies the heat equation*

$$\frac{\partial u}{\partial t} = \frac{\partial^2 u}{\partial x^2} \tag{5.97}$$

for $t \in (0, t]$ *and* $x \in (0, l)$ *then the function attains its extremum points either at the initial moment or at* $x = 0$ *or* $x = l$.

Indeed let us denote by M the maximum value of $u(x, t)$ for $t = 0$ or for $x = 0$ or $x = l$. We shall proceed by reductio ad absurdum. We suppose that at a point (x_0, t_0), $(0 < x_0 < l, 0 < t_0 \leqslant T)$ the function $u(x, t)$ takes the value

$$u(x_0, t_0) = M + \varepsilon \text{ with } \varepsilon > 0.$$

Since (x_0, t_0) is a maximum point for $u(x, t)$ then

$$\frac{\partial u}{\partial x}(x_0, t_0) = 0 \text{ and } \frac{\partial^2 u}{\partial x^2}(x_0, t_0) \leqslant 0.$$

But the function $u(x_0, t)$ attains its maximum value for $t = t_0$ and thus

$$\frac{\partial u}{\partial t}(x_0, t_0) \geqslant 0.$$

But from equation (5.97) it follows that this is possible only if both sides are null. We shall prove that there is a point (x_1, t_1) where

$$\frac{\partial^2 u}{\partial x^2} \leqslant 0 \text{ and } \frac{\partial u}{\partial t} > 0.$$

For this purpose let K be a constant and set

$$v(x, t) = u(x, t) + k(t - t_0).$$

Obviously

$$v(x_0, t_0) = u(x_0, t_0) = M + \varepsilon$$

and

$$k(t_0 - t) \leqslant kT.$$

Let us consider k so that $kT < \frac{\varepsilon}{2}$ that is $k < \frac{\varepsilon}{2T}$. In this case

$$v(x, t) \leqslant M + \frac{\varepsilon}{2} \text{ for } t = 0 \text{ or } x = 0, \text{ or } x = l. \tag{5.98}$$

As the function $v(x, t)$ is continuous it attains its maximum at a point (x_1, t_1). We have

$$v(x_1, t_1) \geqslant v(x_0, t_0) = M + \varepsilon.$$

But from equation (5.97) it follows that $t_1 > 0$ and $0 < x_1 < l$.

We have the following relationship

$$\frac{\partial^2 v}{\partial x^2}(x_1, t_1) = \frac{\partial^2 u}{\partial x^2}(x_1, t_1) \leqslant 0$$

and

$$\frac{\partial v}{\partial t}(x_1, t_1) = \frac{\partial u}{\partial t}(x_1, t_1) - k \geqslant 0$$

whence it follows that

$$\frac{\partial u}{\partial t}(x_1, t_1) \geqslant k > 0.$$

Consequently at the point (x_1, t_1) equation (5.97) cannot be satisfied as its two sides have opposite signs.

Remark. The function $u(x, t) =$ constant satisfies the heat equation and attains its maximum (or minimum) value at every point. This does not contradict the theorem because from the statement of the theorem it follows that if an extremum of the function is attained in the interior of the domain, then it has to be attained for $t = 0$ or $x = 0$ or $x = l$.

Remark. The physical sense of this theorem which is called the principle of the maximum is obvious. If the temperature at the initial moment or on the boundary does not exceed a certain figure M we cannot have a higher temperature than M.

From the above theorem the following theorem of uniqueness results:

Theorem. *If the solution* $u(x, t)$ *of equation* (5.80) *with the initial condition* (5.84) *has the property that there is a number* M *so that* $|u(x, t)| < M$ *for* $-\infty < x < \infty$ *and* $t \geqslant 0$ *then it is unique.*

Let us suppose that equation (5.80) with the condition given by formula (5.84) has two bounded solutions, and let u_1 and u_2 be these solutions. Then

$$v(x, t) = u_1(x, t) - u_2(x, t)$$

satisfies equation (5.80) and we have

$$|v(x, t)| \leqslant |u_1(x, t)| + |u_2(x, t)| < 2M$$

and

$$v(x, 0) = 0.$$

In order to use the principle of the maximum value let us consider the domain $|x| < L$ for $L \to \infty$.

Given the function

$$V(x, t) = \frac{4M}{L^2}\left(\frac{x^2}{2} + a^2 t\right)$$

this function is continuous, satisfies the heat equation and the following conditions

$$V(x, 0) \geqslant |v(x, 0)| = 0,$$

$$V(\pm L, t) \geqslant 2M \geqslant v(\pm L, t).$$

Using the principle of the maximum value for $|x| \leqslant L$ we have

$$-\frac{4M}{L^2}\left(\frac{x^2}{2} + a^2 t\right) \leqslant v(x, t) \leqslant \frac{4M}{L^2}\left(\frac{x^2}{2} + a^2 t\right).$$

Fixing (x, t), for $L \to \infty$ we obtain

$$v(x, t) = 0.$$

Corollary. *Let us suppose that $f(x)$ is continuous and uniformly bounded for every x. Then $u(x, t)$ given by formula* (5.91) *represents the unique solution of the heat equation with the initial condition* (5.84).

5.4 EQUATIONS OF ELLIPTIC TYPE

5.4.1 The Riemann-Graves integral

We shall present here the notion of Riemann's integral of a function with values in a Banach space. Let us consider a set $E \subset R^m$ which is closed, bounded and measurable in Jordan's sense (Ref 37). Let $E = \bigcup_{k=1}^{n} E_k$ be a finite partition of E in measurable Jordan sets.

Let us denote by ω_d the greater diameter of the sets belonging to the previous partition, with $\nu(E_k)$ the jordanian measure of the set E_k, $(k = 1, \ldots, m)$. We can prove (Ref. 37) that if $x: E \to B$, where B is a Banach space [Vol. 1, Section 3.3], is continuous, then for every $\theta_k \in E_k$

$$\lim_{\substack{n\to\infty \\ \omega_d \to 0}} \sum_{k=1}^{n} x(\theta_k)\ \nu(E_k)$$

exists. This limit is denoted by $\int_E x(t)\, \mathrm{d}t$ and is called the integral in Riemann-Graves' sense of the function x on the set E.

5.4.2 Harmonic functions; definitions, properties

Let E be a bounded, sufficiently regular domain of the space R^m. Under these conditions, if u is a function with values in a Banach space X and v a real function, both belonging to the class C^2 on E, then one can immediately establish the formula

$$\int_E v\Delta u \ \mathrm{d}\omega + \int_E \left(\sum_{i=1}^{m} \frac{\partial v}{\partial x_i}\ \frac{\partial u}{\partial x_i}\right) \mathrm{d}\omega = \int_{\mathrm{Fr}E} v\, \frac{\partial u}{\mathrm{d}n_i}\, \mathrm{d}\sigma, \tag{5.99}$$

where $\dfrac{\mathrm{d}}{\mathrm{d}n_i}$ means the interior normal derivative and Δ is the Laplace operator. Formula (5.99) is called the first formula of Green*. Substituting u by v in this formula and substracting the new formula from formula (5.99) we obtain

$$\int_E (u\Delta v - v\Delta u)\ \mathrm{d}\omega = \int_{\mathrm{Fr}E} \left(u \frac{\mathrm{d}v}{\mathrm{d}n_i} - v \frac{\mathrm{d}u}{\mathrm{d}n_i}\right) \mathrm{d}\sigma. \tag{5.99'}$$

The above formula is called the second formula of Green. If $v = 1$ then the previous formula becomes

$$\int_E \Delta u\ \mathrm{d}\omega = \int_{\mathrm{Fr}\,E} \frac{\mathrm{d}u}{\mathrm{d}n_i}\, \mathrm{d}\sigma. \tag{5.99''}$$

* We denote by FrE the boundary of E.

In the particular case when u is a function with real values, the three above formulae have been established in Vol. 1, Section 1.3.4.

Let $G \subset R^m$ be an open set, X be a Banach space and let the map be $x: \overline{G} \to X$.

Definition. The continuous map $x : \overline{G} \to X$ is a harmonic function in G if:

(1) $x \in C^2$ in G,

(2) $\Delta x = 0_X$ in G, that is $\Delta x(t) = 0_X$ for every $t \in G$.

Proposition 1. *If the map $x: \overline{G} \to X$ is harmonic in G then for every x^* from the dual X^* of the space X, the numerical map $x^*(x)$ is harmonic in G.*

Indeed, let $u = x^*(x)$. We have

$$\frac{1}{h}[u(t + e_k h) - u(t)] = \frac{1}{h}[x^*(x(t + e_k h)) - x^*(x(t))] =$$

$$= x^* \left[\frac{x(t + e_k h) - x(t)}{h}\right]$$

where $e_1 = (1, 0, \ldots, 0, 0)$, $e_2 = (0, 1, 0, \ldots, 0)$, $\ldots$, $e_m = (0, 0, \ldots, 0, 1)$.

Hence

$$\frac{\partial u}{\partial t_k} = x^* \left(\frac{\partial x}{\partial t_k}\right)$$

and thus

$$\Delta u = x^*(\Delta x) = 0.$$

Remark. In theorem 2 we shall prove that a continuous function $x: \overline{G} \to X$ for which $x^*(x)$ is harmonic in G for every $x^* \in X^*$ is harmonic.

Let us consider a point $t \in G$ and let us denote by $S(t, r)$ the ball with the centre at t and with radius r. Let us denote by $\omega_m(r)$ and $\sigma_m(r)$ the measure of the sets $S(t, r)$ and $\mathrm{Fr}\, S(t; r)$.

We shall consider the following formulae

$$\mu_0(x, t; r) = \frac{1}{\sigma_m(r)} \int_{\mathrm{Fr}.S(t;\, r)} x(t')\, d\sigma(t'), \tag{5.100}$$

$$\mu_1(x, t; r) = \frac{1}{\omega_m(t)} \int_{S(t;\, r)} x(t')\, d\omega(t') \tag{5.101}$$

The function μ_0 is the peripheric mean of x on the considered sphere. The function μ_1 is the mean of the values of x on the ball. Writing $\sigma_m(1) = \sigma_m$, $\omega_m(1) = \omega_m$ we have

$$\sigma_m(r) = r^{m-1}\sigma_m, \quad \omega_m(r) = r^m \omega_m,$$

$$\omega_m(r) = \int_0^r \sigma_m(\rho)\,\mathrm{d}\rho = \frac{r}{m}\,\sigma_m(r).$$

Between the two means μ_1 and μ_0 there is as a consequence of Fubini's theorem the following relationship

$$\mu_1(x,\ t;\ r) = \frac{m}{r^m}\int_0^r \rho^{m-1}\mu_0(x,\ t;\ \rho)\,\mathrm{d}\rho. \tag{5.102}$$

After these preliminaries we can give the proposition.

Proposition 2. *Let $x : G \to X$. If $x \in C^1$ and*

$$\int_{\mathrm{Fr}.\,S} \frac{\mathrm{d}x(t)}{\mathrm{d}n}\,\mathrm{d}\sigma(t) = 0_X \tag{5.103}$$

for every sphere $S \subset G$, then x is harmonic in G.

Indeed, from formula (5.99″) it follows that

$$\int_S \Delta x\,\mathrm{d}\omega = 0_X$$

for every sphere $S \subset G$. We deduce that

$$\int_S x^*(\Delta x)\,\mathrm{d}\omega = 0$$

for every $x^* \in X^*$ and $S \subset G$ and hence $x^*(\Delta x) = 0$. Since x^* is arbitrary it follows that $\Delta x = 0_X$ that is $\Delta x(t) = 0_X$ for every $t \in G$.

Hence x is harmonic in G.

Proposition 3. *If the continuous function $x : \overline{G} \to X$ verifies one of the relationships*

$$x(t) = \mu_0(x;\ t;\ r), \quad x(t) = \mu_1(x;\ t;\ r) \tag{5.104}$$

for every $t \in G$ and $S(t;\ r) \subset G$, then it verifies both relationships.

The proof of the theorem is immediate. Let us denote by $\mathcal{G}$ the class of the continuous functions on $\overline{G}$ which verifies one of the previous relationships and with $\mathcal{H}$ the class of the continuous functions on $\overline{G}$ and harmonic in G. From the following considerations we shall have $\mathcal{G} = \mathcal{H}$.

Proposition 4. *If $x \in \mathcal{H}$ then $x \in \mathcal{G}$ (that is $\mathcal{H} \subset \mathcal{G}$).*
Indeed, from formula (5.103) it follows that

$$\int_{\mathrm{Fr}.\, S(t;\, r)} \frac{\mathrm{d}x(t)}{\mathrm{d}n} \mathrm{d}\sigma(t) = 0_X$$

for every $S(t;\, r) \subset G$. This relationship can be written

$$\frac{\mathrm{d}}{\mathrm{d}r} \int_{\mathrm{Fr}.\, S(t;\, 1)} x(t')\, \mathrm{d}\sigma_0(t') = 0_X,$$

$\mathrm{d}\sigma_0$ being the element of area of the unit sphere. We deduce

$$\int_{\mathrm{Fr}.\, S(t;\, 1)} x(t')\, \mathrm{d}\sigma_0(t') = c.$$

Hence

$$\mu_0(x;\, t;\, r) = c.$$

If $r \to 0$, then $\mu_0(x;\, t;\, r) \to x(t)$, thus $x \in \mathcal{G}$ and the proposition is established.

Proposition 5. *If $x \in C^1$ and*

$$\int_{\mathrm{Fr}.\, S} \frac{\mathrm{d}x}{\mathrm{d}n} \mathrm{d}\sigma = 0_X$$

for every sphere $S \subset G$, then $x \in \mathcal{G}$.
This proposition is an immediate consequence of the calculus we made for proposition 4.

Lemma. *If $x \in \mathcal{G}$ then $x \in C^\infty$.*
Indeed, let us consider the function $\alpha\colon [0,\, \infty) \to R$ defined by the relationships

$$\alpha(r) = \begin{cases} e^{\frac{1}{r^2} - \frac{1}{r_0^2}} & \text{for } r \in [0,\, r_0], \\ 0 & \text{for } r > r_0. \end{cases}$$

One can immediately verify that α is indefinitely differentiable at every point $r \in [0, \infty)$. We shall suppose (possibly by multiplying α by a scalar) that in addition the function α verifies the condition

$$\sigma_m \int_0^{r_0} r^{m-1}\alpha(r)\, dr = 1.$$

Let us put $d(t, t') = \overline{tt'}$ *

$$y(t) = \int_G \alpha(\overline{tt'})\, x(t')\, d\omega(t')$$

and denote by G_{r_0} the set of the points t which satisfy the condition $d(t, \complement G) > r_0$ where $d(t, \complement G)$ is the distance from the point t to the set $\complement G$. We can easily notice that on the open set G_{r_0} we have $y \in C^\infty$.

By convertion to the polar coordinates in R^m and taking into account the definition of the function the previous relationship can be written as

$$y(t) = \sigma_m \int_0^{r_0} r^{m-1}\alpha(r)\, \mu_0(x;\ t;\ 1)\, dr =$$

$$= \sigma_m x(t) \int_0^{r_0} r^{m-1}\, \alpha(r)\, dr = x(t).$$

Hence $x \in C^\infty$ in G_{r_0}.

Theorem 1. *The continuous map $x\colon \overline{G} \to X$ is harmonic in G if, and only if, $x \in \mathcal{G}$.*

In order to prove the theorem it is sufficient to show that $\mathcal{G} \subset \mathcal{H}$. But if $x \in \mathcal{G}$ then based on the previous lemma we have $x \in C^\infty$. From the relationship

$$x(t) = \mu_0(x;\ t;\ r)$$

we deduce, by differentiating with respect to r, that

$$\int_S \frac{dx}{dr}\, d\omega = 0_X$$

---

* $d(t, t')$ represents the distance from t to t'.

for every sphere $S \subset G_{r_0}$ and thus, from proposition 2, it follows that x is harmonic in G_{r_0}, for every $r_0 > 0$, that is, x is harmonic in H since

$$G = \bigcup_{r_0 > 0} G_{r_0}.$$

The theorem proved above has important applications in the theory of the harmonic functions. One of them concerns the converse of proposition 1.

Theorem 2. *Let x: $\overline{G} \to X$ be a continuous function. If $x^*(x)$ is harmonic for every $x^* \in X^*$, then x is harmonic.*

Indeed, if $x^*(x)$ is harmonic, then, writing $y = x^*(x)$, we have, for example

$$y(t) = \mu_0(y;\ t;\ r)$$

for every $t \in G$ and $r < d(t, \complement G)$. The previous relationship can be also written

$$x^*(x(t) - \mu_0(x;\ t;\ r)) = 0$$

whence as x^* is arbitrary in X^*, it follows that

$$x(t) = \mu_0(x;\ t;\ r)$$

hence the function x is harmonic according to theorem 1.

5.4.3 Liouville's theorem

Let $x : \overline{G} \to X$, where G is an open set in R^m and X a Banach space.

Let us suppose that x is continuous in $\overline{G}$ and harmonic in G. Then $\dfrac{\partial x}{\partial t_k}$ is also harmonic in G. Thus, for every $t \in G$ and for $r < d(t, \complement G)$ otherwise arbitrary, we shall have

$$\frac{\partial x(t)}{\partial t_k} = \frac{1}{\omega_m(r)} \int_{S(t;\,r)} \frac{\partial x(t')}{\partial t_k}\, \mathrm{d}\omega_m(t')$$

or

$$\frac{\partial x(t)}{\partial t_k} = \frac{1}{\omega_m(r)} \int_{\mathrm{Fr.}S\,(t,r)} x(t')\ \mathrm{d}\sigma_m(t').$$

From the hypothesis, there is a positive number M so that $\| x(t) \| < M$, for every $t \in \overline{G}$. We deduce that

$$\left\| \frac{\partial x(t)}{\partial t_k} \right\| \leqslant \frac{\sigma_m(r)}{\omega_m(r)} M < \frac{Mm}{r}. \tag{5.105}$$

Liouville's theorem. *If $x : R^m \to X$ is harmonic and bounded then x is a constant function.*

Since $\| x(t) \| < M$ for every $t \in R^m$, from inequalities (5.105) it follows that for $r \to \infty$

$$\frac{\partial x(t)}{\partial t_k} = 0_X, \; k = 1, 2, \ldots, m,$$

hence

$$x(t) = \text{constant}.$$

5.4.4 Green's formula

Let $E \subset R^m$ be a domain sufficiently regular, let $u: E \to X$ and $v: E \to R$ be harmonic functions in E. Then from formula (5.99′) it follows that

$$\int_{\text{Fr}.E} \left(u \frac{dv}{dn_i} - v \frac{du}{dn_i} \right) d\sigma = 0. \tag{5.106}$$

Let $t \in \mathring{E}$ and * let

$$v(t, t') = \begin{cases} \ln \dfrac{1}{\| t' - t \|}, & m = 2, \\ \dfrac{1}{\| t' - t \|^{m-2}}, & m > 2. \end{cases} \tag{5.107}$$

We know that the function v is harmonic on the set $G = \mathring{E} - \{t\}$. Let us consider a sphere $S(t; \varepsilon) \subset \mathring{E}$. On the set $E - S(t; \varepsilon)$ the function v is harmonic. Applying formula (5.106) we obtain

$$\int_{\text{Fr}.E} \left(u \frac{dv}{dn_i} - v \frac{du}{dn_i} \right) d\sigma = \int_{\text{Fr}.S(t;\varepsilon)} \left(u \frac{dv}{dn_i} - v \frac{du}{dn_i} \right) d\sigma.$$

---

* We write $\mathring{E} = \text{int } E$.

But on Fr.$S(t;\varepsilon)$

$$\frac{dv}{dn_i} = \begin{cases} -\dfrac{1}{\varepsilon} \text{ if } m = 2, \\ -\dfrac{1}{m-2}\dfrac{1}{\varepsilon^{m-1}} \text{ if } m > 2. \end{cases}$$

From theorem 1 in Section 5.4.2 it follows that we can apply the formulae (5.104) to the function u and thus we obtain

$$\int_{\text{Fr}.S(t;\varepsilon)} u\frac{dv}{dn_i}\,d\sigma = \begin{cases} -\sigma_2 u(t), \text{ if } m = 2, \\ -\sigma_m (m-2) \text{ if } m > 2, \end{cases}$$

and

$$\lim_{\varepsilon\to 0} \int_{\text{Fr}.S(t;\varepsilon)} v\frac{du}{dn_i}\,d\sigma = 0.$$

Consequently writing $r(t') = \|t' - t\|$, it follows that

$$u(t) = \begin{cases} \dfrac{1}{\sigma_2}\displaystyle\int_{\text{Fr}.E}\left(u\frac{d\ln\frac{1}{r}}{dn_i} - \ln\frac{1}{r}\frac{du}{dn_i}\right)d\sigma \text{ for } m = 2, \\ \dfrac{1}{(m-2)\sigma_m}\displaystyle\int_{\text{Fr}.E}\left(u\frac{d\frac{1}{r^{m-2}}}{dn_i} - \frac{1}{r^{m-2}}\frac{du}{dn_i}\right)d\sigma, \text{ for } m>2. \end{cases} \tag{5.108}$$

This formula is called Green's formula for the harmonic functions.

5.4.5 Dirichlet's problem

From the mean-value formula for the harmonic functions

$$x(t) = \mu_1(x;\ t;\ r)$$

we deduce

$$\|x(t)\| \leqslant \mu_1(\|x\|;\ t;\ r).$$

Remark. It follows that the function $\| x \|: E \to R^+$ * can not attain its upper bound at an interior point.

In these conditions let x and y be two harmonic functions, defined on E which take the same values on the boundary of E. Then the function $z = x - y$ is harmonic in $\mathring{E}$ and null on Fr.E. But based on the above remark, $\| z \|$ attains its upper bound on Fr.E. Thus $z = 0$ in E that is $x = y$. In other words:

There is at most a function defined and continuous on E, harmonic in $\mathring{E}$ *which on* Fr.E *takes given values.*

The problem of the existence of such a function is called *Dirichlet's problem.*

We shall prove that, in the case when E is a domain with the sufficiently regular boundary, this problem can be reduced to a particular Dirichlet problem for the real harmonic functions.

Indeed, let us suppose that we know haw to determine the real function $g: E \to R$, harmonic in $\mathring{E}$ whose restriction to Fr. E is equal to the restriction of the function v to Fr.E. By applying formula (5.106) to the functions u and g and substracting it from relationship (5.108) we obtain

$$u(t) = \begin{cases} \dfrac{1}{\sigma_2} \displaystyle\int_{\text{Fr.}E} u \frac{dG}{dn_i} d\sigma(t'), & m = 2, \\ \dfrac{1}{(m-2)\sigma_m} \displaystyle\int_{\text{Fr.}E} u \frac{dG}{dn_i} d\sigma(t'), & m > 2, \end{cases} \tag{5.109}$$

where we wrote

$$G(t, t') = v(t, t') - g(t, t').$$

Thus Dirichlet's problem for the harmonic functions with values in a Banach space is reduced to Dirichlet's problem for the real harmonic functions. The function G is called *Green's function* with respect to the set E.

Remark. By obtaining formula (5.109) we have not yet solved Dirichlet's problem. We still have to verify that if in this formula we substitute in the right-hand side, the function by a function f: Fr. $E \to X$, then the function u thus obtained tends to f, as the point t tends (in a way which we have to specify) to Fr. E.

In any case, if Dirichlet's problem for E has a solution, then it is necessarily given by the formula (5.109).

* We denoted by R^+ the set of the positive real numbers.

5.4.6 Dirichlet's problem for the sphere

In several particular cases, Green's function can be determined by simple geometrical considerations. This is for example the case of the sphere. Let us consider the sphere $S(t_0;\ R) \subset R^m$, $(m > 2)$. If t is an interior point of this sphere and t_1 is its harmonic conjugate with respect to the sphere, it is known that for every point z on the boundary of the sphere we have denoting by

$$d(z,\ t) = r, \quad d(z,\ t_1) = r_1,$$

$$\frac{r_1}{r} = \frac{R}{l} \tag{5.110}$$

l being the distance of the point t to the centre of the sphere.
From relationship (5.110) we deduce that the function

$$g(t,\ t') = \frac{R}{l}\,\frac{1}{r_1^{\,m-2}}$$

is harmonic in the sphere $S(t_0;\ R)$ and coincides with the function $v = \dfrac{1}{r^{m-2}}$ on the boundary of the sphere.
Thus Green's function for the sphere $S(t_0,\ R)$ is in the case where $m > 2$,

$$G(t,\ t') = \frac{1}{r^{m-2}} - \frac{R}{l}\,\frac{1}{r_1^{m-2}}\,. \tag{5.111}$$

In the case of the plane $(m = 2)$ we shall obtain

$$g\,(t,\ t') = \ln\left(\frac{R}{l}\cdot\frac{1}{r_1}\right)$$

whence

$$G(t,\ t') = \ln\frac{1}{r} - \ln\left(\frac{R}{l}\cdot\frac{1}{r_1}\right). \tag{5.112}$$

Substituting the right-hand side of formula (5.111) into formula (5.109), we obtain (we will not go into the detailes of the calculus we used) the following formula

$$x(t) = \frac{R^2 - l^2}{R\sigma_m}\int_{\mathrm{Fr.}S}\frac{x(t')\,d\sigma(t')}{r^m}\,. \tag{5.113}$$

This formula is called Poisson's formula. We shall now prove that this formula solves Dirichlet's problem for the sphere.

Let $u : \mathrm{Fr.}\, S \to R$ be a continuous function. We shall consider the function

$$x(t) = \frac{R^2 - l^2}{R\sigma_m} \int\limits_{\mathrm{Fr}.S} \frac{u(t')\, d\sigma(t')}{r^m} . \tag{5.114}$$

Theorem. (1) *The function $x(t)$ given by formula* (5.114) *is harmonic in the interior of the sphere S.* (2) *The function $x(t)$ tends to $u(t')$ for every point $t' \in \mathrm{Fr}.S$, as the point t tends from the interior of the sphere to the point $t' \in \mathrm{Fr}.S$.*

(1) For simplicity's sake we shall suppose $t_0 = 0$ and $t' = (0, 0, \ldots, 0, R)$. It will be sufficient to prove that the function

$$\varphi(t) = \frac{R^2 - l^2}{r^m} = \frac{R^2 - \sum\limits_{k=1}^{m} t_k^2}{\left[\sum\limits_{k=1}^{m} t_k^2 + (R - t_m)^2\right]^{\frac{m}{2}}}$$

is harmonic. But we can write

$$\varphi(t) = \frac{2\, R^2 - 2\, Rt_m - \sum\limits_{k=1}^{m-1} t_k^2 - (R - t_m)^2}{\left[\sum\limits_{k=1}^{m-1} t_k^2 + (R - t_m)^2\right]^{\frac{m}{2}}} = \frac{1}{r^{m-2}} + \frac{2R}{m-2} \frac{\partial \dfrac{1}{r^{m-2}}}{\partial t_m},$$

a formula which underlines the harmonicity of the function φ. It is obvious that the reasoning we made does not depend on the chosen particular position of the point t', a position which can be always realized by a change of the axis. Such a change leaves the Laplacian of a function invariant.

In order to study the function x, let us consider a sphere Σ with the center in the point t and the radius small enough in order to have $\Sigma \subset S$. On the compact $\overline{\Sigma} \cup \mathrm{Fr}.\, S$ the function $\dfrac{\partial \varphi}{\partial t_k}$ is continuous. We write

$$x_k(t) = \frac{1}{R\sigma_m} \int\limits_{\mathrm{Fr}.S} \frac{\partial \varphi}{\partial t_k}\, x(t')\, d\sigma(t').$$

If $|h|$ is less than the radius of the sphere Σ, we can write

$$\left\| \frac{x(t+e_k h) - x(t)}{h} - x_k(t) \right\| \leqslant$$

$$\leqslant \frac{1}{R\sigma_m} \int_{\mathrm{Fr}.S} \left| \frac{\varphi(t+e_k h) - \varphi(t)}{h} - \frac{\partial \varphi}{\partial t_k} \right| \| x(t') \| \, d\sigma(t')$$

or, applying to the integral the mean-value formula we have

$$\left\| \frac{x(t+e_k h) - x(t)}{h} - x_k(t) \right\| \leqslant$$

$$\leqslant \frac{1}{R\sigma_m} \int_{\mathrm{Fr}.S} \left| \frac{\partial \varphi(t+\theta e_k h)}{\partial t_k} - \frac{\partial \varphi(t)}{\partial t_k} \right| \| x(t') \| \, d\sigma(t').$$

As the function $\frac{\partial \varphi}{\partial t_k}$ is continuous on $\bar{\Sigma} \cup \mathrm{Fr}.S$, we shall be able to take the number $h_0 > 0$ less than the radius of the sphere $\bar{\Sigma}$, so that for $|h| < h_0$ the difference under the integral should be less than the number $\frac{\varepsilon}{M}$, where M is an upper bound for the numerical set $\{\| u(t') \|\}$. For $|h| < h_0$ we shall obtain

$$\left\| \frac{x(t+e_k h) - x(t)}{h} - x_k(t) \right\| < \varepsilon$$

whence

$$\frac{\partial x(t)}{\partial t_k} = x_k(t) = \frac{1}{R\sigma_m} \int_{\mathrm{Fr}.S} u(t') \frac{\partial \varphi}{\partial t_k} \, d\sigma(t').$$

Similarly

$$\frac{\partial^2 x(t)}{\partial t_k^2} = \frac{1}{R\sigma_m} \int_{\mathrm{Fr}.S} u(t') \frac{\partial^2 \varphi}{\partial t_k^2} \, d\sigma(t').$$

Hence

$$\Delta x(t) = \frac{1}{R\sigma_m} \int_{\mathrm{Fr}.S} u(t') \Delta \varphi \; d\sigma(t') = 0.$$

therefore x is a harmonic function in S.

(2) Let us now prove that

$$\lim_{t \to t_0'} x(t) = u(t_0'), \quad (t_0' \in \mathrm{Fr}.S).$$

The function u is continuous at the point t'. Thus as $\varepsilon > 0$ is given there is a spheric segment c with the centre at the point t_0, and with solid angle $2\delta_\varepsilon$, so that, for every pair of points $t_1', t_2' \in c$, we have $\| u(t_2') - u(t_1') \| < \frac{\varepsilon}{2}$. We shall write $C = \mathrm{Fr}.S - c$.

We have

$$x(t) = \frac{R^2 - l^2}{R\sigma_m} \int_C \frac{u(t')\, d\sigma(t')}{r^m} + \frac{R^2 - l^2}{R\sigma_m} \int_c \frac{u(t')\, d\sigma(t')}{r^m}.$$

If we notice that the (unique) harmonic function in S, which at every point of the boundary of S tends to 1, is the constant function equal to 1, we deduce that

$$1 = \frac{R^2 - l^2}{R\sigma_m} \int_{\mathrm{Fr}.S} \frac{d\sigma(t')}{r^m}$$

and hence that

$$u(t_0') = \frac{R^2 - l^2}{R\sigma_m} \int_{\mathrm{Fr}.S} \frac{u(t_0')\, d\sigma(t')}{r^m}.$$

Thus

$$x(t) - u(t_0') = \frac{R^2 - l^2}{R\sigma_m} \int_{\mathrm{Fr}.S} \frac{u(t') - u(t_0')}{r^m}\, d\sigma(t') =$$

$$= \frac{R^2 - l^2}{R\sigma_m} \int_c \frac{u(t') - u(t_0')}{r^m}\, d\sigma(t') + \frac{R^2 - l^2}{R\sigma_m} \int_C \frac{u(t') - u(t_0')}{r^m}\, d\sigma(t').$$

But

$$\frac{R^2 - l^2}{R\sigma_m} \left\| \int_c \frac{u(t') - u(t_0')}{r^m}\, d\sigma(t') \right\| \leqslant \frac{(R^2 - l^2)\,\varepsilon}{2R\sigma_m} \int_c \frac{d\sigma(t')}{r^m} \leqslant$$

$$\leqslant \frac{(R^2 - l^2)\,\varepsilon}{2R\sigma_m} \int_{\mathrm{Fr}.S} \frac{d\sigma(t')}{r^m} = \frac{\varepsilon}{2}.$$

Let K be the cone with the summit at t_0 and the generators leaning upon the spherical segment c and let K_0 be a cone with the same summit and the generators leaning upon a spherical segment $c_0 \subset c$ with the centre at t_0'.

If $t \in \overline{K}_0$ the function $\psi: \overline{K}_0 \to R$ defined by the relationship

$$\psi(t) = \frac{R^2 - l^2}{R\sigma_m} \int_C \frac{\| u(t') - u(t'_0) \|}{r^m} \, d\sigma(t')$$

is continuous and null for $t \in c_0$, that is for $l = R$. But we can write $\overline{K}_0 = [0, R] \times c_0$, which means that we represent the point t by the pair (l, t') where l is the distance of t to the centre of the sphere S and t', is the point where the line passing through the points t_0 and t intersects the spherical segment c_0. Thus we can write

$$\psi(t) = \chi(l, t'),$$

where the function χ is continuous on $[0, R] \times c_0 = \overline{K}_0$ null for $l = R$, for every $t' \in c_0$. But χ is a continuous function of l, uniformly with respect to $t' \in c_0$, therefore there exists $R_0 < R$, so that $t' \in c_0$, $l > R_0$ should imply that $\psi(t) < \frac{\varepsilon}{2}$.

The domain defined by the relationships

$$t' \in c_0, \quad l > R_0$$

is $D_0 = K_0 - K_0 \cap S(t_0; R_0)$. Hence $t \in D_0$ implies that $\| x(t) - u(t') \| < \varepsilon$ which is equivalent to

$$\lim_{t \to t'} x(t) = u(t').$$

5.4.7 Poisson's equation

We shall write again a definition already given in (Chap. 2) in the form which we shall need later.

Definition. A function $u : E \subset R^m \to R$ *is called an elementary solution of the equation* $\Delta u = 0$ *if* u *is a solution of the equation*

$$\Delta u = \delta_0$$

where δ_0 *is Dirac's function.*

Remark. An elementary solution is not uniquely determined. Indeed we can add to this solution, a solution w of Laplace's equation. In particular Green's function is an elementary solution of Laplace's equation.

Let us now consider the non-homogeneous equation

$$\Delta u = f(x). \tag{5.115}$$

This equation is called *Poisson's equation.*

From the definition we gave to an elementary solution we can expect that a particular solution of this equation should be of the form

$$u(x) = \frac{1}{(m-2)\,\omega_m} \int_E f(x') \frac{1}{r^{m-2}}\, d\omega(x'), \quad (m > 3). \tag{5.116}$$

Indeed, by differentiating formally under the integral sign and by taking into account the above remark we have

$$\Delta u(x) = \frac{1}{m-2} \int_E \frac{1}{\omega_m} \Delta \frac{1}{r^{m-2}} f(x')\, d\omega(x') =$$

$$= \frac{1}{m-2} \int_E \delta_0(x', x)\, f(x')\, d\omega(x') = f(x).$$

In the following our purpose will be to justify the above calculus.

Proposition. *If f is a function of class C^1 on $E \cup$ Fr.E then u is of class C^1 on* Fr.E *and of class C^2 in E and satisfies Poisson's equation* (5.115).

First we shall verify that the first derivatives of u can be obtained by differentiating in formula (5.116) under the integral sign. To this purpose let us define the function

$$\varphi_\varepsilon(r) = \begin{cases} \dfrac{1}{2}\left(\dfrac{r^2}{\varepsilon^2} - 1\right) + \dfrac{1}{\varepsilon^{m-2}}, & r \leqslant \varepsilon. \\[2ex] \dfrac{1}{r^{m-2}}, & r > \varepsilon \end{cases} \tag{5.117}$$

and write

$$u_\varepsilon(x) = \frac{1}{(m-2)\,\omega_m} \int_E \varphi_\varepsilon(r)\, f(x')\, d\omega(x'). \tag{5.118}$$

Let us notice that the function $\varphi_\varepsilon(r)$ differs from an elementary solution $\frac{1}{r^{m-2}}$ only in a sphere S_ε with the centre in x and of radius ε.

It is easy to verify that f_ε is of class C^1 in $E \cup$ Fr.E. We have

$$|u - u_\varepsilon| \leqslant \frac{1}{(m-2)\,\omega_m} \int_E \left| \frac{1}{r^{m-2}} - \varphi_\varepsilon \right| |f(x')|\, d\omega(x') \leqslant$$

$$\leqslant \frac{1}{\omega_m (m-2)} \int_{S_\varepsilon} |f| \left(\frac{1}{r^{m-2}} + \frac{1}{2}\frac{r^2}{\varepsilon^2} + \frac{1}{2} + \frac{1}{\varepsilon^{m-2}} \right) d\omega(x') \leqslant$$

$$\leqslant \frac{M}{(m-2)\,\omega_m} \int_{\text{Fr.}S_0} d\sigma_0 \int_0^\varepsilon \left(r^{m-1} \frac{1}{r^{m-2}} + \frac{1}{2} \frac{r^{m+1}}{\varepsilon^2} + \frac{1}{2} r^{m-1} + \frac{r^{m-1}}{\varepsilon^{m-2}} \right) dr,$$

where M is a bound of the function f, S_0 is the sphere with its centre in x and of radius 1 and $d\sigma_0$ is the element of area of the unit sphere. From the above inequalities it follows that u_ε converges uniformly to u in $E \cup \text{Fr}.E$.

Since f is of class C^1 in $E \cup \text{Fr}.E$ we can differentiate in formula (5.118) under the integral sign and we obtain

$$\frac{\partial u_\varepsilon}{\partial x_i} = \frac{1}{(m-2)\,\omega_m} \int_E \frac{\partial \varphi_\varepsilon}{\partial x_i} f(x')\, d\omega(x'). \tag{5.119}$$

Let us consider the function

$$\Phi(x) = \frac{1}{(m-2)\,\omega_m} \int_E f(x') \frac{x_i - x_i}{r^m}\, d\omega(x'). \tag{5.120}$$

We have

$$\left| \frac{\partial u_\varepsilon}{\partial x_i} - \Phi(x) \right| \leqslant \frac{1}{(m-2)\omega_m} \int_E \left| \left(\frac{\partial \varphi_\varepsilon}{\partial x_i} - \frac{x_i' - x_i}{r^m} \right) f(x') \right| d\omega(x') \leqslant$$

$$\leqslant \frac{1}{(m-2)\,\omega_m} \int_{S_\varepsilon} \left| \frac{\partial \varphi_\varepsilon}{\partial x_i} - \frac{x_i' - x_i}{r^m} \right| |f|\, d\omega(x') \leqslant$$

$$\leqslant \frac{M}{(m-2)\,\omega_m} \int_{\text{Fr}.S_0} d\sigma_0 \int_0^\varepsilon (r^{m-4} \,|\, \alpha_i \,|\, r + |\, \alpha_i \,|)\, dr \leqslant$$

$$\leqslant \frac{M}{(m-2)\,\omega_m} \int_{\text{Fr}.S_0} d\sigma_0 \int_0^\varepsilon (r^m + r^{m-1})\, dr =$$

$$= \frac{M}{m-2} \left(\frac{\varepsilon^{m+1}}{m+1} + \frac{\varepsilon^m}{m} \right),$$

where we wrote $x_i' - x_i = \alpha_i r$. From the above evaluation if follows that $\frac{\partial u_\varepsilon}{\partial x_i}$ tends to $\Phi(x)$ uniformly in $E \cup \text{Fr}.E$. But $u_\varepsilon \to u$ thus $\frac{\partial u}{\partial x_i}$ exists, is continuous in $E \cup \text{Fr}.E$ and

$$\frac{\partial u}{\partial x_i} = \Phi(x).$$

Consequently

$$\frac{\partial u}{\partial x_i} = \frac{1}{(m-2)\,\omega_m} \int_E f(x') \frac{\partial \frac{1}{r^{m-2}}}{\partial x_i} \, \mathrm{d}\omega\,(x') \tag{5.121}$$

or

$$\frac{\partial u}{\partial x_i} = -\frac{1}{(m-2)\,\omega_m} \int_E f(x') \frac{\partial \frac{1}{r^{m-2}}}{\partial x'_i} \, \mathrm{d}\omega\,(x'). \tag{5.122}$$

As we have

$$\int_E \frac{\partial}{\partial x'_i}\left(f \frac{1}{r^{m-2}}\right) \mathrm{d}\omega(x') = \int_E \frac{\partial f}{\partial x'_i} \cdot \frac{1}{r^{m-2}} \, \mathrm{d}\omega(x') + \int_E f \frac{\partial \frac{1}{r^{m-2}}}{\partial x'_i} \, \mathrm{d}\omega(x'),$$

it follows that relationship (5.122) can be written

$$\frac{\partial u}{\partial x_i} = \frac{1}{(m-2)\omega_m} \int_E \frac{\partial f}{\partial x'_i} \frac{1}{r^{m-2}} \, \mathrm{d}\omega(x') - \frac{1}{(m-2)\omega_m} \int_E \frac{\partial}{\partial x'_i}\left(f \frac{1}{r^{m-2}}\right) \mathrm{d}\omega(x').$$

Applying the theorem of divergence to the second term of the above relationship we obtain

$$\frac{\partial u}{\partial x_i} = \frac{1}{(m-2)\omega_m} \int_E \frac{\partial f}{\partial x'_i} \frac{1}{r^{m-2}} \, \mathrm{d}\omega(x') - \frac{1}{(m-2)\omega_m} \int_{\mathrm{Fr}.E} f \cdot \frac{1}{r^{m-2}} \frac{\mathrm{d}x_i}{\mathrm{d}n} \, \mathrm{d}\omega(x').$$

Taking into account the hypothesis and reasoning in the same way as for the function u, we notice that the above expression can be differentiated under the integral sign and we obtain the following expression

$$\frac{\partial^2 u}{\partial x_i^2} = \frac{1}{(m-2)\omega_m} \int_E \frac{\partial f}{\partial x'_i} \frac{\partial \frac{1}{r^{m-2}}}{\partial x_i} \, \mathrm{d}\omega(x') -$$

$$- \frac{1}{\omega_m(m-2)} \int_{\mathrm{Fr}.E} f \frac{\partial \frac{1}{r^{m-2}}}{\partial x_i} \frac{\mathrm{d}x'_i}{\mathrm{d}n} \, \mathrm{d}\sigma = -\frac{1}{(m-2)\omega_m} \int_E \frac{\partial f}{\partial x'_i} \frac{\partial \frac{1}{r^{m-2}}}{\partial x'_i} \, \mathrm{d}\omega(x') +$$

$$+ \frac{1}{(m-2)\omega_m} \int_{\mathrm{Fr}.E} f \frac{\partial \frac{1}{r^{m-2}}}{\partial x'_i} \frac{\mathrm{d}x_i}{\mathrm{d}n} \, \mathrm{d}\sigma.$$

Hence

$$\Delta u = -\frac{1}{(m-2)\omega_m}\int_E \left(\sum_{i=1}^{m} \frac{\partial f}{\partial x_i'} \frac{\partial \frac{1}{r^{m-2}}}{\partial x_i'}\right) d\omega(x') +$$

$$+\frac{1}{(m-2)\omega_m}\int_{\mathrm{Fr.}\, E} f \frac{d\frac{1}{r^{m-2}}}{dn} d\sigma,$$

whence we deduce that Δ exists and is continuous in E. We still have to prove that $\Delta u = f$. To this purpose we shall write the above relationship in the following form

$$\Delta u = \lim_{\varepsilon \to 0}\left[-\frac{1}{(m-2)\omega_m}\int_{E-S_\varepsilon} \sum \frac{\partial f}{\partial x_i} \frac{\partial \frac{1}{r^{m-2}}}{\partial x_i} d\omega + \right.$$

$$\left. + \frac{1}{(m-2)\omega_m}\int_{\mathrm{Fr.}\, E} f \frac{d\frac{1}{r^{m-2}}}{dn} d\sigma\right] \tag{5.123}$$

where S_ε is a sphere with its centre in x and of radius ε. Applying Green's first formula (5.99) to the first term of the relationship (5.123) we obtain

$$\Delta u = \lim_{\varepsilon \to 0}\left| \frac{1}{(m-2)\omega_m}\int_{E-S_\varepsilon} f\Delta \frac{1}{r^{m-2}} d\omega(x') - \frac{1}{(m-2)\omega_m}\int_{\mathrm{Fr.}\, E} f \frac{d\frac{1}{r^{m-2}}}{dn} d\sigma + \right.$$

$$+ \frac{1}{(m-2)\omega_m}\int_{\mathrm{Fr.}\, S_\varepsilon} f \frac{d\frac{1}{r^{m-2}}}{dn} d\sigma + \frac{1}{(m-2)\omega_m}\int_{\mathrm{Fr.}\, E} f \frac{d\frac{1}{r^{m-2}}}{dn} d\sigma =$$

$$= \lim_{\varepsilon \to 0} \frac{1}{(m-2)\omega_m}\int_{\mathrm{Fr.}\, S_\varepsilon} f \frac{d\frac{1}{r^{m-2}}}{dn} d\sigma.$$

But

$$\lim_{\varepsilon \to 0} \frac{1}{(m-2)\omega_m}\int_{\mathrm{Fr.}\, S_\varepsilon} f \frac{d\frac{1}{r^{m-2}}}{dn} d\sigma = \lim_{\varepsilon \to 0} \frac{1}{(m-2)\omega_m}\int_{\mathrm{Fr.}\, S_\varepsilon} f d\sigma_0$$

where $d\sigma_0$ is the element of the solid angle in the n-dimensional space ($d\sigma = r^{m-1}d\sigma_0$). Thus

$$\Delta u = f.$$

Remark. One can easily notice that the solution u of Poisson's equation which on the boundary of the domain E, is equal to a continuous function $h(x)$, is given by $u = u' + u''$, where u' is the solution of Poisson's equation and u'' is the solution of Laplace's equation with the condition $u''(x) = h(x) - u'(x)$ on Fr.E. It is easy to see that u is unique. Indeed, if there would be two solutions u_1 and u_2 then the function $v = u_1 - u_2$ would be harmonic in E and zero on Fr.E. But from proposition 3, Section 1.3.4, Vol. 1 which can be extended with no difficulty to the space R^m and which is also called the principle of the maximum, if follows that $v = 0$ or $u_1 = u_2$.

5.5 EXERCISES

1. Determine the solution of the equation

$$2\frac{\partial^2 u}{\partial x^2} - 7\frac{\partial^2 u}{\partial x \partial y} + 3\frac{\partial^2 u}{\partial y^2} = 0$$

which satisfies the conditions

$$u(x, y)\Big|_{x=0} = y^3, \quad \frac{\partial u}{\partial x}\Big|_{x=0} = y.$$

2. A homogeneous chord of length l, fixed at the extremities $x = 0$ and $x = l$, vibrates under a perturbatory exterior force $F(x, t)$, distributed continuously along the whole chord. Determine the displacement of the chord from the position of equilibrium taking into account it undergoes free oscillations, produced by the initial displacement and speed.

3. Determine the general solution of the equation

$$x^2\frac{\partial^2 u}{\partial x^2} - 2xy\frac{\partial^2 u}{\partial x \partial y} + y^2\frac{\partial^2 u}{\partial y^2} + x\frac{\partial u}{\partial x} + y\frac{\partial u}{\partial y} = 0.$$

4. Solve Poisson's equation

$$\Delta u = k, \ (k = \text{constant})$$

in the rectangle $0 \leqslant x \leqslant a, -\frac{b}{2} \leqslant y \leqslant \frac{b}{2}$, knowing that we have on the boundary of the rectangle $u = 0$.

6

Probability Theory and Mathematical Statistics

6.1 BOOLE ALGEBRAS

6.1.1 Boole algebras

Definition 1. *A non-empty set* $\mathfrak{a}$ *in which the operations* $\cup$ *(union),* $\cap$ *(intersection) and* $\complement$ *(the complement) are defined and which satisfies the following axioms is called a Boole algebra.*

(i) $A \cup B = B \cup A$, $A \cap B = B \cap A$, for every A, $B \in \mathfrak{a}$ (commutativity);
(ii) $A \cup (B \cup C) = (A \cup B) \cup C$, $A \cap (B \cap C) = (A \cap B) \cap C$, for every A, B, $C \in \mathfrak{a}$, (associativity);
(iii) $(A \cap B) \cup A = A$, $A \cap (A \cup B) = A$, (absorbtion);
(iv) $A \cap (B \cup C) = (A \cap B) \cup (A \cap C)$, $A \cup (B \cap C) = (A \cup B) \cap (A \cup C)$ (distributivity);
(v) $(A \cap \complement A) \cup B = B$, $(A \cup \complement A) \cap B = B$ (complementarity).

One can immediately verify that the set $\mathfrak{P}(E)$ of all parts of a non-empty set forms a Boole algebra.

Definition 2. *If* A, $B \in \mathfrak{a}$, *we say that* A *is included in* B *(it is denoted by* $A \subset B$ *or* $B \supset A$*) if*

$$A \cap B = A. \tag{6.1}$$

From the above definition and the absorbtion axiom it follows that relationship (6.1) is equivalent to the relationship

$$A \cup B = B. \tag{6.1'}$$

Indeed, if we consider relationship (6.1′), from the second equality of the axiom (iii) it follows that

$$A \cap B = A.$$

If we now write the first relationship from the axiom (iii) in the form

$$(B \cap A) \cup B = B$$

and replace relationship (6.1) written in the form $B \cap A = A$, we obtain

$$A \cup B = B.$$

Let us now prove some properties.

Proposition 1. *For every $A \in \mathfrak{a}$ we have*

$$A \cup A = A, \qquad A \cap A = A. \tag{6.2}$$

Indeed, applying the commutativity to the first relationship from axiom (iii) we can write

$$A = A \cup (A \cap B).$$

Applying the distributivity axiom we obtain

$$A = A \cup (A \cap B) = (A \cup A) \cap (A \cup B) = [A \cap (A \cup B)] \cup [A \cap (A \cup B)],$$

and hence, taking into account the second relationship from axiom (iii)

$$A = A \cup A.$$

By analogy, the second relationship (6.2) can be also proved.

Proposition 2. *If $A \subset B$ and $B \subset A$, then $A = B$.*

Indeed, from $B \subset A$ and relationship (6.1′) it follows that

$$A = A \cup B.$$

On the other hand from $A \subset B$ and equality (6.1) we obtain

$$A = A \cap B$$

which substituted in the above relationship, yields, if we take into account axiom (iii), the following equalities

$$A = (A \cap B) \cup B = B.$$

Proposition 3. *If $A, B, C \subset \mathfrak{a}$ and if $A \subset B$, then we have*

$$A \cup C \subset B \cup C, \qquad A \cap C \subset B \cap C.$$

From the inclusion $A \subset B$ it follows that relationships (6.1) and (6.1′) are fulfilled if we apply the axiom of associativity and proposition 1. We obtain

$$(A \cup C) \cup (B \cup C) = (A \cup B) \cup (C \cup C) = B \cup C$$

hence based on definition 2 it follows that

$$A \cup C \subset B \cup C.$$

By analogy the second inclusion can be also proved.

Proposition 4. *In a Boole algebra there are two elements which are denoted by $\wedge$ and $\vee$ and which are called zero element and total element, so that for every $A \in \mathfrak{a}$ we have*

$$A \cap \complement A = \wedge, \qquad A \cup \complement A = \vee. \tag{6.3}$$

If $A, B \in \mathfrak{a}$ from definition 2 and from the first relationship of axiom (v) it follows that

$$A \cap \complement A \subset B. \tag{6.4}$$

From the second relationship of axiom (v) we obtain

$$B \subset A \cup \complement A. \tag{6.4'}$$

Substituting into relationship (6.4) B for $B \cap \complement B$ and into relationship (6.4′) B for $B \cup \complement B$, we obtain

$$A \cap \complement A \subset B \cap \complement B, \qquad B \cup \complement B \subset A \cup \complement A. \tag{6.5}$$

Substituting B for A and A for B, we obtain

$$B \cap \complement B \subset A \cap \complement A, \qquad A \cup \complement A \subset B \cup \complement B. \tag{6.5'}$$

From relationships (6.5) and (6.5′) and from proposition 2, we deduce that

$$A \cap \complement A = B \cap \complement B, \qquad A \cup \complement A = B \cup \complement B.$$

Proposition 5. *For every $A \in \mathfrak{a}$ we have*

$$A \cap \vee = A, \qquad A \cup \vee = \vee, \qquad A \cup \wedge = A, \qquad A \cap \wedge = \wedge. \quad (6.6)$$

These relationships follow immediately by taking formulae (6.3) into account in axiom (v).

From relationships (6.6) and definition 2, we deduce proposition 5′.

Proposition 5′. *For every $A \in \mathfrak{a}$ we have*

$$\wedge \subset A, \qquad A \subset \vee.$$

Proposition 6. *If $A \cap B = \wedge$ and $A \cup B = \vee$ then $B = \complement A$.*

From formulae (6.6) and (6.3) we obtain

$$B = \wedge \cup B = (A \cap \complement A) \cup B$$

and taking into account the axiom of distributivity

$$B = (A \cup B) \cap (\complement A \cup B) = \vee \cap (\complement A \cup B) = \complement A \cup B,$$

from which, based on relationship (6.1′) it follows that $B \subset \complement A$.

By analogy we have

$$B = \vee \cap B = (A \cup \complement A) \cap B = (A \cap B) \cup (\complement A \cap B) = \wedge \cup (\complement A \cap B) = \complement A \cap B$$

and based on relationship (6.1) it follows that $\complement A \subset B$. Hence $B = \complement A$.

Proposition 7. *(De Morgan's formula.) For every $A, B \in \mathfrak{a}$, we have the relationships*

$$\complement(A \cup B) = \complement A \cap \complement B, \qquad \complement(A \cap B) = \complement A \cup \complement B.$$

In order to prove this formula let us write $C = \complement A \cap \complement B$. Taking into account the distributivity and associativity we deduce

$$(A \cup B) \cap C = (A \cup B) \cap (\complement A \cap \complement B) = [A \cap (\complement A \cap \complement B)] \cup [(B \cap (\complement A \cap \complement B)] =$$

$$= [(A \cap \complement A) \cap \complement B] \cup [(B \cap \complement B) \cap \complement A],$$

from which, considering relationships (6.3) and (6.6), we have

$$(A \cup B) \cap C = (\wedge \cap \complement B) \cup (\wedge \cap \complement A) = \wedge \cup \wedge = \wedge .$$

By analogy

$$(A \cup B) \cup C = (A \cup B) \cup (\complement A \cap \complement B) = [(A \cup B) \cup \complement A] \cap [(A \cup B) \cup \complement B] =$$

$$= [(A \cup \complement A) \cup B] \cap [A \cup (B \cup \complement B)] = (\vee \cup B) \cap (A \cup \vee) = \vee \cap \vee = \vee .$$

Thus

$$(A \cup B) \cap C = \wedge , \qquad (A \cup B) \cup C = \vee$$

from which, based on proposition 6, it follows that

$$C = \complement(A \cup B),$$

that is

$$\complement(A \cup B) = \complement A \cap \complement B.$$

In the same way the second De Morgan's formula can also be proved.

Definition 3. Two elements $A, B \in \mathfrak{a}$ are called disjoint elements if $A \cap B = \wedge$.

Definition 4. If A, $B \in \mathfrak{a}$, the element $A - B = A \cap \complement B$ is called their difference.

Definition 5. An element $A \in \mathfrak{a}$, $A \neq \wedge$ is called an atom if for every $B \in \mathfrak{a}$ the inclusion $B \subset A$ implies $B = \wedge$ or $B = A$.

From this definition it results that, if A is an atom, for every $B \in \mathfrak{a}$ we have $A \subset B$ or $A \cap B = \wedge$.

Definition 6. Let $\mathfrak{a}$ be a Boole algebra and $\mathfrak{m}$ a non-empty family of elements from $\mathfrak{a}$. The element $\bigcup_{A \in \mathfrak{m}} A$ which satisfies the following conditions is called union of the elements $A \in \mathfrak{m}$.

(i) $A \subset \bigcup_{A \in \mathfrak{m}} A$ for every $A \in \mathfrak{m}$;

(ii) if $A \subset B$ for every $A \in \mathfrak{m}$ then $\bigcup_{A \in \mathfrak{m}} A \subset B$.

The element $\bigcap_{A \in \mathfrak{m}} A$ which satisfies the following conditions is called intersection of the elements $A \in \mathfrak{m}$

(i) $\bigcap_{A \in \mathfrak{m}} A \subset A$ for every $A \in \mathfrak{m}$;

(ii) if $B \subset A$ for every $A \in \mathfrak{m}$, then $B \subset \bigcap_{A \in \mathfrak{m}} A$.

If $\mathfrak{m}$ is a sequence of elements $A_1, A_2, \ldots, A_n, \ldots$, we shall denote the union and the intersection respectively of elements from $\mathfrak{m}$ by

$$\bigcup_i A_i, \bigcap_i A_i.$$

Definition 7. A Boole algebra $\mathfrak{a}$ *is called σ-algebra if for every sequence of elements* $\{A_i\} \subset \mathfrak{a}$ *we have* $\bigcup_i A_i \in \mathfrak{a}$.

It can be verified immediately that the infinite unions and the infinite intersections are commutative and associative, and the union is distributive with respect to the intersection and the intersection with respect to the union and that De Morgan's formulae are valid.

6.1.2 Algebra of parts

In the following we shall denote by Ω an arbitrary set, and with $\mathfrak{P}(\Omega)$ the set of all the parts of the set Ω.

Definition 8. A non-empty family $\mathcal{K} \subset \mathfrak{P}(\Omega)$ *is called an algebra of parts if it has the properties:*

(i) if $A \in \mathcal{K}$ then $\complement A \in \mathcal{K}$;

(ii) if $A, B \in \mathcal{K}$ then $A \cup B \in \mathcal{K}$.

We shall prove now some properties of an algebra of parts.

Proposition 8. *If* $\varnothing$ *is an empty set, then* $\varnothing \in \mathcal{K}$, $\Omega \in \mathcal{K}$.

As the algebra $\mathcal{K}$ is non-empty, there is at least one element $A \in \mathcal{K}$. Based on the properties (i) and (ii) from definition 8, it follows that $A \cup \complement A = \Omega \in \mathcal{K}$, hence we deduce $\varnothing = \complement \Omega \in \mathcal{K}$.

Proposition 9. *If $A_i \in \mathcal{K}$, $(i = 1, \ldots, n)$, then $\bigcap_{i=1}^{n} A_i \in \mathcal{K}$.*

Based on De Morgan's formula (proposition 7) we have

$$\bigcap_{i=1}^{n} A_i = \complement\left(\bigcap_{i=1}^{n} \complement A_i\right)$$

and applying, the properties (i) and (ii) a finite number of times from definition 8 it follows $\bigcap_{i=1}^{n} A_i \in \mathcal{K}$.

Proposition 10. *If A, $B \in \mathcal{K}$ then $A - B \in \mathcal{K}$.*

We have $A - B = A \cap \complement B$ and from proposition 9 it follows that $A - B \in \mathcal{K}$.

Definition 9. A non-empty family $\mathcal{K} \subset \mathfrak{P}(\Omega)$ is called a σ-algebra if it has the properties:

(i) if $A \in \mathcal{K}$ then $\complement A \in \mathcal{K}$;

(ii) if $A_i \in \mathcal{K}$ $(i = 1, \ldots, n, \ldots)$, then $\bigcup_i A_i \in \mathcal{K}$.

As in the case of an algebra of parts proposition 11 can be proved.

Proposition 11. *We have $\varnothing \in \mathcal{K}$, $\Omega \in \mathcal{K}$. If $A_i \in \mathcal{K}$, $(i = 1, \ldots, n, \ldots)$ then $\bigcap_i A_i \in \mathcal{K}$. If A, $B \in \mathcal{K}$ then $A - B \in \mathcal{K}$.*

6.2 THE PROBABILITY SPACE

6.2.1 The space of events

The basic idea in the theory of probabilities is the event. Of course, this is interesting for us from the point of view of its occurence or non-occurrence in certain given conditions.

Definition 1. If A is an arbitrary event, we call the contrary event of A or non A the event which occurs when and only when A does not occur and we denote it by $\complement A$.

Definition 2. *If A and B are two arbitrary events, the event or A or B or the union of the events A and B is the event which occurs when and only when at least one of the events A or B occurs. We denote it by* $A \cup B$.

Definition 3. *The event and A and B or the intersection of the events A and B is the event which occurs when and only when both events A and B occur. We denote it by* $A \cap B$.

Hence it follows that the set of events forms a Boole algebra with respect to the operations: $\complement$, $\cup$, $\cap$.

The certain event which we shall denote by Ω and the impossible event which we shall denote by $\varnothing$ correspond to the total element $\vee$ and the zero element $\wedge$ of the Boole algebra.

Definition 4. *The difference of the events A and B is the event which occurs when, and only when, A occurs but B does not. We shall denote it by* $A - B$. *The symmetric difference of the events A and B is the event* $A \Delta B = (A - B) \cup (B - A)$.

Definition 5. *If the occurence of the event A necessarily implies the occurrence of the event B then we say that A implies B and we denote it by* $A \subseteq B$. *If* $A \subseteq B$ *and* $B \subseteq A$ *the events A and B are equivalent and we write* $A = B$. *If* $A \subseteq B$ *but* $A \neq B$ *we write* $A \subset B$.

Definition 6. *If* $A \cap B = \varnothing$ *we say that the events A and B are incompatible (or mutually exclusive).*

Definition 7. *A set* Ω *such that* $\mathfrak{K} \subset \mathfrak{P}(\Omega)$ *is a σ-algebra is called a space of events. We shall denote it by* $\{\Omega, \mathfrak{K}\}$. *If the set* Ω *is finite, the space is called a finite space of events.*

Definition 8. *A system of events* $A_k \in \mathfrak{K} (k = 1, \ldots, n)$ *is a complete system of events if* $A_k \neq \varnothing$, $A_j \cap A_k = \varnothing$, $(j \neq k)$, $\bigcup_{k=1}^{n} A_k = \Omega$.

Definition 9. *An event* $A \in \mathfrak{K}$ *is called a compound event if there are two events* $B, C \in \mathfrak{K}$ *different from A so that we have* $A = B \cup C$. *An event which is not a compound event and which is not the impossible one is called an elementary event.*

Let as now prove some properties of the elementary events.

Proposition 1. *If $A \in \mathcal{K}$ is an elementary event and $B \in \mathcal{K}$ an arbitrary event the relationship $B \subseteq A$ implies $B = \varnothing$ or $B = A$.*

Indeed let us suppose that $B \neq \varnothing$ and $B \neq A$. Then writing $C = A - B$, from relationships $B \subset A$ and $B \neq \varnothing$ we have $C \neq A$ and $A = B \cup C$ which is not possible since A is an elementary event.

Proposition 2. *The necessary and sufficient condition for the event $A \neq \varnothing$ to be elementary is that no event $B \neq \varnothing$ exists so that $B \subset A$.*

Indeed, let us suppose that there is an event $B \neq \varnothing$ and $B \subset A$. Hence it follows that $B \neq A$. But

$$A = A \cap (B \cup \complement B) = (A \cap B) \cup (A \cap \complement B)$$

and from $B \subset A$ we deduce that $A \cap B = B$, thus

$$A = B \cup (A \cap \complement B).$$

If we write $C = A \cap \complement B$, we have $C \neq A$, since otherwise we would have

$$B = A \cap B = C \cap B = (A \cap \complement B) \cap B = A \cap (\complement B \cap B) = A \cap \varnothing = \varnothing.$$

But the relationship $A = B \cup C$ with $B \neq A$, $C \neq A$ is absurd, since A is elementary.

By analogy proposition 3 can be proved.

Proposition 3. *The necessary and sufficient condition for the event $A \neq \varnothing$ to be elementary, is that for every event B we have*

$$A \cap B = \varnothing \text{ or } A \cap B = A.$$

Proposition 4. *Every two distinct elementary events are mutually exclusive events.*

Let A_1 and A_2 be two elementary events. Let us suppose that $A_1 \cap A_2 \neq \neq \varnothing$. As A_1 is an elementary event, from the above assumption and proposition 3 it follows that

$$A_1 = A_1 \cap A_2.$$

By analogy, as A_2 is elementary it follows that

$$A_2 = A_1 \cap A_2.$$

From these two relationships we deduce that $A_1 = A_2$, which is absurd. That means that $A_1 \cap A_2 = \varnothing$.

Proposition 5. *In a finite space of events, given every compound event $B \in \mathcal{K}$, there is an elementary event A so that $A \subset B$.*

As B is a compound event, from proposition 2 it follows that there is an event $A_1 \neq \varnothing$ with $A_1 \subset B$. If A_1 is elementary, the property is proved. If A_1 is not elementary, there is an event $A_2 \neq \varnothing$ with $A_2 \subset A_1$ whence $A_2 \subset B$. If A_2 is elementary, the property is proved. If A_2 is not elementary, the operation goes on. But as the space is finite, after a finite number of steps the operation must come to an end. Thus there is an elementary event $A_n \neq \varnothing$ so that $A_n \subset A_{n-1} \subset \ldots \subset A_1 \subset B$.

Proposition 6. *Every event of a finite space of events can be written in a unique form as an union of elementary events.*

Let B be a compound event. From proposition 5, there is an elementary event, A_1 with $A_1 \subset B$. Writing $B_1 = B - A_1$, it follows that $B = A_1 \cup B_1$. If B_1 is an elementary event, the first part of the theorem is proved. If B_1 is a compound event from proposition 5, it follows that there is an event A_2 with $A_2 \subset B_1$. Writing $B_2 = B_1 - A_2$ it follows that $B_1 = A_2 \cup B_2$, that is $B = A_1 \cup A_2 \cup B_2$. If B_2 is elementary the first part of the theorem is proved. If B_2 is a compound event the procedure continues. But, taking into account that the space of events is finite, the procedure cannot continue indefinitely, thus after a finite number of steps, we obtain

$$B = A_1 \cup A_2 \cup \ldots \cup A_r, \tag{6.7}$$

with A_h, $(h = 1, \ldots, r)$ elementary events.

Let us now suppose that the decomposition given in formula (6.7) is not unique hence

$$B = A_1 \cup A_2 \cup \ldots \cup A_r = A'_1 \cup A'_2 \cup \ldots \cup A'_s, \tag{6.8}$$

where, for example,

$$A_1 \neq A'_k, \ (k = 1, \ldots, s). \tag{6.9}$$

From formula (6.8) it follows that

$$A_1 \cap (A_1 \cup A_2 \cup \ldots \cup A_r) = A_1 \cap (A'_1 \cup A'_2 \cup \ldots \cup A'_s)$$

that is, taking into account proposition 4 and relationship (6.9) it follows that

$$A_1 = \varnothing$$

which is absurd.

Proposition 7. *In a finite space of events, the certain event is the union of all elementary events.*

From proposition 6, we have

$$\Omega = A_1 \cup A_2 \cup \ldots \cup A_m. \tag{6.10}$$

Let us suppose that in the space of events there is one more elementary event $A_n \neq A_i$, $(i = 1, \ldots, m)$. Then from formula (6.10) it follows that

$$A_n \cap \Omega = A_n \cap (A_1 \cup A_2 \cup \ldots \cup A_m)$$

i.e. taking into account proposition 4,

$$A_n = \varnothing$$

which is absurd.

6.2.2 The probability space

Let us consider a space of events $(\Omega, \mathcal{K})$.

Definition 10. (*Kolmogorov*). *A function P defined on $\mathcal{K}$ with real values, is called a probability on $\mathcal{K}$, if it satisfies the conditions:*

(i) $P(A) \geqslant 0$, *for every* $A \in \mathcal{K}$;

(ii) $P(\Omega) = 1$;

(iii) *If* $A, B \in \mathcal{K}$ *and* $A \cap B = \varnothing$, *then* $P(A \cup B) = P(A) + P(B)$.

The triple $\{\Omega, \mathcal{K}, \mathfrak{P}\}$ is called a probability space.

Definition 11. A function P defined in $\mathcal{K}$ with real values, is called a completely additive probability in $\mathcal{K}$ if it satisfies the conditions:

(i) $P(A) \geqslant 0$, *for every* $A \in \mathcal{K}$;

(ii) $P(\Omega) = 1$;

(iii) *If* $A_i \in \mathcal{K}$, $i \in I$, *where I is at most a countable set of indices and* $A_i \cap A_j = \varnothing$, $(i \neq j)$, *then* $P\left(\bigcup_{i \in I} A_i\right) = \sum_{i \in I} P(A_i)$.

Let us suppose that $\{\Omega, \mathcal{K}\}$ is a finite space of events whose elementary events we shall denote by $\omega_1, \ldots, \omega_n$. Then from formula (6.10) and definition 10 it follows that

$$P(\omega_i) \geqslant 0, \ (i = 1, \ldots, n), \ \sum_{i=1}^{n} P(\omega_i) = P(\Omega) = 1. \tag{6.11}$$

If

$$P(\omega_1) = \ldots = P(\omega_n)$$

we shall say that the elementary events are equally probable. In this case from formula (6.11) it follows that

$$P(\omega_i) = \frac{1}{n}, \quad (i = 1, \ldots, n). \tag{6.12}$$

Let there be $A = \omega_{i_1} \cup \omega_{i_2} \cup \ldots \cup \omega_{i_m}$ an arbitrary event of the space. Then, from formula (6.12) we deduce that

$$P(A) = \frac{m}{n}.$$

Thus we have the following proposition.

Proposition 8. *In a finite space of events whose elementary events are equally probable, the probability of an arbitrary event is equal to the ratio between the number of elementary events favorable to the given event and the total number of elementary events of the space (the frequency of the given event).*

Let us prove some properties which are deduced from definitions 9 and 10.

Proposition 9. *If $A \in \mathcal{K}$ then*

$$P(\complement A) = 1 - P(A). \tag{6.13}$$

By taking into account that $A \cup \complement A = \Omega$ and $A \cap \complement A = \varnothing$, from axioms 2 and 3 and definition 9 it follows that

$$1 = P(\Omega) = P(A) + (P\complement A),$$

whence we deduce formula (6.13).

Proposition 10. *We have*

$$(P\varnothing) = 0. \tag{6.14}$$

Indeed, we have $\complement\, \varnothing = \Omega$ thus, applying relationship (6.13) and axiom 2 definition 9 we deduce that

$$P(\varnothing) = 1 - P(\Omega) = 1 - 1 = 0.$$

Proposition 11. *If the events $A_1, A_2, \ldots A_m \in \mathcal{K}$ are mutually exclusive then*

$$P(A_1 \cup A_2 \cup \ldots \cup A_m) = P(A_1) + P(A_2) + \ldots + P(A_m). \tag{6.15}$$

In order to prove this proposition we shall use the method of complete induction. From axiom 3 of definition 10 the property is true for $m = 2$. We shall suppose that formula (6.15) is true for $m - 1$. By hypothesis we have

$$(A_1 \cup A_2) \cap A_k = (A_1 \cap A_k) \cup (A_2 \cap A_k) = \varnothing,\ (k = 3, \ldots, m),$$

thus

$$P[(A_1 \cup A_2) \cup A_3 \cup \ldots \cup A_m] = P(A_1 \cup A_2) + P(A_3) + \ldots +$$
$$+ P(A_m) = P(A_1) + P(A_2) + P(A_3) + \ldots + P(A_m).$$

Proposition 12. *If $A_1, \ldots, A_n$ forms a complete system of events of the space $\{\Omega, \mathcal{K}\}$, then*

$$P(A_1) + \ldots + P(A_n) = 1. \tag{6.16}$$

Indeed, based on definition 8 we have

$$A_1 \cup \ldots \cup A_n = \Omega,\ A_i \cap A_j = \varnothing,\ (i \neq j,\ i, j = 1, \ldots, n)$$

thus, taking into account that $P(\Omega) = 1$, from formula (6.15) relationship (6.16) follows.

Proposition 13. *If $B \subseteq A$, then $P(B) \leqslant P(A)$.*

Indeed, if $B \subseteq A$, denoting by $C = A \cap \complement B$, we have

$$A = B \cup C, \quad B \cap C = \varnothing,$$

thus based on axiom 3 of definition 10, it follows that

$$P(A) = P(B) + P(C).$$

But $P(C) \geqslant 0$, thus

$$P(A) \geqslant P(B).$$

Proposition 14. *If* $A, B \in \mathcal{K}$ *then*

$$P(A \cup B) = P(A) + P(B) - P(A \cap B), \quad (6.17)$$

$$P(B - A) = P(B) - P(A \cap B), \quad (6.18$$

$$P(A \Delta B) = P(A) + P(B) - 2\,P(A \cap B). \quad (6.19)$$

We have

$$A \cup (\complement A \cap B) = (A \cup \complement A) \cap (A \cup B) = \Omega \cap (A \cup B) = A \cup B$$

$$A \cap (\complement A \cap B) = \varnothing \cap B = \varnothing$$

thus based on the axiom 3 of definition 10 we deduce that

$$P(A \cup B) = P[A \cup (\complement A \cap B)] = P(A) + P(\complement A \cap B). \quad (6.20)$$

On the other hand we have

$$B = \Omega \cap B = (A \cup \complement A) \cap B = (A \cap B) \cup (\complement A \cap B)$$

and

$$(A \cap B) \cap (\complement A \cap B) = \varnothing \cap B = \varnothing,$$

thus

$$P(B) = P(A \cap B) \cup (\complement A \cap B) = P(A \cap B) + P(\complement A \cap B). \quad (6.21)$$

From formulae (6.20) and (6.21) the relationship (6.19) follows. We have

$$(B - A) \cup (A \cap B) = (B \cup \complement A) \cup (B \cap A) = B \cap (A \cup \complement A) = B \cap \Omega = B$$

and

$$(B - A) \cup (A \cap B) = (B \cap \complement A) \cup (A \cap B) = \varnothing \cap B = \varnothing$$

thus,

$$P(B) = P[(B - A) \cup (A \cap B)] = P(B - A) + P(A \cap B).$$

We have

$$A \Delta B = (A - B) \cup (B - A)$$

and

$$(A - B) \cap (B - A) = (A \cap \complement B) \cap (B \cap \complement A) = \varnothing,$$

thus,

$$P(A \Delta B) = P(A - B) + P(B - A)$$

and by taking into account formula (6.18), we obtain formula (6.19).

Proposition 15. *If $A \subseteq B$, then*

$$P(B - A) = P(B) - P(A). \tag{6.22}$$

From $A \subseteq B$ it follows that $A \cap B = A$ which, substituted in formula (6.19), yields relationship (6.22).

Proposition 16. *If $\{\Omega, \mathcal{K}, P\}$ is a probability space and $A_i \in \mathcal{K}$ $(i = 1,2, \ldots, n, \ldots)$, then*

$$P\left(\bigcup_{i=1}^{\infty} A_i\right) \leqslant \sum_{i=1}^{\infty} P(A_i).$$

In order to prove this inequality let us consider the events

$$B_1 = A_1$$

$$B_2 = A_2 - A_1$$

$$\ldots\ldots\ldots\ldots\ldots\ldots$$

$$B_n = A_n - \bigcup_{j=1}^{n-1} A_j$$

$$\ldots\ldots\ldots\ldots\ldots\ldots$$

Let us prove that the events of the sequence B_i, $(i = 1, \ldots, n, \ldots)$ are mutually exclusive. We have

$$B_l \cap B_{l+k} = \left((A_l - \bigcup_{j=1}^{l-1} A_j\right) \cap \left(A_{l+k} - \bigcup_{h=1}^{l+k-1} A_h\right) =$$

$$= \left[A_l \cap \complement\left(\bigcup_{j=1}^{l-1} A_j\right)\right] \cap \left[A_{l+k} \cap \complement\left(\bigcup_{h=1}^{l+k-1} A_h\right)\right] = \tag{6.23}$$

$$= \left[A_l \cap \left(\bigcap_{j=1}^{l-1} \complement A_j\right)\right] \cap \left[A_{l+k} \cap \left(\bigcap_{h=1}^{l+k-1} A_h\right)\right] = \varnothing$$

for every natural number l and k. Taking into account that $B_n \subseteq A_n$ it follows that

$$\bigcup_{i=1}^{\infty} B_i \subseteq \bigcup_{i=1}^{\infty} A_i. \tag{6.24}$$

Let there be $\omega \in \bigcup_{i=1}^{\infty} A_i$ and n_0 the smallest natural number for which $\omega \in A_{n_0}$. Then $\omega \in B_{n_0} \subset \bigcup_{i=1}^{\infty} B_i$ and thus

$$\bigcup_{i=1}^{\infty} A_i \subseteq \bigcup_{i=1}^{\infty} B_i. \tag{6.25}$$

From formulae (6.24) and (6.25) it follows that

$$\bigcup_{i=1}^{\infty} B_i = \bigcup_{i=1}^{\infty} A_i$$

and thus

$$P\left(\bigcup_{i=1}^{\infty} A_i\right) = P\left(\bigcup_{i=1}^{\infty} B_i\right) = \sum_{i=1}^{\infty} P(B_i).$$

But taking into account that $B_i \subseteq A_i$, from proposition 13 it follows that

$$P(B_i) \leqslant P(A_i), \qquad (i = 1, \ldots, n, \ldots)$$

and

$$P\left(\bigcup_{i=1}^{\infty} A_i\right) \leqslant \sum_{i=1}^{\infty} P(A_i).$$

Proposition 17. *(Boole's inequality). If $\{\Omega, \mathcal{K}, P\}$ is a probability space and $A_i \in \mathcal{K}$, $(i \in I)$ at most a countable set of events, then*

$$P\left(\bigcap_{i \in I} A_i\right) \geqslant 1 - \sum_{i \in I} P(\complement A_i). \tag{6.26}$$

For the set of events $A_i (i \in I)$ we have De Morgan's formula

$$\complement\left(\bigcap_{i \in I} A_i\right) = \bigcup_{i \in I} \complement A_i,$$

that is

$$\bigcap_{i \in I} A_i = \complement\left(\bigcup_{i \in I} \complement A_i\right)$$

and thus

$$P\left(\bigcap_{i \in I} A_i\right) = P\left[\complement\left(\bigcup_{i \in I} \complement A_i\right)\right]$$

or based on formula (6.13)

$$P\left(\bigcap_{i \in I} A_i\right) = 1 - P\left(\bigcup_{i \in I} \complement A_i\right). \tag{6.27}$$

From formula (6.23) we deduce that

$$P\left(\bigcup_{i \in I} \complement A_i\right) \leqslant \sum_{i \in I} P(\complement A_i),$$

which, substituted in formula (6.27) yields the inequality (6.26).

If $I = \{1, 2, \ldots, n\}$ taking into account that

$$P(\complement A_i) = 1 - P(A_i), \ (i = 1, \ldots, n),$$

it follows that

$$\sum_{i=1}^{n} P(\complement A_i) = n - \sum_{i=1}^{n} P(A_i)$$

and inequality (6.26) becomes

$$P(A_1 \cap A_2 \cap \ldots \cap A_n) \geqslant P(A_1) + P(A_2) + \ldots + P(A_n) - (n-1). \tag{6.26'}$$

Proposition 18. *If* $A_i \in \mathcal{K}$ $(i = 1, \ldots, n)$ *then*

$$P(A_1 \cup A_2 \cup \ldots \cup A_n) = \sum_{k=1}^{n} (-1)^{k-1} S_k^{(n)}, \tag{6.28}$$

where

$$S_k^{(n)} = \sum_{1 \leqslant i_1 \leqslant i_2 \leqslant \ldots \leqslant i_k \leqslant n} P(A_{i_1} \cap A_{i_2} \cap \ldots \cap A_{i_k}).$$

We shall prove this relationship by a complete induction. For $n = 2$, formula (6.28) becomes

$$P(A_1 \cup A_2) = P(A_1) + P(A_2) - P(A_1 \cap A_2)$$

which is identical to relationship (6.17).

Let us suppose that formula (6.28) is true for $n - 1$, that is

$$P(A_1 \cup A_2 \cup \ldots \cup A_{n-1}) = \sum_{k=1}^{n-1} (-1)^{k-1} S_k^{(n-1)}. \tag{6.29}$$

Based on the above relationship we have

$$P(A_1 \cup A_2 \cup \ldots \cup A_n) = P[(A_1 \cup A_2 \cup \ldots \cup A_{n-1}) \cup A_n] =$$
$$= P(A_1 \cup A_2 \cup \ldots \cup A_{n-1}) + P(A_n) - P[(A_1 \cup \ldots \cup A_{n-1}) \cap A_n] =$$
$$= \sum_{k=1}^{n-1} (-1)^{k-1} S_k^{(n-1)} +$$
$$+ P(A_n) - P[(A_1 \cap A_n) \cup (A_2 \cap A_n) \cup \ldots \cup (A_{n-1} \cap A_n)].$$

From formula (6.29) it follows that

$$P[(A_1 \cap A_n) \cup \ldots \cup (A_{n-1} \cap A_n)] = \sum_{k=1}^{n-1} (-1)^{k-1} T_k^{(n-1)},$$

where

$$T_k^{(n-1)} = \sum_{1 \leqslant i_1 \leqslant i_2 \leqslant \ldots \leqslant i_k \leqslant n-1} P[(A_{i_1} \cap A_n) \cap \ldots \cap (A_{i_k} \cap A_n)],$$

that is

$$T_k^{(n-1)} = \sum_{1 \leqslant i_1 \leqslant i_2 \leqslant \ldots \leqslant i_k \leqslant n-1} P(A_{i_1} \cap A_{i_2} \cap \ldots \cap A_{i_k} \cap A_n).$$

Thus

$$P(A_1 \cup A_2 \cup \ldots \cup A_n) =$$
$$= \sum_{k=1}^{n-1} (-1)^{k-1} \Big[\sum_{1 \leqslant i_1 \leqslant i_2 \leqslant \ldots \leqslant i_k \leqslant n-1} P(A_{i_1} \cap A_{i_2} \cap \ldots \cap A_{i_k}) \Big] +$$
$$+ P(A_n) - \sum_{k=1}^{n-1} (-1)^{k-1} \Big[\sum_{1 \leqslant i_1 \leqslant i_2 \leqslant \ldots \leqslant i_k \leqslant n-1} P(A_{i_1} \cap A_{i_2} \cap \ldots \cap A_{i_k} \cap A_n) \Big].$$

By expanding the right-hand side sums of this equality, we obtain

$$P(A_1 \cup A_2 \cup \ldots \cup A_n) = P(A_1) + P(A_2) + \ldots + P(A_{n-1}) +$$
$$+ P(A_n) - P(A_1 \cap A_2) - P(A_1 \cap A_3) - \ldots - P(A_{n-1} \cap A_n) + \ldots +$$
$$+ (-1)^{n-1} P(A_1 \cap A_2 \cap \ldots \cap A_n),$$

that is, relationship (6.28).

6.2.3 Conditioned probabilities. Independent events

Definition 12. If $\{\Omega, \mathcal{K}, P\}$ *is a probability space and* $A, B \in \mathcal{K}$ *with* $P(A) \neq 0$, *the following expression is called the probability of the event B conditioned by the event A,*

$$P_A(B) = P(B/A) = \frac{P(A \cap B)}{P(A)}. \tag{6.30}$$

Let us prove now some properties of the conditioned probabilities.

Proposition 19. *The triple* $\{\Omega, \mathcal{K}, P_A\}$ *is a probability space.*

In order to prove this property let us show that conditions (i), (ii), and (iii) from definition 11 are verified.

Taking into account that

$$P(A \cap B) \geqslant 0, \; P(A) > 0,$$

it follows from formula (6.30) that

$$P_A(B) \geqslant 0.$$

Taking into account that $A \cap \Omega = A$, we can write

$$P_A(\Omega) = \frac{P(A \cap \Omega)}{P(A)} = \frac{P(A)}{P(A)} = 1.$$

Let there be $A_i \in K$, $(i \in I)$ at most a countable set of mutually exclusive events, that is $A_i \cap A_j = \varnothing$, $(i, j \in I)$.

Then we have

$$(A \cap A_i) \cap (A \cap A_j) = A \cap A_i \cap A_j = \varnothing$$

and thus

$$P_A(\bigcup_{i \in I} A_i) = \frac{P[A \cap (\bigcup_{i \in I} A_i)]}{P(A)} = \frac{P[\bigcup_{i \in I} (A \cap A_i)]}{P(A)} = \frac{\sum_{i \in I} P(A \cap A_i)}{P(A)} = \sum_{i \in I} P_A(A_i).$$

Proposition 20. *If* $P(A) \neq 0$, $P(B) \neq 0$, *then*

$$P(A)\; P_A(B) = P(B)\; P_B(A). \tag{6.31}$$

Indeed, from formula (6.30) it follows that

$$P(A \cap B) = P(A)\; P_A(B). \tag{6.32}$$

By analogy, if $P(B) \neq 0$, we have

$$P_B(A) = \frac{P(A \cap B)}{P(B)}$$

whence

$$P(A \cap B) = P(B)\, P_B(A).$$

From this and from equation (6.32) formula (6.31) is obtained.

Proposition 21. *(The formula of total probability.) If $A_i \in \mathcal{K}$ $(i \in I)$ is at most a countable set of mutually exclusive events, with $\bigcup_{i \in I} A_i = \Omega$ and $P(A_i) \neq 0$, $(i \in I)$. Then*

$$P(A) = \sum_{i \in I} P(A_i)\, P_{A_i}(A). \tag{6.33}$$

Taking into account formula (6.32) it follows that

$$P(A_i)\, P_{A_i}(A) = P(A \cap A_i),$$

whence

$$\sum_{i \in I} P(A_i)\, P_{A_i}(A) = \sum_{i \in I} P(A \cap A_i). \tag{6.34}$$

But since the events A_i are mutually exclusive, if follows that

$$\sum_{i \in I} P(A \cap A_i) = P\left[\bigcup_{i \in I} (A \cap A_i)\right] = P\left[A \cap (\bigcup_{i \in I} A_i)\right] = P(A \cap \Omega) = P(A),$$

which substituted in formula (6.34) yields formula (6.33).

Proposition 22. *(The multiplication formula of the probabilities.) If $A_i \in \mathcal{K}$ $(i = 1, \ldots, n)$ and $P(A_1 \cap A_2 \cap \ldots \cap A_{n-1}) \neq 0$, then*

$$P(A_1 \cap A_2 \cap \ldots \cap A_n) =$$
$$= P(A_1)\, P(A_2/A_1)\, P(A_3/A_1 \cap A_2) \ldots (P(A_n/A_1 \cap \ldots \cap A_{n-1})). \tag{6.35}$$

From hypothesis $P(A_1 \cap A_2 \cap \ldots \cap A_{n-1}) \neq 0$ it follows that $P(A_1 \cap A_2 \cap \ldots \cap A_{n-1}) > 0$. On the other hand, we have

$$A_1 \cap A_2 \cap \ldots \cap A_{n-1} \subseteq A_1 \cap A_2 \cap \ldots \cap A_{n-2} \subseteq \ldots \subseteq A_1 \cap A_2 \subseteq A_1$$

and based on proposition 13 it follows that

$$0 < P(A_1 \cap A_2 \cap \dots \cap A_{n-1}) \leqslant P(A_1 \cap A_2 \cap \dots \cap A_{n-2}) \leqslant \dots \leqslant$$
$$\leqslant P(A_1 \cap A_2) \leqslant P(A_1),$$

thus

$$P(A_1) > 0,\ P(A_1 \cap A_2) > 0, \dots, P(A_1 \cap A_2 \cap \dots \cap A_{n-1}) > 0,$$

whence we deduce that the conditioned probabilities

$$P(A_2/A_1),\ P(A_3/A_1 \cap A_2), \dots, P(A_n/A_1 \cap \dots \cap A_{n-1})$$

are defined.

We shall prove formula (6.35) by a complete induction. For $n = 2$ formula (6.35) becomes identical to formula (6.32). We shall suppose that $P(A_1 \cap \dots \cap A_{n-1}) = P(A_1)\, P(A_2/A_1) \dots P(A_{n-1}/A_1 \cap \dots \cap A_{n-2})$. Thus we have

$$P(A_1 \cap \dots \cap A_n) = P[(A_1 \cap \dots \cap A_{n-1}) \cap A_n] =$$
$$= P(A_1 \cap \dots \cap A_{n-1})\, P(A_n/A_1 \cap \dots \cap A_{n-1}) =$$
$$= P(A_1)\, P(A_2/A_1) \dots P(A_{n-1}/A_1 \cap \dots \cap A_{n-2})\, P(A_n/A_1 \cap \dots \cap A_{n-1})$$

i.e. formula (6.35).

Proposition 23. (*Bayes' formula.*) *If $A_i \in \mathcal{K}$ is at most a countable set of mutually exclusive events and if $\bigcup_{i \in I} A_i = \Omega$ and $P(A_i) \neq 0$, $(i \in I)$, then*

$$P_A(A_j) = \frac{P(A_j)\ P_{A_j}(A)}{\sum_{i \in I} P(A_i)\ P_{A_i}(A)}, \quad (j \in I). \tag{6.36}$$

We have

$$P_A(A_j) = \frac{P(A_j \cap A)}{P(A)}, \ (j \in I).$$

By substituting relationships (6.32) and (6.33) in the above formula we obtain formula (6.36).

Definition 13. *The events $A_1, A_2 \in \mathcal{K}$ are independent if*

$$P(A_1 \cap A_2) = P(A_1)\, P(A_2). \tag{6.37}$$

Proposition 24. *If the events $A_1, A_2 \in \mathcal{K}$ are independent then the following pairs are also independent: A_1 and $\complement A_2$, A_2 and $\complement A_1$, $\complement A_1$ and $\complement A_2$.*

We have

$$A_1 = A_1 \cap \Omega = A_1 \cap (A_2 \cup \complement A_2) = (A_1 \cap A_2) \cup (A_1 \cap \complement A_2)$$

and

$$(A_1 \cap A_2) \cap (A_1 \cap \complement A_2) = A_1 \cap A_2 \cap \complement A_2 = \varnothing,$$

thus

$$P(A_1) = P(A_1 \cap A_2) + P(A_1 \cap \complement A_2),$$

that is

$$P(A_1 \cap \complement A_2) = P(A_1) - P(A_1 \cap A_2),$$

or, taking into account formula (6.37)

$$P(A_1 \cap \complement A_2) = P(A_1) - P(A_1)\, P(A_2) = P(A_1)\,[1 - P(A_2)].$$

From relationship (6.13) it follows that

$$P(A_1 \cap \complement A_2) = P(A_1)\; P(\complement A_2)$$

hence the events A and $\complement A_2$ are independent.

It can be proved by analogy for the other pairs of events.

Definition 14. *The events $A_i \in \mathcal{K}$, $(i = 1, \ldots, n)$ are independent for every $1 \leqslant i_1 \leqslant \ldots \leqslant i_s \leqslant n$, $(s = 2, \ldots, n)$ if we have*

$$P(A_{i_1} \cap \ldots \cap A_{i_s}) = P(A_{i_1}) \ldots P(A_{i_s}).$$

Definition 15. *Let $(A_1, A_2, \ldots, A_m)$ and $(B_1, B_2, \ldots, B_n)$ be two complete systems of events of the space $\{\Omega, \mathcal{K}, P\}$. These two systems are independent if*

$$P(A_j \cap B_k) = P(A_j)\, P(B_k), \; (j = 1, \ldots, m;\; k = 1, \ldots, n).$$

6.3 DISCRETE RANDOM VARIABLES

6.3.1 Discrete random variables

Definition 1. *If* $\{\Omega, \mathcal{K}, P\}$ *is a probability space and* $A_i \in \mathcal{K}$, $(i \in I)$ *a finite or countable system of mutually exclusive events for which* $P\left(\bigcup_{i \in I} A_i\right) = \sum_{i \in I} P(A_i) = 1$, *the system* $\{A_i\}$ *is called a complete system of events and the numerical system* $p_i = P(A_i)$, $(i \in I)$ *is called the distribution of the space* $\{\Omega, \mathcal{K}, P\}$.

Definition 2. *A function* $\xi = \xi(\omega)$, *defined on the set of the elementary events* $\omega \in \Omega$ *with real values is called a discrete random variable if it has the following properties:*

(i) *It takes the values* x_i, $(i \in I)$ *where* I *is at most a countable set*

(ii) $$\{\omega : \xi(\omega) = x_i\} \in \mathcal{K}, \ (i \in I).$$

A discrete random variable for which the set I is finite, is called a *simple random variable.* The simplest discrete random variable is the *indicator* of an arbitrary event $A \in \mathcal{K}$

$$\xi_A = \begin{cases} 1, & \text{if } A \text{ occurs,} \\ 0, & \text{if } A \text{ does not occur.} \end{cases}$$

Let ξ be a discrete random variable which has the distinct values x_n, $(n = 1, 2, \ldots)$. To this random variable we can assign a complete system of events $\{A_n\}$ that is A_n is the event $\xi = x_n$. Since the values x_n are distinct, it follows that $A_n \cap A_m = \varnothing$ for $n \neq m$ and $\bigcup_{n=1}^{\infty} A_n = \Omega$, thus

$$P\left(\bigcup_{n=1}^{\infty} A_n\right) = \sum_{n=1}^{\infty} P(A_n) = 1.$$

Conversely, to every complete system of events $\{A_n\}$ we can assign at least a discrete random variable, that is, $\xi = n$ if A_n occurs.

Definition 3. *Let* ξ *be a random variable. The function* $F(x)$ *which is defined for every real* x *by*

$$F(x) = P(\xi < x) \tag{6.38}$$

is called the distribution function of the random variable ξ.

If x_n, $(n = 1, 2, \ldots)$ are discrete random variables ξ and $p_n = P(\xi = x_n)$ is the distribution of this random variable, then from formula (6.38) it follows that

$$F(x) = \sum_{x_k < x} p_k, \tag{6.38'}$$

the summation being extended to all the values of k for which $x_k < x$.

Definition 4. *Let ξ and η be two random variables defined by*

$$\xi(\omega) = x_n, \text{ for } \omega \in A_n, \quad (n = 1, 2, \ldots),$$
$$\eta(\omega) = y_n, \text{ for } \omega \in B_m, \quad (m = 1, 2, \ldots),$$

where $\{A_n\}$ and $\{B_m\}$ *are two complete systems of events.*

We say that the random variables ξ and η are independent if

$$P(A_n \cap B_m) = P(A_n)\, P(B_m), \tag{6.39}$$

that is if the two complete systems of events are independent.

Definition 5. *Let ξ and η be two random variables defined by*

$$\xi(\omega) = x_n, \textit{ for } \omega \in A_n, \quad (n = 1, 2, \ldots,),$$
$$\eta(\omega) = y_m, \textit{ for } \omega \in B_m, \quad (m = 1, 2, \ldots,),$$

where $\{A_n\}$ and $\{B_m\}$ are two complete systems of events and $g(x, y)$ a real function of two real variables.

The random variable $\zeta = g(\xi, \eta)$ has the values $\zeta_{nm} = g(x_n, y_m)$ as it is defined by the complete system of events $C_{nm} = \{\xi = x_n, \eta = y_m\}$.

Definition 6. *Let ξ be a random variable having the values x_n and the distribution $p_n = P(\xi = x_n)$. If the series $\sum_n p_n x_n$ is absolutely convergent, the expression*

$$E[\xi] = \sum_n p_n x_n$$

is called the mean value of the random variable or its mathematical expectation.

Let us prove some properties of the mean value of a random variable.

Proposition 1. *Let ξ and η be two discrete random variables. If $E[\xi]$ and $E[\eta]$ exist, there exists also $E[\xi + \eta]$ and we have*

$$E[\xi + \eta] = E[\xi] + E[\eta]. \tag{6.40}$$

Let $\{A_i\}$ and $\{B_j\}$ be two complete systems of events $x_i = \xi(\omega)$, $\omega \in A_i$, $(i = 1, 2, \ldots)$ and $y_j = \eta(\omega)$, $\omega \in B_j$, $(j = 1,2, \ldots)$.

Let us denote by C_{ij} the event $\xi = x_i$ and $\eta = y_j$. As $\{C_{ij}\}$ forms a complete system of events, we have

$$\bigcup_i C_{ij} = B_j, \quad \bigcup_j C_{ij} = A_i, \tag{6.41}$$

whence

$$\sum_i P(C_{ij}) = P(B_j), \quad \sum_j P(C_{ij}) = P(A_i), \tag{6.42}$$

thus

$$E[\xi + \eta] = \sum_i \sum_j (x_i + y_j) P(C_{ij}) = \sum_i \sum_j x_i P(C_{ij}) +$$

$$+ \sum_i \sum_j y_j P(C_{ij}) = \sum_i x_i \Big(\sum_j P(C_{ij})\Big) + \sum_j y_j \Big(\sum_i P(C_{ij})\Big),$$

or taking into account formula (6.42), we obtain

$$E[\xi + \eta] = \sum_i x_i P(A_i) + \sum_j y_j P(B_j). \tag{6.43}$$

If we take into account that $E[\xi]$ and $E[\eta]$ exist, it follows that the right-hand side series in formula (6.43) exists and that means that $E[\xi + \eta]$ also exists. Also from formula (6.43) we obtain formula (6.40).

By complete induction we have proposition 2.

Proposition 2. *Let ξ_k, $(k = 1, \ldots, n)$ be n discrete random variables. If $E[\xi_k]$, $k \leqslant n$, exist, then $E[\xi_1 + \ldots + \xi_n]$ exists and we have*

$$E[\xi_1 + \ldots + \xi_n] = E[\xi_1] + \ldots + E[\xi_n].$$

Proposition 3. *Let ξ be a discrete random variable and c a constant. If $E[\xi]$ exists, then $E[c\,\xi]$ exists and we have*

$$E[c\,\xi] = cE[\xi].$$

Indeed, from definition 6 it follows that

$$E\{c\,\xi] = \sum_n p_n(cx_n) = c \sum_n p_n x_n = cE[\xi].$$

From propositions 2 and 3 we deduce proposition 4.

Proposition 4. *Let ξ_k, $(k = 1, \ldots, n)$ be n discrete random variables and let c_k be n constants. If $E[\xi_k]$, $k \leqslant 1$, exist, then $E\left[\sum_{k=1}^{n} c_k \xi_k\right]$ exists and we have*

$$E\left[\sum_{k}^{n} c_k \xi_k\right] = \sum_{k=1}^{n} c_k E[\xi_k]. \tag{6.41'}$$

Definition 7. *If ξ is a discrete random variable and $E[\xi]$ exists, $\eta = \xi - E[\xi]$ is called the standard deviation of the random variable ξ.*

By means of the above definition let us prove proposition 5.

Proposition 5. *If the standard deviation of a random variable exists, then its mean value is zero.*

Indeed let $\zeta =$ constant be a random variable. Then

$$E[\zeta] = \sum_{n} p_n \zeta = \zeta \sum_{n} p_n.$$

But $p_n = P(A_n)$ and $\{A_n\}$ is a complete system of events, hence $\sum_{n} p_n = 1$, that is

$$E[\zeta] = \zeta.$$

Since for every random variable ξ whose mean value exists, we have $E[\xi] =$ constant, it follows that

$$E[E[\xi]] = E[\xi]$$

and by taking into account formula (6.41′) we deduce that

$$E[\eta] = E[\xi - E[\xi]] = E[\xi] - E[E[\xi]] = 0.$$

Proposition 6. *(Schwarz's inequality). Let there be ξ and η two discrete random variables, so that $E[\xi^2]$ and $E[\eta^2]$ exist. Then we have*

$$|E[\xi\eta]| \leqslant \sqrt{E[\xi^2]\, E[\eta^2]} \tag{6.42'}$$

In order to prove this proposition let us consider the random variable $\zeta_\lambda = (\xi - \lambda\eta)^2$ where λ is a real parameter.

From formula (6.41′) we have

$$E[\zeta_\lambda] = E[\xi^2] - 2\lambda\, E[\xi\, \eta] + \lambda^2\, E[\eta^2].$$

But $\zeta_\lambda \geqslant 0$, thus from definition 6 we deduce that $E[\zeta_\lambda] \geqslant 0$, for every value of λ. It follows that

$$E[\xi^2] - 2\lambda E[\xi\eta] + \lambda^2 E[\eta^2] \geqslant 0$$

whence we obtain relationship (6.42′).

Proposition 7. *If ξ and η are two independent discrete random variables and if $E[\xi]$ and $E[\eta]$ exist then $E[\xi\eta]$ exists and we have*

$$E[\xi\eta] = E[\xi]\,E[\eta]. \tag{6.43'}$$

Indeed, if we denote by C_{ij} the event $\xi = x_i$ and $\eta = y_j$, we have

$$E[\xi\eta] = \sum_i \sum_j x_j y_j P(C_{ij}).$$

But as ξ and η are independent, from definition 4 we deduce that

$$P(C_{ij}) = P(\xi = x_i)\,P(\eta = y_j)$$

thus

$$E[\xi\eta] = \left[\sum_i x_i P(\xi = x_i)\right]\left[\sum_j y_j P(\eta = y_j)\right] = E[\xi]\,E[\eta].$$

Definition 8. *Let ξ be a discrete random variable with values x_k, $(k \in I)$, with the distribution $P(\xi = x_k)$ and let A be an event for which $P(A) \neq 0$. If the series $\sum_{k \in I} x_k P_A(\xi = x_k)$ is absolutely convergent, the expression*

$$E[\xi/A] = \sum_{k \in I} x_k P_A(\xi = x_k) \tag{6.44}$$

is called the conditioned mean value of ξ.

With this definition let us prove proposition 8.

Proposition 8. *If $\{A_n\}$, $(n = 1, 2, \ldots)$ is a complete system of events and ξ a discret random variable for which $E[\xi]$ exists then*

$$E[\xi] = \sum_n P(A_n)\,E[\xi/A_n]. \tag{6.45}$$

Indeed if we take into account that $P(A) \neq 0$, from formula (6.33) we deduce that

$$P(\xi = x_k) = \sum_n P_{A_n}(\xi = x_k)\,P(A_n).$$

Taking into account this relationship in the definition of the mean value $E[\xi]$, we obtain

$$E[\xi] = \sum_{k \in I} x_k P(\xi = x_k) = \sum_{k \in I} x_k \left[\sum_n P_{A_n}(\xi = x_k) P(A_n) \right] =$$

$$= \sum_n P(A_n) \left[\sum_{k \in I} x_k P_{A_n}(\xi = x_k) \right],$$

and from formula (6.44), we obtain

$$E[\xi] = \sum_n P(A_n) E[\xi \mid A_n].$$

Definition 9. *Let ξ be a discrete random variable and p a natural number. If the mean value of ξ^p exists then this mean value is called the moment of order p of the random variable ξ and is denoted by*

$$\alpha_p[\xi] = E[\xi^p] = \sum_k x_k^p p_k.$$

The mean value of the random variable $|\xi|^p$ is called the absolute moment of order p and is denoted by

$$\beta_p[\xi] = E[|\xi|^p] = \sum_k |x_k|^p p_k.$$

Definition 10. *Given a discrete random variable ξ, the moment of order p of the standard deviation is called the centred moment of the order p and is denoted by*

$$\mu_p[\xi] = \alpha_p[\xi - E[\xi]].$$

Definition 11. *Given a discrete random variable ξ the centred moment of the second order is called the variance of the random variable ξ and is denoted by $D^2(\xi)$ or by σ^2, that is*

$$D^2(\xi) = \mu_2[\xi].$$

The number $D(\xi) = \sigma = \sqrt{\mu_2[\xi]}$ is called the mean square deviation of the random variable ξ.

Proposition 9. *We have*

$$D^2(\xi) = E[\xi^2] - (E[\xi])^2. \tag{6.46}$$

From definitions 10 and 11 it follows that

$$D^2(\xi) = \alpha_2[\xi - E[\xi]] = E[(\xi - E[\xi])^2] = E[\xi^2 - 2E[\xi]\,\xi + (E[\xi])^2].$$

But $E[\xi]$ is a constant and from definition 6 it follows that the mean value of a constant is equal to the constant, that is

$$E[(E[\xi])^2] = (E[\xi])^2,$$

or by taking into account formula (6.41′) we deduce that

$$D^2(\xi) = E[\xi^2] - 2E[\xi]\,E[\xi] + (E[\xi])^2 = E[\xi^2] - (E[\xi])^2.$$

Proposition 10. *If $\eta = a\,\xi + b$ where a and b are constants, then*

$$D(\eta) = |a|\,D(\xi).$$

Indeed, from formula (6.41′) it follows that

$$E[\eta] = aE[\xi] + b,\ E[\eta^2] = a^2\,E[\xi^2] + 2abE[\xi] + b^2$$

and by taking into account formula (6.46) it follows that

$$D^2(\eta) = a^2\,D^2(\xi) \tag{6.47}$$

hence

$$D(\eta) = |a|\,D(\xi).$$

In particular, if $b = 0$, formula (6.47) becomes

$$D^2(a\ \xi) = a^2\,D^2(\xi). \tag{6.47′}$$

Proposition 11. *Let $\xi_1, \ldots, \xi_n$, be n discrete random variables mutually independent and let $a_1, \ldots, a_n$ be n constants. Then*

$$D^2\left(\sum_{k=1}^{n} a_k\,\xi_k\right) = \sum_{k=1}^{n} a_k^2\,D^2(\xi_k). \tag{6.48}$$

From formula (6.46) we have

$$D^2\left(\sum_{k=1}^{n} a_k\,\xi_k\right) = E\left[\left(\sum_{k=1}^{n} a_k\,\xi_k\right)^2\right] - \left(E\left[\sum_{k=1}^{n} a_k\,\xi_k\right]\right)^2.$$

But taking into account formula (6.41′) it follows that

$$E\left[\sum_{k=1}^{n} a_k \xi_k\right] = \sum_{k=1}^{n} a_k E[\xi_k]$$

thus

$$D^2\left(\sum_{k=1}^{n} a_k \xi_k\right) = E\left[\left(\sum_{k=1}^{n} a_k \xi_k\right)^2\right] - \left(\sum_{k=1}^{n} a_k E[\xi_k]\right)^2 =$$

$$= E\left[\sum_{k=1}^{n} a_k^2 \xi_k^2 + 2\sum_{h<k} a_h a_k \xi_h \xi_k\right] - \sum_{k=1}^{n} a_k^2 (E[\xi_k])^2 -$$

$$-2\sum_{h<k} a_h a_k E[\xi_h] E[\xi_k] = \sum_{k=1}^{n} a_k^2 E[\xi_k^2] + 2\sum_{h<k} a_h a_k E[\xi_h \xi_k] -$$

$$-\sum_{k=1}^{n} a_k^2 (E[\xi_k])^2 - 2\sum_{h<k} a_h a_k E[\xi_h] E[\xi_k].$$

If we take into account that the random variables ξ_k $(k = 1, \ldots, n)$ are mutually independent, formula (6.43′) yields

$$E[\xi_h \xi_k] = E[\xi_h] E[\xi_k]$$

hence

$$D^2\left(\sum_{k=1}^{n} a_k \xi_k\right) = \sum_{k=1}^{n} a_k^2 E[\xi_k^2] + 2\sum_{h<k} a_h a_k E[\xi_h] E[\xi_k] -$$

$$-\sum_{k=1}^{n} a_k^2 (E[\xi_k])^2 - 2\sum_{h<k} a_h a_k E[\xi_h] E[\xi_k] = \sum_{k=1}^{n} a_k^2 [E[\xi_k^2] - (E[\xi_k])^2]$$

or, by taking into account formula (6.46) we obtain formula (6.48).

Definition 12. Let ξ and η be two discrete random variables whose mean values exist. The expression

$$\lambda_{\xi,\eta} = E[(\xi - E[\xi])(\eta - E[\eta])]$$

is called the covariance of the random variables ξ and η and the expression

$$\rho_{\xi,\eta} = \frac{\lambda_{\xi,\eta}}{D(\xi)\,D(\eta)}$$

is called the correlation coefficient of the random variables ξ and η.

Definition 13. Two discrete random variables ξ *and* η *are not correlated if* $\rho_{\xi,\eta} = 0$.

Proposition 12. *If* ξ *and* η *are two discrete random variables whose mean values exist, then*

$$\lambda_{\xi,\eta} = E[\xi\eta] - E[\xi]E[\eta]. \tag{6.49}$$

Indeed, from formula (6.41′) and from the fact that $E[\xi]$ and $E[\eta]$ are constant, we have

$$\lambda_{\xi,\eta} = E[\xi\eta - E[\eta]\xi - E[\xi]\eta + E[\xi]\, E[\eta]] = E[\xi\eta] - E[\xi]\, E[\eta] -$$
$$- E[\xi]E[\eta] + E[\xi]E[\eta] = E[\xi\eta] - E[\xi]\, E[\eta].$$

From formula (6.49) and proposition 7 proposition 13 follows.

Proposition 13. *Two independent discrete random variables are not correlated.*

The converse of this property is not true, as we can see from several examples.

Proposition 14. *For every two discrete random variables* ξ *and* η *whose mean values exist, we have*

$$\rho^2_{\xi,\eta} \leqslant 1. \tag{6.50}$$

In order to prove this property let us consider two arbitrary real constants a and b and let us form the expression

$$g(a,\ b) = E[\{\, a(\xi - E[\xi]) + b(\eta - E[\eta])\}^2], \tag{6.51}$$

which is non-negative if we take into account definition 6. Expanding the right-hand side expression, from definitions 11 and 12 we obtain

$$g(a,\ b) = a^2D^2(\xi) + 2ab\ \lambda_{\xi,\eta} + b^2D^2(\eta) \geqslant 0. \tag{6.51′}$$

It follows that the discriminant of the square form $g\,(a,\ b)$ is non-negative, that is

$$D^2(\xi)D^2(\eta) - \lambda^2_{\xi,\eta} \geqslant 0,$$

hence

$$\rho^2_{\xi,\eta} = \frac{\lambda^2_{\xi,\eta}}{D^2(\xi)\, D^2(\eta)} \leqslant 1.$$

roposition 15. *If $\rho^2_{\xi,\eta} = 1$, then between the discrete random variables and η there is a linear relation.*

Indeed, if in (6.51′) we take $a = -\rho_{\xi,\eta}\, D(\eta)$, $b = D(\xi)$, we obtain

$$g(-\rho_{\xi,\eta} D(\eta),\ D(\xi)) = \rho^2_{\xi,\eta}\, D^2(\xi)\, D^2(\eta) - 2\rho^2_{\xi,\eta}\, D^2(\xi)\, D^2(\eta) + D^2(\xi)\, D^2(\eta).$$

But $\rho_{\xi,\eta} = \pm 1$, whence by taking into account (6.51) we deduce

$$E[\{\mp D(\eta)(\xi - E[\xi]) + D(\xi)(\eta - E[\eta])\}^2] = 0.$$

Taking into account definition 6, it follows that

$$\mp D(\eta)(\xi - E[\xi]) + D(\eta)(\eta - E[\eta]) = 0,$$

ch is a linear relation between ξ and η.

2 Classical distributions

us consider an urn in which there are $a + b$ balls, a white balls and b k balls. Such an urn is called Bernoulli's urn. We denote by A the nt which consists of drawing a white ball. Then $\complement A$ is the event which sists of drawing a black ball. We have

$$P(A) = \frac{a}{a+b}, \quad P(\complement A) = \frac{b}{a+b}.$$

If we consider $P(A) = p$, $P(\complement A) = q$, it follows $p > 0$, $q > 0$, $p + q = 1$.

We shall draw n times a ball from the urn and put back the drawn ball h time. These operations are thus independent. Let us denote by B the nt which consists of obtaining k white balls by n repeated operations. follows that the events $\{B_k\}$, $(k = 0, 1, \ldots, n)$ form a complete system events. Taking into account that the order of drawing these k balls is no interest it follows that

$$W_k = P(B_k) = \binom{n}{k} p^k q^{n-k}, \quad (k = 0, 1, \ldots, n). \tag{6.52}$$

Definition 14. The distribution determined by the probabilities (6.52) *is called the binomial distribution of order n and parameter p and the random variable*

$$\xi\begin{pmatrix} 0, & 1, & 2, & ,\dots, n \\ q^n, & \binom{n}{1}pq^{n-1}, & \binom{n}{2}p^2q^{n-2}, & \dots, p^n \end{pmatrix}; \quad (q = 1 - p)$$

is called the binomial random variable.

Let us now prove the following proposition.

Proposition 16. *If ξ is a binomial random variable of order n and parameter p, then*

$$E[\xi] = np, \quad D^2(\xi) = npq.$$

Indeed, from definition 5 we have

$$E[\xi] = \sum_{k=1}^{n} k\binom{n}{k}p^kq^{n-k}.$$

In order to calculate the sum of the right-hand side, of this equality let us consider the identity

$$(pt + q)^n = \sum_{k=0}^{n}\binom{n}{k}p^kt^kq^{n-k}$$

and let us differentiate it term by term with respect to t. We obtain

$$np(pt + q)^{n-1} = \sum_{k=1}^{n} k\binom{n}{k}p^kt^{k-1}q^{n-k}. \tag{6.53}$$

By taking $t = 1$ in the above equality and by taking into account that $p + q = 1$ we deduce

$$\sum_{k=1}^{n} k\binom{n}{k}p^kq^{n-k} = np,$$

thus

$$E[\xi] = np.$$

In order to prove the second relationship let us calculate the moment of the second-order of the random variable ξ. We have

$$E[\xi^2] = \sum_{k=1}^{n} k^2\binom{n}{k}p^kq^{n-k}.$$

om formula (6.53) we deduce

$$npt\,(pt+q)^{n-1} = \sum_{k=1}^{n} k \binom{n}{k} p^k q^{n-k} t^k.$$

Differentiating this relationship term by term with respect to t and taking $t = 1$ we have

$$\sum_{k=1}^{n} k^2 \binom{n}{k} p^k q^{n-k} = np + n(n-1)p^2,$$

thus

$$E[\xi^2] = np + n(n-1)p^2.$$

Substituting the values of $E[\xi]$ and $E[\xi^2]$, in formula (6.46), we obtain

$$D^2(\xi) = np + n(n-1)\,p - n^2p^2 = np(1-p) = npq.$$

Let us consider an urn which contains n_1 balls of colour c_1, n_2 balls of colour $c_2, \ldots, n_r$ balls of colour c_r. Let us denote by A_l the event consisting of drawing a ball of colour c_l, $(l = 1, \ldots, r)$ and set $p_l = P(A_l)$. It is obvious that we have $p_l = \dfrac{n_l}{n_1 + n_2 + \ldots + n_r}$ and thus $p_1 + \ldots + p_r = 1$. Finally let us denote by $B_{k_1k_2 \ldots k_r}$ the event which consists of obtaining, after n such operations and after putting back the n drawn ball every time into the urn, k_1 balls of colour c_1, k_2 balls of colour $c_2, \ldots, k_r$ balls of colour c_r, where $k_1 + k_2 + \ldots + k_r = n$.

Taking into account that the order of drawing the balls is unimportant, we obtain

$$W_{k_1k_2 \ldots k_r} = P(B_{k_1k_2 \ldots k_r}) = \frac{n!}{k_1!\, k_2! \ldots k_r!} p_1^{k_1} p_2^{k_2} \ldots p_r^{k_r} \qquad (6.54)$$

Definition 15. The distribution determined by the probabilities (6.54) *is called the multinomial* (*polynomial*) *distribution and the corresponding variable is called the multinomial* (*polynomial*) *random variable.*

Let us consider an urn in which there are a white balls and b black balls. We shall draw n times a ball from the urn without putting back the drawn ball.

Denoting by C_k the event which consists of obtaining k white balls from these n operations the events $\{C_k\}$ $(\max\,(0, n-b) \leqslant k \leqslant \min\,(n, a))$ form a complete system of events. In order to determine the probability

$P(C_k)$ let us determine the total number of elementary events of the space and the number of elementary events favorable to C_k. The total number of elementary events of the space is equal to the number of combinations which can be made with the total number of balls from the urn taken, with respect to the number of the drawn balls i.e. $\binom{a+b}{n}$. The second number is given by the number of groups which can be formed with k white balls and $n-k$ black balls. With the a white balls we can form $\binom{a}{k}$ groups which associated with $\binom{b}{n-k}$ groups which can be formed with the b black balls yields the number $\binom{a}{k}\binom{b}{n-k}$. Thus

$$P(C_k) = \frac{\binom{a}{k}\binom{b}{n-k}}{\binom{a+b}{n}}. \tag{6.55}$$

Definition 16. The distribution determined by the probabilities (6.55) *is called the hypergeometric distribution and the random variable is called the hypergeometric random variable.*

By means of a simple calculus we can prove the following proposition.

Proposition 17. *The mean value and the variance of a hypergeometrical random variable are*

$$E[\xi] = \frac{na}{a+b}, \quad D^2[\xi] = \frac{nab\,(a+b-n)}{(a+b)^2\,(a+b-1)}.$$

Let us consider again the binomial distribution and suppose that

$$np = \lambda \text{ (constant).} \tag{6.56}$$

We want to determine the value of the probability W_k as $n \to \infty$. By taking into account the expression (6.52) of W_k and relationship (6.56) we can write

$$W_k = \frac{n(n-1)\dots(n-k+1)}{k!}\,p^k\,(1-p)^{n-k} =$$

$$= \frac{n(n-1)\dots(n-k+1)}{k!}\,\frac{\lambda^k}{n^k}\cdot\left(1-\frac{\lambda}{n}\right)^{n-k}.$$

But

$$\lim_{n\to\infty} \frac{n(n-1)\dots(n-k+1)}{n^k} = 1, \lim_{n\to\infty}\left(1-\frac{\lambda}{n}\right)^{n-k} = e^{-\lambda},$$

thus

$$\lim_{n\to\infty} W_k = \frac{\lambda^k}{k!}e^{-\lambda}, \quad (k = 0, 1, 2, \dots).$$

If we denote by

$$P_k = \frac{\lambda^k}{k!}e^{-\lambda} \tag{6.57}$$

we have

$$\sum_{k=0}^{\infty} P_k = e^{-\lambda}\sum_{k=0}^{\infty}\frac{\lambda^k}{k!} = e^{-\lambda}e^{\lambda} = 1.$$

Definition 17. The distribution determined by the probabilities (6.57) *is called Poisson's distribution of parameter* λ *and the random variable*

$$\xi\begin{pmatrix} 0, & 1, & 2, & \dots, & k, & \dots \\ e^{-\lambda}, & \frac{\lambda}{1!}e^{-\lambda}, & \frac{\lambda^2}{2!}e^{-\lambda}, & \dots, & \frac{\lambda^k}{k!}e^{-\lambda}, & \dots \end{pmatrix}$$

is called Poisson's random variable.

By means of the above definition let us prove the following proposition.

Proposition 18. *If* ξ *is a Poisson random variable of parameter* λ*, then*

$$E[\xi] = \lambda, \quad D^2(\xi) = \lambda.$$

Indeed, taking into account definition 6 we deduce that

$$E[\xi] = \sum_{k=0}^{\infty} k\frac{\lambda^k}{k!}e^{-\lambda} = \lambda\left(\sum_{k=1}^{\infty}\frac{\lambda^{k-1}}{(k-1)!}\right)e^{-\lambda} = \lambda e^{\lambda}e^{-\lambda} = \lambda.$$

From this equality and from definition 9, it follows that

$$E[\xi^2] = \sum_{k=0}^{\infty} k^2\frac{\lambda^k}{k!}e^{-\lambda} = \sum_{k=0}^{\infty}(k^2-k+k)\frac{\lambda^k}{k!}e^{-\lambda} =$$

$$= \sum_{k=0}^{\infty} k(k-1)\frac{\lambda^k}{k!}e^{-\lambda} + \sum_{k=0}^{\infty} k\frac{\lambda^k}{k!}e^{-\lambda} =$$

$$= \lambda^2\left(\sum_{k=2}^{\infty}\frac{\lambda^{k-2}}{(k-2)!}\right)e^{-\lambda} + \lambda = \lambda^2 e^{\lambda}e^{-\lambda} + \lambda = \lambda^2 + \lambda.$$

Using this equality in formula (6.46) we obtain

$$D^2(\xi) = E[\xi^2] - (E[\xi])^2 = \lambda^2 + \lambda - \lambda^2 = \lambda.$$

6.4 RANDOM VARIABLES

6.4.1 Random variables

Definition 1. *Let* $\{\Omega, \mathcal{K}, P\}$ *be a probability space. A real function* $\xi = \xi(\omega)$, *defined on the set of the elementary events* $\omega \in \Omega$ *is called a random variable (r.v.) if all the level sets* $A_a = \{\omega; \xi(\omega) < a\}$ *belong to the space* $\mathcal{K}$.

Proposition 1. *If* ξ *is a random variable then*

$$\{\omega;\ \xi(\omega) \leqslant a\} \in \mathcal{K}, \tag{6.58}$$

$$\{\omega;\ \xi(\omega) > a\} \in \mathcal{K}, \tag{6.58'}$$

$$\{\omega;\ \xi(\omega) \geqslant a\} \in \mathcal{K}, \tag{6.58''}$$

for every real a.

We have

$$\{\omega;\ \xi(\omega) \leqslant a\} = \bigcap_{n=1}^{\infty} \left\{\omega;\ \xi(\omega) < a + \frac{1}{n}\right\}.$$

If we take into account definition 1 it follows that

$$\left\{\omega;\ \xi(\omega) < a + \frac{1}{n}\right\} \in \mathcal{K}$$

for every real a and natural n. Thus

$$\bigcap_{n=1}^{\infty} \left\{\omega;\ \xi(\omega) < a + \frac{1}{n}\right\} \in \mathcal{K}$$

whence we deduce relationship (6.58).

Also, we have

$$\{\omega;\ \xi(\omega) > a\} = \complement\{\omega;\ \xi(\omega) \leqslant a\},$$

$$\{\omega;\ \xi(\omega) \geqslant a\} = \complement\{\omega;\ \xi(\omega) < a\},$$

and taking into account definition 1 and relationship (6.58) we have relationships (6.58′) and (6.58″).

Proposition 2. *If* ξ *and* η *are two random variables and* k *a real number, then*

$$A_1 = \{\omega;\ \xi(\omega) < \eta(\omega) + k\} \in \mathcal{K}, \tag{6.59}$$

$$A_2 = \{\omega;\ \xi(\omega) \leqslant \eta(\omega) + k\} \in \mathcal{K}, \tag{6.59'}$$

$$A_3 = \{\omega;\ \xi(\omega) = \eta(\omega) + k\} \in \mathcal{K}. \tag{6.59''}$$

Let us suppose that $\omega_0 \in A_1$, that is let us suppose that ω_0 is an elementary event for which $\xi(\omega_0) < \eta(\omega_0) + k$. Then there is a rational number $\frac{m_0}{n_0}$ with $n_0 > 0$, so that we obtain

$$\xi(\omega_0) < \frac{m_0}{n_0} < \eta(\omega_0) + k,$$

from which it follows that

$$\omega_0 \in \left\{\omega;\ \xi(\omega) < \frac{m_0}{n_0}\right\} \cap \left\{\omega;\ \eta(\omega) + \frac{m_0}{n_0} - k\right\},$$

that is

$$\omega_0 \in \bigcup_{m=-\infty}^{+\infty} \left[\bigcup_{n=0}^{\infty} \left(\left\{\omega;\ \xi(\omega) < \frac{m}{n}\right\} \cap \left\{\omega;\ \eta(\omega) > \frac{m}{n} - k\right\}\right)\right].$$

If ω_0 is an event of this set of events, then there is at least a rational number $\frac{m_0}{n_0}$ so that

$$\omega_0 \in \left\{\omega;\ \xi(\omega_0) < \frac{m_0}{n_0}\right\} \cap \left\{\omega;\ \eta(\omega_0) > \frac{m_0}{n_0} - k\right\},$$

whence we deduce that

$$\xi(\omega_0) < \frac{m_0}{n_0} \quad \text{and} \quad \eta(\omega_0) > \frac{m_0}{n_0} - k,$$

that is

$$\xi(\omega_0) < \eta(\omega_0) + k$$

and thus $\omega_0 \in A_1$. It follows that

$$A_1 = \bigcup_{m=-\infty}^{+\infty} \left[\bigcup_{n=0}^{\infty} \left(\left\{\omega;\ \xi(\omega) < \frac{m}{n}\right\} \cap \left\{\omega;\ \eta(\omega) > \frac{m}{n} - k\right\}\right)\right].$$

Since $\xi(\omega)$ and $\eta(\omega)$ are random variables from proposition 1 we deduce that $A_1 \in \mathcal{K}$.

Also we have

$$A_2 = \complement\{\omega;\ \eta(\omega) < \xi(\omega) - k\}$$

and thus $A_2 \in \mathcal{K}$ and

$$A_3 = A_1 \cap \complement A_2$$

whence it follows that $A_3 \in \mathcal{K}$.

Proposition 3. *If ξ is an random variable, and $k \neq 0$ a real number, then $\xi + k$, $k\xi$, $|\xi|$, ξ^2 and $\frac{1}{\xi}$, $(\xi \neq 0)$ are also random variables.*

Taking into account that ξ is a random variable from the relationship

$$\{\omega;\ \xi(\omega) + k < a\} = \{\omega;\ \xi(\omega) < a - k\} \in \mathcal{K},$$

it follows that $\xi + a$ is also a r.v.
From relationship

$$\{\omega;\ k\xi(\omega) < a\} = \begin{cases} \left\{\omega;\ \xi(\omega) > \frac{a}{k}\right\} \in \mathcal{K}, \text{ if } k < 0, \\ \left\{\omega;\ \xi(\omega) < \frac{a}{k}\right\} \in \mathcal{K}, \text{ if } k > 0, \end{cases}$$

we deduce that $k\xi$ is a r.v. Taking into account that

$$\{\omega;\ |\xi(\omega)| < a\} = \{\omega;\ \xi(\omega) < a\} \cap \{\omega;\ \xi(\omega) > -a\} \in \mathcal{K},$$

it follows that $|\xi|$ is a r.v.
From relationship

$$\{\omega;\ \xi^2(\omega) < a\} = \{\omega;\ |\xi(\omega)| < \sqrt{a}\} \in \mathcal{K},$$

we deduce that ξ^2 is a r.v.
Taking into account that, if $\xi \neq 0$ we have

$$\left\{\omega;\ \frac{1}{\xi(\omega)} < a\right\} = \begin{cases} \{\omega;\ \xi(\omega) < 0\} \cap \left\{\omega;\ \xi(\omega) > \frac{1}{a}\right\} \in \mathcal{K}, \text{ if } a < 0, \\ \{\omega;\ \xi(\omega) < 0\} \in \mathcal{K}, \text{ if } a = 0, \\ \{\omega;\ \xi(\omega) < 0\} \cup \left[\{\omega;\ \xi(\omega) > 0\} \cap \left\{\omega;\ \xi(\omega) > \frac{1}{a}\right\}\right] \in \mathcal{K}, \\ \qquad \text{if } a > 0, \end{cases}$$

it follows that $\frac{1}{\xi}$ is a r.v. if $\xi \neq 0$.

Proposition 4. *If ξ and η are two r.v. then $\xi \pm \eta$, $\xi\eta$ and $\frac{\xi}{\eta}$ $(\eta \neq 0)$, are r.v.*

Based on relationship (6.59) we have

$$\{\omega;\ \xi(\omega) - \eta(\omega) < a\} = \{\omega;\ \xi(\omega) < \eta(\omega) + a\} \in \mathcal{K}$$

and thus $\xi - \eta$ is a r.v.

In order to prove that $\xi + \eta$ is a r.v. we write

$$\xi + \eta = \xi - (-\eta)$$

and from proposition 3 we have $-\eta = (-1)\eta$ is a r.v.
From formula

$$\xi\eta = \frac{1}{4}[(\xi + \eta)^2 - (\xi - \eta)^2]$$

and since from proposition 3, we have that $(\xi + \eta)^2$ and $(\xi - \eta)^2$ are r.v. it follows that $\xi\eta$ is a r.v.

From proposition 3 it follows that $\frac{1}{\eta}$ is a r.v. and thus

$$\frac{\xi}{\eta} = \xi \cdot \frac{1}{\eta}$$

is a r.v.

Definition 2. A r.v. which takes a finite number of finite values is called a simple r.v.

Taking into account definition 1, it follows that $\xi(\omega)$ is a simple r.v. if:

(i) it takes a finite number of finite values $x_1, \ldots, x_m$,

(ii) $\Omega_i = \{\omega;\ \xi(\omega) = x_i\} \in \mathcal{K}$, $(i = 1, \ldots, m)$.

If we denote by χ_{Ω_i} the indicator of the event Ω_i, that is

$$\chi_{\Omega_i} = \begin{cases} 1, & \text{if } \omega \in \Omega_i, \\ 0, & \text{if } \omega \notin \Omega_i, \end{cases}$$

then we can write the r.v. ξ in the form

$$\xi = \sum_{i=1}^{m} x_i\, \chi_{\Omega_i}. \tag{6.60}$$

In the following we shall denote by S the set of the simple r.v. It is obvious that, if $\xi, \eta \in S$ and if k is a real number, then

$$|\xi| \in S,\ k\xi \in S,\ \xi^2 \in S,\ \xi + \eta \in S,\ \xi\eta \in S.$$

Let us now prove the following proposition.

Proposition 5. *If ξ is a non-negative r.v. then there is an increasing sequence $\{\xi_n\}$ of non-negative simple r.v. which is convergent to ξ.*

Let us denote by

$$\xi_n(\omega) = \begin{cases} \dfrac{i-1}{2^n}, & \text{if } \dfrac{i-1}{2^n} < \xi(\omega) < \dfrac{i}{2^n}, \qquad (i = 1, 2, 3, \ldots, 2^n), \\ n, & \text{if } \xi(\omega) \geqslant n, \end{cases}$$

for every $n = 1, 2, \ldots$

Whence it follows that $\xi_n \in S$ and that $\xi_n \geqslant 0$.

If $\xi(\omega) < \infty$ from the definition of $\xi_n(\omega)$ we deduce that

$$|\xi(\omega) - \xi_n(\omega)| < \frac{1}{2^n}$$

and thus $\lim\limits_{n\to\infty} \xi_n(\omega) = \xi(\omega)$.

If $\xi(\omega) = \infty$, then for every natural number n we have $\xi_n = n$ and in this case $\lim\limits_{n\to\infty} \xi_n(\omega) = \xi(\omega)$.

Writing

$$\xi^+ = \sup(\xi, 0), \quad \xi^- = -\inf(\xi, 0),$$

if ξ is not non-negative ξ^+ and ξ^- are non-negative random variables and we have

$$\xi = \xi^+ - \xi^-.$$

It follows that every r. v. ξ is the limit of a convergent sequence $\{\xi_n\}$ of simple r. v. (not necessarily increasing).

Definition 3. The random variables $\xi_1, \ldots, \xi_n$ are called independent random variables if, for every system of real numbers $a_1, \ldots, a_n$, we have

$$P(\xi_1 < a_1, \ldots, \xi_n < a_n) = P(\xi_1 < a_1) \ldots P(\xi_n < a_n). \tag{6.61}$$

Definition 4. A vector $\xi(\xi_1, \ldots, \xi_n)$ whose components ξ_i, $(i = 1, \ldots, n)$, are random variables is called a n-dimensional random vector (or a n-dimensional r.v.).

Definition 5. If ξ_1 *and* ξ_2 *are two real r.v., then* $\xi = \xi_1 + i\xi_2$, $(i = \sqrt{-1})$ *is called a complex r.v.*

It is obvious that the complex r.v. can be put in a one-to-one correspondance with the bidimensional r.v. (ξ_1, ξ_2).

6.4.2 The distribution function

Definition 6. Given a r.v. ξ, *the function*

$$F(x) = P(A_x) = P(\xi < x)$$

is called the distribution function of the r.v. ξ.

Let us prove some properties of this function.

Proposition 6. *The function* $F(x)$ *is monotonously increasing.*

Let x_1 x_2 be two real numbers and $A_{x_1} = \{\omega;\ \xi(\omega) < x_1\}$, $A_{x_2} = \{\omega;\ \xi(\omega) < x_2\}$ their corresponding level sets. We have $A_{x_1} \subseteq A_{x_2}$ and thus based on proposition 13, Section 6.2.2,

$$P(A_{x_1}) \leqslant P(A_{x_2}),$$

i.e. by taking into account definition 6, we obtain

$$F(x_1) \leqslant F(x_2),$$

thus the distribution function is monotonously increasing.

Proposition 7. *The function* $F(x)$ *is continuous to the left.*

To prove this property we must prove that

$$\lim_{h\to 0,\ h>0} F(x - h) = F(x). \tag{6.62}$$

Let us consider a fixed real number x and let

$$x_1 < x_2 < \ldots < x_n < \ldots$$

be an increasing sequence of real numbers convergent to x. Let us consider the events

$$A = \{\omega;\ \xi(\omega) < x\},$$

$$A_1 = \{\omega;\ \xi(\omega) < x_1\},$$

$$B_n = \{\omega;\ x_n \leqslant \xi(\omega) < x_{n+1}\}, \qquad (n = 1, 2, \ldots).$$

We have

$$A = A_1 \cup B_1 \cup B_2 \cup \ldots,$$

where the events of the right-hand side are mutually exclusive, thus

$$P(A) = P(A_1) + P(B_1) + P(B_2) + \ldots \tag{6.63}$$

On the other hand

$$P(A) = F(x), \quad P(A_1) = F(x_1)$$

and

$$P(B_n) = P(\{\omega;\ \xi(\omega) < x_{n+1}\}) - P(\{\omega;\ \xi(\omega) < x_n\}) =$$

$$= F(x_{n+1}) - F(x_n), \quad (n = 1, 2, \ldots).$$

Using this formula, in formula (6.63) we obtain

$$F(x) = F(x_1) + F(x_2) - F(x_1) + F(x_3) - F(x_2) + \ldots$$

$$\ldots + F(x_{n+1}) - F(x_n) + \ldots$$

thus

$$F(x) = \lim_{n \to \infty} F(x_n)$$

that is

$$F(x) = F(x - 0).$$

The relationship is equivalent to formula (6.62).

Proposition 8. *Every distribution function has the properties*

$$F(-\infty) = 0, \ F(+\infty) = 1.$$

In order to prove the first relationship let us consider the decreasing sequence of real numbers

$$x_1 > x_2 > \ldots > x_n > \ldots$$

which tends to $-\infty$ and the events

$$A = \{\omega;\ \xi(\omega) < x_1\},$$

$$B_n = \{\omega;\ x_n \leqslant \xi(\omega) < x_{n-1}\}, \ (n = 1, 2, \ldots).$$

The events B_n, $(n = 1, 2, \ldots)$ are mutually exclusive and we have

$$A = \bigcup_{n=1}^{\infty} B_n$$

thus

$$P(A) = \sum_{n=1}^{\infty} P(B_n),$$

that is

$$F(x_1) = [F(x_1) - F(x_2)] + [F(x_2) - F(x_3)] + \ldots$$
$$\ldots + [F(x_{n-1}) - F(x_n)] + \ldots$$

or

$$F(x_1) = F(x_1) - \lim_{n\to\infty} F(x_n).$$

Considering the fact that the sequence x_n tends to $-\infty$, we deduce that

$$\lim_{n\to\infty} F(x_n) = F(-\infty) = 0.$$

In order to prove the second relation let us consider the increasing sequence of real numbers

$$x_1 < x_2 < \ldots < x_n < \ldots$$

which tends to $+\infty$ and the events

$$\Omega = \{\omega;\ \xi(\omega) \leqslant \infty\},$$
$$D_0 = \{\omega;\ \xi(\omega) < x_1\},$$
$$D_n = \{\omega;\ x_n \leqslant \xi(\omega) < x_{n+1}\},\ (n = 1, 2, \ldots).$$

But the events D_0, D_1, D_2, ... are mutually exclusive and

$$\Omega = \bigcup_{j=0}^{\infty} D_j,$$

thus

$$P(\Omega) = \sum_{j=0}^{\infty} P(D_j)$$

that is

$$1 = F(x_1) + [F(x_2) - F(x_1)] + \dots + [F(x_{n+1}) - F(x_n)] + \dots$$

or

$$1 = \lim_{n\to\infty} F(x_n).$$

Taking into account that the sequence $\{x_n\}$ tends to ∞, we have

$$\lim_{n\to\infty} F(x_n) = F(+\infty) = 1.$$

We shall now prove a theorem which represents the converse of the propositions 6, 7 and 8.

Proposition 9. *Every monotonously increasing function $F(x)$ continuous to the left which has the values $F(-\infty) = 0$, $F(+\infty) = 1$ is a distribution function for at least one r.v. defined in a conveniently chosen probability space.*

Let us suppose that F is a function which has the above properties and let $x = G(y)$ be an inverse of the function $y = F(x)$.

We shall consider as set Ω the interval (0, 1), as space $\mathcal{K}$, the σ-algebra of the measurable subsets of this interval and as probability $P(A)$; $(A \in K)$ the Lebesgue measure of the set A. Then for the r.v. $\eta(y) = G(y)$, we have

$$P(\eta < x) = P[G(y) < x] = P[y < F(x)] = F(x).$$

Thus $F(x)$ is a distribution function of the r.v. η.

Proposition 10. *A distribution function $F(x)$ which is continuous at every point is uniformly continuous on R.*

If $x \in R$ then we have

$$0 = F(-\infty) \leqslant F(x) \leqslant F(+\infty) = 1$$

thus, we can determine two real numbers a and b with $a < b$ so that for $x < a$ we have

$$0 \leqslant F(x) \leqslant \frac{\varepsilon}{2}$$

and for $x > b$ we have

$$1 - \frac{\varepsilon}{2} \leqslant F(x) \leqslant 1.$$

Let us now consider two points x_1, $x_2 \in [a, b]$. As the function F is continuous in R, it is uniformly continuous on an arbitrary closed interval and hence also on $[a, b]$. It follows that we can determine a number $\eta = \eta(\varepsilon) > 0$ so that, as soon as $| x_1 - x_2 | < \eta$, we have

$$| F(x_1) - F(x_2) | < \frac{\varepsilon}{2} \cdot$$

Whence we deduce that if $x_1, x_2 \in R$ and $| x_1 - x_2 | < \delta = \min (\eta, b - a)$ we have $| F(x_1) - F(x_2) | < \varepsilon$, that is $F(x)$ is uniformly continuous in R.

Proposition 11. *If ξ is a r.v. and $F(x)$ is its distribution function then for every real x we have*

$$P(\xi = x) = F(x + 0) - F(x). \tag{6.64}$$

Let us consider a decreasing sequence of real numbers

$$x_1 > x_2 > \ldots > x_n > \ldots$$

which tends to the real number x and set

$$A = \{\omega;\ \xi(\omega) \geqslant x\},$$

$$A_1 = \{\omega;\ \xi(\omega) \geqslant x_1\},$$

$$B_n = \{\omega;\ x_{n+1} \leqslant \xi(\omega) < x_n\},\ (n = 1, 2, \ldots),$$

$$C = \{\omega;\ \xi(\omega) = x\}.$$

We have

$$A = C \cup A_1 \cup \left(\bigcup_{n=1}^{\infty} B_n \right),$$

and the events of the right-hand side of the equality are mutually exclusive. Thus

$$P(A) = P(C) + P(A_1) + \sum_{n=1}^{\infty} P(B_n). \tag{6.65}$$

If we take into account that

$$P(\complement A) = P(\{\omega;\ \xi(\omega) < x\}) = F(x)$$

we deduce that

$$P(A) = 1 - F(x).$$

By analogy

$$P(A_1) = 1 - F(x_1).$$

We also have

$$P(B_n) = P(x_{n+1} \leqslant \xi < x_n) = P(\xi < x_n) - P(\xi < x_{n+1}) =$$
$$= F(x_n) - F(x_{n+1}).$$

Substituting these relationships into formula (6.65) we obtain

$$1 - F(x) = P(C) + 1 - F(x_1) + \sum_{n=1}^{\infty} [F(x_n) - F(x_{n+1})]$$

whence it follows that

$$1 - F(x) = P(C) + 1 - \lim_{n\to\infty} F(x_n).$$

If we take into account that the sequence $\{x_n\}$ is decreasing and convergent to x, we deduce that

$$\lim_{n\to\infty} F(x_n) = F(x + 0),$$

that is

$$P(C) = F(x + 0) - F(x)$$

which, taking into account the expression of C, coincides with formula (6.65).

Proposition 12. *If ξ is a r.v. whose distribution function is $F(x)$ and $a < b$ are two real numbers, then we have*

$$P(a \leqslant \xi < b) = F(b) - F(a), \tag{6.66}$$
$$P(a < \xi < b) = F(b) - F(a + 0), \tag{6.66'}$$
$$P(a < \xi \leqslant b) = F(b + 0) - F(a + 0), \tag{6.66''}$$
$$P(a \leqslant \xi \leqslant b) = F(b + 0) - F(a). \tag{6.66'''}$$

We have

$$\{\omega;\ \xi(\omega) < a\} \cup \{\omega;\ a \leqslant \xi(\omega) < b\} = \{\omega;\ \xi(\omega) < b\}.$$

As the events from the left-hand side are mutually exclusive it follows that:

$$P(\xi < a) + P(a \leqslant \xi < b) = P(\xi < b)$$

whence it follows that

$$P(a \leqslant \xi < b) = P(\xi < b) - P(\xi < a) = F(b) - F(a).$$

In order to prove relationship (6.66′) we notice that

$$P(a < \xi < b) = P(a \leqslant \xi < b) - P(\xi = a),$$

and hence, taking into account formulae (6.66) and (6.65), formula (6.66′) follows.

We also have

$$P(a < \xi \leqslant b) = P(a < \xi < b) + P(\xi = b)$$

and if we substitute in this formula the relationships (6.66′) and (6.65) we obtain formula (6.66″).

By taking into account that

$$P(a \leqslant \xi \leqslant b) = P(a \leqslant \xi < b) + P(\xi = b),$$

from formulae (6.66) and (6.65), we obtain formula (6.66‴).

Definition 7. The distribution function corresponding to a discrete random variable is called a distribution function of a discrete type.

If ξ is a discrete random variable whose values are $x_1, x_2, \ldots$ with the corresponding probabilities $p_1, p_2, \ldots$ then from definitions 6 and 7 it follows that the distribution function of this random variable is

$$F(x) = \sum_{x_n < x} p_n .$$

Definition 8. Let ξ be a r.v. whose distribution function is $F(x)$. If there is an integrable function $f(u)$ so that we have

$$F(x) = \int_{-\infty}^{x} f(u) \, \mathrm{d}u, \tag{6.67}$$

then $F(x)$ is called a distribution function of the continuous type and the r.v. ξ is called the r.v. of the continuous type. The function $f(u)$ is called the density of distribution or density of probability.

From formula (6.67) it follows that $f(u) \geqslant 0$ and

$$\int_{-\infty}^{+\infty} f(u) \, \mathrm{d}u = 1.$$

Also from formula (6.67) we deduce that

$$f(x) = F'(x) = \lim_{\Delta x \to 0} \frac{F(x + \Delta x) - F(x)}{\Delta x} = \lim_{\Delta x \to 0} \frac{P(x \leqslant \xi \leqslant x + \Delta x)}{\Delta x} \tag{6.68}$$

whence it follows that

$$P(x \leqslant \xi < x + \mathrm{d}x) = f(x)\ \mathrm{d}x. \tag{6.68'}$$

Definition 9. Let ξ be a r.v. whose distribution function is the continuous and strictly increasing function $F(x)$. We call the value of x for which $F(x) = q$, q-cvartile of the r.v. ξ. In the case when $q = \frac{1}{2}$, the q-cvartile is called the median.

Definition 10. The real function

$$F(x_1, \ldots, x_n) = P(\xi_1 < x_1, \ldots, \xi_n < x_n)$$

is called the distribution function of the n-dimensional r.v.

The following property can be proved in the same way as in the one dimensional case.

Proposition 13. *The function $F(x_1, \ldots, x_n)$ is monotonously increasing and continuous to the left with respect to every argument, it is zero if at least one of the variables $x_1, \ldots, x_n$ is equal to $-\infty$ and is equal to 1 if $x_k = \infty$, $(k = 1, \ldots, n)$.*

Definition 11. Let $\zeta(\xi_1, \ldots, \xi_n)$ be a n-dimensional r.v. whose distribution function is $F(x_1, \ldots, x_n)$. If there is an integrable function $f(u_1, \ldots, u_n)$ so that

$$F(x_1, \ldots, x_n) = \int_{-\infty}^{x_1} \cdots \int_{-\infty}^{x_n} f(u_1, \ldots, u_n)\ \mathrm{d}u_1 \ldots \mathrm{d}u_n$$

the function f is called the density of distribution or density of probability of the r.v. ζ.

6.4.3 Stieltjes' integral

Let $[a, b]$ be an interval and let f and g be two functions defined on $[a, b]$. Let us denote by Δ the partition

$$a = x_0 < x_1 < \ldots < x_{n-1} < x_n = b$$

of the interval $[a, b]$ and with $\nu(\Delta)$ the norm of this partition, i.e.

$$\nu(\Delta) = \max\,(x_i - x_{i-1}),\ (i = 1, \ldots, n).$$

We shall choose an arbitrary point y_i in every interval $[x_{i-1}, x_i]$ and we hall write

$$\sigma = \sigma(d, f, g, y_i) = \sum_{i=1}^{n} f(y_i)\,[g(x_i) - g(x_{i-1})]. \tag{6.69}$$

)efinition 12. *If there is a real number I with the property that for every > 0 there is a number $\eta = \eta(\varepsilon) > 0$ so that we have*

$$|\,\sigma - I\,| < \varepsilon$$

or every partition Δ with $\nu(\Delta) < \eta$ the points y_i being arbitrary, then the umber I is called Stieltjes' integral of the function $f(x)$ with respect to $g(x)$ nd is denoted by

$$I = \int_a^b f(x)\,\mathrm{d}g(x) = \int_a^b f\,\mathrm{d}g.$$

In case where $g(x) = x$ Stieltjes' integral is reduced to Riemann's integral.

If we consider the above definition we can immediately prove the ollowing property:

'roposition 14. *If the functions f_1 and f_2 are Stieltjes integrable with espect to the function g in $[a, b]$, and α_1 and α_2 are real constants, then the unction $\alpha_1 f_1 + \alpha_2 f_2$ is also Stieltjes integrable with respect to g in $[a, b]$ and 'e have*

$$\int_a^b (\alpha_1 f_1 + \alpha_2 f_2)\,\mathrm{d}g = \alpha_1 \int_a^b f_1 \mathrm{d}g + \alpha_2 \int_a^b f_2\,\mathrm{d}g.$$

'roposition 15. *If the function f is Stieltjes integrable with respect to the unctions g_1 and g_2 in the interval $[a, b]$ and α_1 and α_2 are real constants then is Stieltjes integrable with respect to the function $\alpha_1 g_1 + \alpha_2 g_2$ on the interval $\imath$, b] and we have*

$$\int_a^b f\,\mathrm{d}(\alpha_1 g_1 + \alpha_2 g_2) = \alpha_1 \int_a^b f\,\mathrm{d}g_1 + \alpha_2 \int_a^b f\,\mathrm{d}g_2.$$

Proposition 16. *If $a < c < b$ and the function f is Stieltjes integrabl with respect to g on the interval $[a, b]$ then f is integrable with respect to on the intervals $[a, c]$ and $[c, b]$ and we have*

$$\int_a^b f \, dg = \int_a^c f \, dg + \int_c^b f \, dg.$$

Proposition 17. *If the functions f_1 and f_2 are Stieltjes integrable wit respect to g on the interval $[a, b]$ and $f_1 \leqslant f_2$ for every $x \in [a, b]$ then*

$$\int_a^b f_1 \, dg \leqslant \int_a^b f_2 \, dg.$$

Proposition 18. *If the function f is Stieltjes integrable with respect t g in $[a, b]$ then*

$$\left| \int_a^b f \, dg \right| \leqslant \int_a^b |f| \, dg.$$

Proposition 19. *If $\{f_n\}$ is a sequence of functions which are Stieltje integrable with respect to g in the interval $[a, b]$ and convergent in this interva to f, then f is Stieltjes integrable on $[a, b]$ and we have*

$$\lim_{n \to \infty} \int_a^b f_n \, dg = \int_a^b f \, dg.$$

Let us now prove the following proposition.

Proposition 20. *If the function f is continuous on the interval $[a, b]$ an the function g is monotonous on the same interval, then f is Stieltjes integrab with respect to g on $[a, b]$.*

We shall first suppose that g is monotonously increasing on $[a, b]$, thu if $a \leqslant x_{i-1} < x_i \leqslant b$, we have

$$g(x_i) - g(x_{i-1}) \geqslant 0. \tag{6.70}$$

Let us write

$$m_i = \inf_{x \in [x_{i-1}, x_i]} f(x), \quad M_i = \sup_{x \in [x_{i-1}, x_i]} f(x)$$

and let us consider the sums

$$s = \sum_{i=1}^{n} m_i[g(x_i) - g(x_{i-1})], \qquad S = \sum_{i=1}^{n} M_i[g(x_i) - g(x_{i-1})]. \quad (6.71)$$

As the function f is continuous on $[a, b]$ for every $y_i \in [x_{i-1}, x_i]$ we have

$$m_i \leqslant f(y_i) \leqslant M_i. \quad (6.72)$$

From formulae (6.70) and (6.72) we deduce that

$$m_i[g(x_i) - g(x_{i-1})] \leqslant f(y_i)[g(x_i) - g(x_{i-1})] \leqslant M_i[g(x_i) - g(x_{i-1})],$$

whence, taking into account formulae (6.69) and (6.71) it follows that

$$s \leqslant \sigma \leqslant S. \quad (6.73)$$

Denoting by

$$m = \inf_{x \in [a, b]} f(x), \qquad M = \sup_{x \in [a, b]} f(x),$$

we obtain

$$s \geqslant m \sum_{i=1}^{n} [g(x_i) - g(x_{i-1})] = m[g(b) - g(a)],$$

$$S \leqslant M \sum_{i=1}^{n} [g(x_i) - g(x_{i-1})] = M[g(b) - g(a)],$$

whence relationship (6.73) becomes

$$m[g(b) - g(a)] \leqslant s \leqslant \sigma \leqslant S \leqslant M[g(b) - g(a)]. \quad (6.73')$$

Let us now consider all the partitions of the interval $[a, b]$. By taking into account relationships (6.73') it follows that the set $\{s\}$ is upper bounded and the set $\{S\}$ is lower bounded. Let us denote by

$$\bar{I} = \sup_{\Delta} \{s\}, \qquad \underline{I} = \inf_{\Delta} \{S\}.$$

We shall suppose that we formed the partition Δ', by adding to the partition Δ a point $\in z(x_{i-1}, x_i)$. Let us denote by

$$m_i' = \inf_{x \in [x_{i-1}, z]} f(x), \qquad m_i'' = \inf_{x \in [z, x_i]} f(x),$$

$$M_i' = \sup_{x \in [x_{i-1}, z]} f(x), \qquad M_i'' = \sup_{x \in [z, x_i]} f(x).$$

It is obvious that

$$m_i' \geqslant m_i, \; m_i'' \geqslant m_i, \qquad M_i' \leqslant M_i, \; M_i'' \leqslant M_i.$$

If we denote by s' and S' the sums (6.71) corresponding to the partitio Δ', it follows that

$$s' \geqslant s, \qquad S' \leqslant S. \tag{6.74}$$

Let us now consider two partitions Δ_1 and Δ_2 and let us denote by s_1 S_1, s_2, S_2 the sums (6.71) corresponding to the partitions. The partition $\tilde{\Delta}$ which was obtained by the superposition of the partitions Δ_1 and Δ_2 i consequtive to both partitions Δ_1 and Δ_2 and denoting by $\tilde{s}$ and $\tilde{S}$ the sum (6.71) corresponding to $\tilde{\Delta}$ from relationship (6.74) we obtain

$$s_1 \leqslant \tilde{s}, \; S_1 \geqslant \tilde{S}, \quad s_2 \leqslant \tilde{s}, \quad S_2 \geqslant \tilde{S}.$$

But $\tilde{s} \leqslant \tilde{S}$, thus we deduce that

$$s_1 \leqslant \tilde{s} \leqslant \tilde{S} \leqslant S_2, \qquad s_2 \leqslant \tilde{s} \leqslant \tilde{S} \leqslant S_1.$$

It follows that every number from the set $\{s\}$ is less than every numbe from the set $\{S\}$ and that

$$s \leqslant \overline{I} \leqslant \underline{I} \leqslant S.$$

From these inequalities and from inequalities (6.73) we deduce that

$$|\sigma - \overline{I}| \leqslant S - s, \tag{6.75}$$

$$|\sigma - \underline{I}| \leqslant S - s. \tag{6.75'}$$

By taking into account that f is continuous on $[a, b]$ it follows that for ever $\varepsilon > 0$ we can determine a number $\eta = \eta(\varepsilon) > 0$ so that, if $|x' - x''| < \eta$ we have

$$|f(x') - f(x'')| < \varepsilon.$$

It follows that, if $\nu(\Delta) < \eta$ then

$$M_i - m_i < \varepsilon.$$

Thus, if $\nu(\Delta) < \eta$ we have

$$S - s = \sum_{i=1}^{n} (M_i - m_i) \,]g(x_i) - g(x_{i-1})] \leqslant \varepsilon \sum_{i=1}^{n}]g(x_i) - g(x_{i-1})] =$$

$$= \varepsilon [g(b) - g(a)],$$

which substituted in inequalities (6.75) and (6.75′) yields

$$|\sigma - \overline{I}| < \varepsilon\,[g(b) - g(a)],$$

$$|\sigma - \underline{I}| < \varepsilon[g(b) - g(a)].$$

It follows that f is Stieltjes integrable with respect to g in $[a, b]$ and we have

$$\overline{I} = \underline{I} = \int_a^b f \, \mathrm{d}g.$$

If g is monotonously decreasing, then $g_1 = -g$ is monotonously increasing and from proposition 15 it follows that the property is valid also in this case.

From propositions 6 and 20 we have the following proposition.

Proposition 21. *If $\varphi(x)$ is continuous on $[a, b]$ and $F(x)$ is the distribution function of a r.v. ξ then $\varphi(x)$ is Stieltjes integrable with respect to $F(x)$ on $[a, b]$.*

The notion of Stieltjes' integral can be extended also for unbounded intervals.

Definition 13. *If f is Stieltjes integrable with respect to g on $[a, b]$ and if*

$$\lim_{b \to \infty} \int_a^b f \, \mathrm{d}g$$

exists, we say that f is Stieltjes integrable with respect to g on $[a, \infty)$ and we write

$$\lim_{b \to \infty} \int_a^b f \, \mathrm{d}g = \int_a^\infty f \, \mathrm{d}g.$$

6.4.4 Moments of a random variable

Let $\{\Omega, \mathcal{K}, P\}$ be a probability space and ξ a r.v. whose distribution function is $F(x)$.

Proposition 22. *If $\tau(x)$ is a real measurable function so that $\tau[\xi(\omega)]$ is summable with respect to the probability P, then*

$$\int_\Omega \tau[\xi(\omega)] \, \mathrm{d}P(\omega) = \int_{-\infty}^{+\infty} \tau(x) \, \mathrm{d}F(x). \tag{6.76}$$

First we shall prove this relationship in the case where $\tau(x)$ has a finite number of values $a_1, a_2, \ldots, a_n$. Setting

$$A_i = \{x;\ \tau(x) = a_i\},\ (i = 1, \ldots, n)$$

we have

$$\tau = \sum_{i=1}^{n} a_i\, x_{A_i}.$$

In order to prove relationship (6.76) in this case it is sufficient to verify it for $\tau = x_{(a\,,\,b)}$. We have

$$\tau[\xi(\omega)] = \begin{cases} 0, \text{ if } \xi(\omega) \notin [a, b], \\ 1, \text{ if } \xi(\omega) \in [a, b], \end{cases}$$

thus

$$\int_{\Omega} \tau\,[\xi(\omega)]\ \mathrm{d}P(\omega) = P(\{\omega;\ \xi(\omega) \in [a,\ b]\}) = F(b) - F(a) = \int_a^b \mathrm{d}F(x).$$

This means that, in this case, relationship (6.76) is true.

Let us now suppose that $\tau(x)$ is an arbitrary function. Without loss of generality, we can suppose $\tau \geqslant 0$. From proposition 5, there is an increasing sequence of functions $\tau_n(x)$, which have a finite number of values, convergent to $\tau(x)$. The sequence of r.v. $\tau_n\,[\xi(\omega)]$, $(n = 1, 2, \ldots)$ is increasing and converges to the r.v. $\tau[\xi(\omega)]$, which by hypothesis is summable. Thus, from proposition 19 we have

$$\int_{\Omega} \tau[\xi(\omega)]\,\mathrm{d}P(\omega) = \lim_{n\to\infty} \int_{\Omega} \tau_n\,[\xi(\omega)]\ \mathrm{d}P(\omega) =$$

$$= \lim_{n\to\infty} \int_{-\infty}^{+\infty} \tau_n(x)\ \mathrm{d}F(x) = \int_{-\infty}^{+\infty} \tau(x)\ \mathrm{d}F(x).$$

Definition 14. Let ξ_0 be a r.v. whose distribution function is $F(x)$. The expression

$$E[\xi] = \int_{\Omega} \xi(\omega)\ \mathrm{d}P(\omega) = \int_{-\infty}^{+\infty} x\ \mathrm{d}F(x),$$

is called the mean value of the r.v. ξ, and the r.v. $\eta = \xi - E[\xi]$ is called the standard deviation of the r.v. ξ.

Definition 15. *Given a r.v. ξ, if p is a natural number the expression*

$$\alpha_p(\xi) = E[\xi^p] = \int_\Omega \xi^p(\omega)\, dP(\omega) = \int_{-\infty}^{+\infty} x^p\, dF(x) \tag{6.77}$$

is called the moment of order p of the r.v. ξ and the expression

$$\beta_p[\xi] = E[|\xi^p|] = \int_\Omega |\xi(\omega)|^p dP(\omega) = \int_{-\infty}^{+\infty} |x|^p\, dF(x) \tag{6.78}$$

is called the absolute moment of order p of the r.v. ξ.

If the r.v. ξ is of continuous type with the density of distribution $f(x)$ then formula (6.77) becomes

$$\alpha_p(\xi) = E[\xi^p] = \int_{-\infty}^{+\infty} x^p f(x)\, dx. \tag{6.77'}$$

Definition 16. *The expression*

$$\mu_p(\xi) = \alpha_p(\xi - E[\xi])$$

is called the centred moment of order p of the r.v.ξ.

Definition 17. *If ξ is of summable square then the expression*

$$D^2(\xi) = \sigma^2 = \mu_2(\xi)$$

is called the variance of the r.v. ξ and

$$D(\xi) = \sigma = \sqrt{\mu_2[\xi]}$$

is called the mean square deviation of the r.v. ξ.

In the case where ξ is a discrete random variable these definitions become identical with definitions 6, 9, 10 and 11 from Section 6.3.

Proposition 23. *If the r.v. ξ is almost everywhere equal to a then*

$$\alpha_p(\xi) = a^p.$$

Let us write

$$A = \{\omega;\ \xi(\omega) \neq a\}.$$

By taking into account that $\xi = a$ except for a set of null measure, it follows that $|A| = 0$. From definition 15 we deduce that

$$\alpha_p(\xi) = \int_{\Omega} \xi^p(\omega)\, dP(\omega) = \int_{\Omega - A} a^p\, dP(\omega) + \int_{A} \xi^p(\omega)\, dP(\omega).$$

But from the relationship $|A| = 0$ it follows (Vol. 1, proposition 3, Section 5.2.1) that

$$\int_{A} \xi^p(\omega)\, dP(\omega) = 0$$

thus

$$\alpha_p(\xi) = a^p \int_{\Omega - A} dP(\omega) = a^p.$$

Proposition 24. *If the r.v. ξ is almost everywhere equal to a constant a then*

$$D^2(\xi) = 0.$$

From the previous property we have

$$E[\xi] = a$$

and almost everywhere we have $\xi - E[\xi] = 0$, whence we obtain

$$D^2(\xi) = \alpha_2(\xi - E[\xi]) = 0.$$

Using the properties of Lebesgue's integral we can prove immediately that propositions 4—7, and 9—11, Section 6.3, are also valid in the case of every r.v.

The definitions 12—14 and the propositions 12—15 from Section 6.3. are also true in the case of every r.v.

Let us now prove the following very important proposition

Proposition 25. *(Tchebysheff's inequality.) If $a > 0$ then*

$$P(\{\omega;\ |\xi(\omega)| \geqslant a\}) \leqslant \frac{\alpha_2(\xi)}{a^2}. \tag{6.79}$$

To this purpose let us write

$$A = \{\omega;\ |\xi(\omega)| \geqslant a\}.$$

We have

$$\alpha_2(\xi) = E[\xi^2] = \int_{\Omega} \xi^2(\omega)\ \mathrm{d}P(\omega) = \int_{A} \xi^2(\omega)\ \mathrm{d}P(\omega) + \int_{\complement A} \xi^2(\omega)\ \mathrm{d}P(\omega). \tag{6.80}$$

But, by taking into account that we have $|\xi(\omega)| \geqslant a$ on the set A, it follows that

$$\int_{A} \xi^2(\omega)\ \mathrm{d}P(\omega) \geqslant \int_{A} a^2\ \mathrm{d}P(\omega) = a^2 \int_{A} \mathrm{d}P(\omega) = a^2\ P(A).$$

We also have

$$\int_{\complement A} \xi^2(\omega)\ \mathrm{d}P(\omega) \geqslant a$$

and thus equality (6.80) becomes

$$\alpha_2(\xi) \geqslant a^2 P(A),$$

hence

$$P(A) \leqslant \frac{\alpha_2(\xi)}{a^2}$$

which is just inequality (6.79).

6.4.5 Some classical distributions

Definition 18. The function whose density of distribution is

$$f(x) = \begin{cases} \dfrac{1}{b-a}, & \text{if } x \in [a, b], \\ 0, \text{ if } x \notin [a, b] \end{cases}$$

is called the uniform distribution function on $[a, b]$.

The r.v. ξ is called uniform on $[a, b]$ if it has its distribution function uniform on $[a, b]$. From formula (6.67) it follows that

$$F(x) = \int_{-\infty}^{x} f(u)\ \mathrm{d}u.$$

If $x \leqslant a$ we have

$$F(x) = 0$$

since in this case $f(u) = 0$.

If $a < x \leqslant b$, we have

$$F(x) = \int_{-\infty}^{a} f(u)\, du + \int_{a}^{x} f(u)\, du.$$

The first integral is zero because for $u < a$ we have $f(u) = 0$. In the second integral we have $f(u) = \dfrac{1}{b-a}$, thus

$$F(x) = \frac{1}{b-a} \int_{a}^{x} du = \frac{x-a}{b-a}.$$

If $x > b$, we have

$$F(x) = \int_{-\infty}^{a} f(u)\, du + \int_{a}^{b} f(u)\, du + \int_{b}^{x} f(u)\, du.$$

The first and the last integrals are zero. In the second integral we have $f(u) = \dfrac{1}{b-a}$, thus

$$F(x) = \frac{1}{b-a} \int_{a}^{b} du = 1.$$

From this we have the following proposition.

Proposition 26. *An uniform distribution function on* $[a, b]$ *is*

$$F(x) = \begin{cases} 0, \text{ if } x \leqslant a, \\ \dfrac{x-a}{b-a}, \quad \text{if } a < x \leqslant b, \\ 1, \text{ if } x > b. \end{cases}$$

Let us now prove another proposition.

Proposition 27. *If ξ is an uniform r.v. on $[a, b]$ we have*

$$E[\xi] = \frac{a+b}{2}, \quad D(\xi) = \frac{b-a}{2\sqrt{3}}.$$

Indeed, considering definition 14 and proposition 26, we deduce that

$$E[\xi] = \int_{-\infty}^{a} x \, dF(x) + \int_{a}^{b} x \, dF(x) + \int_{b}^{\infty} x \, dF(x) = \frac{1}{b-a}\int_{a}^{b} x \, dx = \frac{a+b}{2}.$$

Also from definition 15 we obtain

$$E[\xi^2] = \int_{-\infty}^{a} x^2 \, dF(x) + \int_{a}^{b} x^2 \, dF(x) + \int_{b}^{\infty} x^2 \, dF(x) =$$

$$= \frac{1}{b-a}\int_{a}^{b} x^2 \, dx = \frac{a^2+ab+b^2}{3}.$$

From the above relationship and from formula (6.46) it follows that

$$D^2(\xi) = \frac{a^2+ab+b^2}{3} - \frac{(a+b)^2}{4} = \frac{(b-a)^2}{12}$$

whence, taking into account that $b > 0$, we obtain

$$D(\xi) = \frac{b-a}{2\sqrt{3}}.$$

Definition 19. *The distribution function whose density of distribution is*

$$f(x; m, \sigma^2) = \frac{1}{\sigma\sqrt{2\pi}} e^{-\frac{(x-m)^2}{2\sigma^2}}$$

is called the normal distribution of parameters m and σ^2.

The r. v. ξ is called the normal r. v. of parameters m, σ^2 if its distribution function is a normal distribution function. We have

$$\int_{-\infty}^{+\infty} f(x; m, \sigma^2) \, dx = \frac{1}{\sigma\sqrt{2\pi}} \int_{-\infty}^{+\infty} e^{-\frac{(x-m)^2}{2\sigma^2}} dx.$$

By making the change of variables $x = m + \sigma t \sqrt{2}$, we obtain

$$\int_{-\infty}^{+\infty} f(x;\, m,\, \sigma^2)\, dx = \frac{1}{\sqrt{\pi}} \int_{-\infty}^{+\infty} e^{-t^2}\, dt = \frac{1}{\sqrt{\pi}} \sqrt{\pi} = 1.$$

Let us now prove the following proposition.

Proposition 28. *If ξ is a normal r.v. of parameters m, σ^2, then*

$$E[\xi] = m,\quad D^2[\xi] = \sigma^2.$$

Indeed we have

$$E[\xi] = \int_{-\infty}^{+\infty} x f(x;\, m,\, \sigma^2)\, dx = \frac{1}{\sigma\sqrt{2\pi}} \int_{-\infty}^{+\infty} x e^{-\frac{(x-m)^2}{2\sigma^2}}\, dx.$$

By the change of variables $x = m + \sigma t \sqrt{2}$, we obtain

$$E[\xi] = \frac{1}{\sqrt{\pi}} \int_{-\infty}^{+\infty} (m + \sigma t \sqrt{2})\, e^{-t^2}\, dt = \frac{m}{\sqrt{\pi}} \int_{-\infty}^{+\infty} e^{-t^2}\, dt +$$

$$+ \frac{\sigma\sqrt{2}}{\sqrt{\pi}} \int_{-\infty}^{+\infty} t e^{-t^2}\, dt = \frac{m}{\sqrt{\pi}} \cdot \sqrt{\pi} - \frac{\sigma}{\sqrt{2\pi}} \int_{-\infty}^{+\infty} e^{-t^2}\, d(-t^2) =$$

$$= m - \frac{\sigma}{\sqrt{2\pi}} e^{-t^2} \Big|_{-\infty}^{\infty} = m.$$

By analogy we have

$$E[\xi^2] = m^2 + \sigma^2, \tag{6.81}$$

thus

$$D^2(\xi) = (E[\xi^2] - E[\xi])^2 = \sigma^2.$$

Proposition 29. *If ξ is a normal r.v. of parameters m, σ^2, then $\eta = \dfrac{\xi - m}{\sigma}$ is a normal r.v. of parameters* 0, 1.

Indeed, we have:

$$E[\eta] = \frac{1}{\sigma} (E[\xi] - m) = \frac{1}{\sigma} (m - m) = 0.$$

Also

$$E[\eta^2] = \frac{1}{\sigma^2}\,(E\,\xi^2 = 2\;m\;E[\xi] + m^2)$$

and taking into account equality (6.81)

$$E[\eta^2] = \frac{1}{\sigma^2}\,(m^2 + \sigma^2 - 2m^2 + m^2) = 1.$$

Proposition 30. *If* $\xi_1, \ldots, \xi_n$ *are normal independent r. v. of the parameters* 0, 1 *then the r. v.*

$$\chi_n^2 = \sum_{i=1}^{n} \xi_i^2$$

has a density of distribution equal to

$$h_n(x) = \begin{cases} \dfrac{1}{2^{\frac{n}{2}}\,\Gamma\left(\dfrac{n}{2}\right)}\, x^{\frac{n}{2}-1}\, e^{-\frac{x}{2}}, & \text{for } x > 0. \\ 0, & \text{for } x \leqslant 0. \end{cases}$$

We shall determine first the density of distribution of the r. v. $\eta_n = \dfrac{\chi_n}{\sqrt{n}}$.

If we denote by $F(y)$ the distribution function of the r.v. η_n it follows that for $y \leqslant 0$ we have $F(y) = 0$. For $y > 0$, $F(y)$ is the probability that the point of co-ordinates $\xi_1(\omega), \ldots, \xi_n(\omega)$ with ω fixed, be in the interior of the sphere

$$\sum_{i=1}^{n} x_i^2 = ny^2.$$

But the r.v. $\xi_1, \ldots, \xi_n$ are independent, thus

$$F(y) = \int_{\sum\limits_{i=1}^{n} x_i^2 < ny^2} \cdots \int \frac{1}{\sqrt{2\pi}}\, e^{-\frac{x_1^2}{2}} \cdots \frac{1}{\sqrt{2\pi}}\, e^{-\frac{x_n^2}{2}}\, \mathrm{d}x_1 \ldots \mathrm{d}x_n =$$

$$= \int_{\sum\limits_{i=1}^{n} x_i^2 < ny^2} \cdots \int \frac{1}{(\sqrt{2\pi})^n}\, e^{-\frac{1}{2}\sum\limits_{i=1}^{n} x_i^2}\, \mathrm{d}x_1 \ldots \mathrm{d}x_n.$$

In order to calculate this integral let us consider the spherical co-ordinates

$$x_1 = \rho \cos \theta_1 \cos \theta_2 \dots \cos \theta_{n-2} \cos \theta_{n-1},$$
$$x_2 = \rho \cos \theta_1 \cos \theta_2 \dots \cos \theta_{n-2} \sin \theta_{n-1},$$
$$\dots\dots\dots\dots\dots\dots\dots\dots\dots\dots$$
$$x_n = \rho \sin \theta_1 .$$

In this way we obtain

$$F(y) = \int\limits_{-\frac{\pi}{2}}^{\frac{\pi}{2}} \dots \int\limits_{-\frac{\pi}{2}}^{\frac{\pi}{2}} \int\limits_{0}^{y\sqrt{n}} \frac{1}{(\sqrt{2\pi})^n} e^{-\frac{\rho^2}{2}} \rho^{n-1} D(\theta_1, \dots, \theta_{n-1})\, d\rho\, d\theta_1 \dots d\theta_{n-1},$$

where $D(\theta_1, \dots, \theta_{n-1})$ is a determinant which depends on $\theta_1, \dots, \theta_{n-1}$ or

$$F(y) = c_n \int\limits_{0}^{y\sqrt{n}} e^{-\frac{\rho^2}{2}} \rho^{n-1}\, d\rho,$$

where we have written

$$c_n = \frac{1}{(\sqrt{2\pi})^n} \int\limits_{-\frac{\pi}{2}}^{\frac{\pi}{2}} \dots \int\limits_{-\frac{\pi}{2}}^{\frac{\pi}{2}} D(\theta_1, \dots, \theta_{n-1})\, d\theta_1 \dots d\theta_{n-1}.$$

In order to determine the value of c_n we take into account that $\lim_{y\to\infty} F(y) = 1$, thus

$$1 = c_n \int\limits_{0}^{\infty} e^{-\frac{\rho^2}{2}} \rho^{n-1}\, d\rho$$

and from the properties of Euler's function Γ (Section 1.1.2) we obtain

$$1 = c_n\, 2^{\frac{n}{2}-1} \Gamma\left(\frac{n}{2}\right).$$

Hence, it follows that

$$c_n = \frac{1}{2^{\frac{n}{2}-1}\Gamma\left(\frac{n}{2}\right)}$$

thus

$$F(y) = \frac{1}{2^{\frac{n}{2}-1}\Gamma\left(\frac{n}{2}\right)} \int_0^{y\sqrt{n}} \rho^{n-1} e^{-\frac{\rho^2}{2}} d\rho.$$

The distribution function of the r.v. χ_n^2 can be obtained from the above relationship by substituting $\sqrt{\frac{x}{n}}$ for y. Thus the distribution function of the r.v. will be

$$H_n(x) = \frac{1}{2^{\frac{n}{2}-1}\Gamma\left(\frac{n}{2}\right)} \int_0^{\sqrt{x}} \rho^{n-1} e^{-\frac{\rho^2}{2}} d\rho$$

or by making the change of variables $\rho = \sqrt{t}$

$$H_n(x) = \frac{1}{2^{\frac{n}{2}}\Gamma\left(\frac{n}{2}\right)} \int_0^{x} t^{\frac{n}{2}-1} e^{-\frac{t}{2}} dt.$$

By taking into account that the density of distribution of the r.v. χ_n^2 is $h_n(x) = H_n'(x)$, we deduce that for $x > 0$

$$h_n(x) = \frac{1}{2^{\frac{n}{2}}\Gamma\left(\frac{n}{2}\right)} x^{\frac{n}{2}-1} e^{-\frac{x}{2}}.$$

Definition 20. The distribution function whose density of distribution is

$$h_n(x) = \begin{cases} \dfrac{1}{2^{\frac{n}{2}}\Gamma\left(\frac{n}{2}\right)} x^{\frac{n}{2}-1} e^{-\frac{x}{2}}, & \text{for } x > 0 \\ 0, & \text{for } x \leqslant 0 \end{cases}$$

is called Pearson's distribution function χ^2 with n degrees of freedom.

Proposition 31. *We have*

$$E[\chi_n^2] = n, \qquad D^2(\chi_n^2) = 2\,n.$$

Indeed

$$E[\chi_n^2] = \int_0^\infty x\ h_n(x)\ dx = \frac{1}{2^{\frac{n}{2}}\Gamma\left(\frac{n}{2}\right)}\int_0^\infty x^{\frac{n}{2}} e^{-\frac{x}{2}}\ dx = \frac{2\Gamma\left(\frac{n}{2}+1\right)}{\Gamma\left(\frac{n}{2}\right)}.$$

But $\Gamma\left(\frac{n}{2}+1\right) = \frac{n}{2}\Gamma\left(\frac{n}{2}\right)$, thus

$$E[\chi_n^2] = n.$$

By analogy

$$E[(\chi_n^2)^2] = \int_0^\infty x^2\ h_n(x)\ dx = \frac{1}{2^{\frac{n}{2}}\Gamma\left(\frac{n}{2}\right)}\int_0^\infty x^{\frac{n}{2}+1} e^{-\frac{x}{2}}\ dx = \frac{2^2\,\Gamma\left(\frac{n}{2}+2\right)}{\Gamma\left(\frac{n}{2}\right)},$$

or, taking into account that $\Gamma\left(\frac{n}{2}+2\right) = \frac{n(n+2)}{4}\ \Gamma\left(\frac{n}{2}\right)$, we have

$$E[(\chi_n^2)^2] = n(n+2).$$

Thus

$$D^2(\chi_n^2) = E[(\chi_n^2)^2] - (E[\chi_n^2])^2 = n(n+2) - n^2 = 2n.$$

6.4.6 Characteristic functions

Definition 21. Given the r.v. ξ whose distribution function is $F(x)$ the function

$$\varphi_\xi(t) = E[e^{it\xi}] = \int_\Omega e^{it\xi(\omega)}\,dP(\omega) = \int_{-\infty}^{+\infty} e^{itx}\,dF(x), \qquad (i = \sqrt{-1}), \quad (6.82)$$

is called the characteristic function of the r.v. ξ.

By taking into account that $|\,e^{itx}\,| = 1$, it follows that the characteristic function exists for every r.v. ξ and is uniquely determined by its distribution function. If ξ is a discrete random function with the distribution

$$\begin{pmatrix} x_1, & x_2, \dots \\ p_1, & p_2, \dots \end{pmatrix}$$

then formula (6.82) becomes

$$\varphi_\xi(t) = \sum_{j=1}^{\infty} e^{itx_j} p_j. \qquad (6.82')$$

If ξ is a r.v. of continuous type with the density of distribution $f(x)$, then formula (6.82) can be written as

$$\varphi_\xi(t) = \int_{-\infty}^{+\infty} e^{itx} f(x)\, \mathrm{d}x. \qquad (6.82'')$$

Thus, in this case $\varphi_\xi(t)$ is a Fourier transform of the function $f(x)$.

Proposition 32. *For every r.v. ξ we have $|\varphi_\xi(t)| \leqslant 1$. For $t = 0$ we have the equality.*

We have

$$|\varphi_\xi(t)| = \left| \int_{-\infty}^{+\infty} e^{itx}\, \mathrm{d}F(x) \right| \leqslant \int_{-\infty}^{+\infty} |e^{itx}|\, \mathrm{d}F(x) = \int_{-\infty}^{+\infty} \mathrm{d}F(x) = 1.$$

Also

$$\varphi_\xi(0) = \int_{-\infty}^{+\infty} \mathrm{d}F(x) = 1.$$

Proposition 33. *The function $\varphi_\xi(t)$ is uniformly continuous.*

Indeed

$$|\varphi_\xi(t_1) - \varphi_\xi(t_2)| = \left| \int_{-\infty}^{+\infty} e^{it_1x}\, \mathrm{d}F(x) - \int_{-\infty}^{+\infty} e^{it_2x}\, \mathrm{d}F(x) \right| =$$

$$= \left| \int_{-\infty}^{+\infty} (e^{it_1x} - e^{it_2x})\, \mathrm{d}F(x) \right| \leqslant \int_{-\infty}^{+\infty} |e^{it_1x} - e^{it_2x}|\, \mathrm{d}F(x) =$$

$$= \int_{|x| \leqslant a} |e^{it_1x} - e^{it_2x}|\, \mathrm{d}F(x) + \int_{|x| > a} |e^{it_1x} - e^{it_2x}|\, \mathrm{d}F(x).$$

But

$$| e^{it_1x} - e^{it_2x} | \leqslant | x | | t_1 - t_2 |$$

and

$$| e^{it_1x} - e^{it_2x} | \leqslant | e^{it_1x} | + | e^{it_2x} | = 2,$$

thus

$$| \varphi_\xi (t_1) - \varphi_\xi (t_2) | \leqslant | t_1 - t_2 | \int\limits_{|x| \leqslant a} | x | \, dF(x) + 2 \int\limits_{|x| > a} dF(x).$$

Given an arbitrary number $\varepsilon > 0$, we shall choose a so that we have

$$P(\{\omega; | \xi(\omega) | > a\}) < \frac{\varepsilon}{2}.$$

Thus

$$2 \int\limits_{|x| > a} dF(x) = 2P(\{\omega; | \xi(\omega) | > a\}) < 2 \cdot \frac{\varepsilon}{2} = \varepsilon.$$

Also, as soon as $| t_1 - t_2 | < \frac{\varepsilon}{a}$, we have

$$| t_1 - t_2 | \int\limits_{|x| \leqslant a} | x | \, dF(x) < \frac{\varepsilon}{a} \cdot a = \varepsilon.$$

It follows that

$$| \varphi_\xi(t_1) - \varphi_\xi(t_2) | < 2 \varepsilon.$$

and hence $\varphi_\xi(t)$ is uniformly continuous.

Proposition 34. *For every real t we have*

$$\varphi_\xi (- t) = \overline{\varphi_\xi (t)}.$$

Indeed

$$\varphi_\xi(- t) = \int_{-\infty}^{+\infty} e^{-itx} \, dF(x) = \int_{-\infty}^{+\infty} \overline{e^{itx}} \, dF(x) = \overline{\int_{-\infty}^{+\infty} e^{itx} \, dF(x)} = \overline{\varphi_\xi (t)}.$$

Proposition 35. *If a and b are real constants and* $\eta = a\xi + b$, *then*

$$\varphi_\eta(t) = e^{ibt}\,\varphi_\xi(at).$$

We have

$$\varphi_\eta(t) = E\,[e^{i(a\xi + b)t}] = e^{ibt}\,E[e^{ia\xi t}] = e^{ibt}\,\varphi_\xi(at).$$

Proposition 36. *If* $\xi_1, \ldots, \xi_n$ *are independent r.v. then*

$$\varphi_{\xi_1 + \ldots + \xi_n}(t) = \varphi_{\xi_1}(t) \ldots \varphi_{\xi_n}(t).$$

We shall prove this for $n = 2$; the general relationship is obtained by a complete induction. We have

$$\varphi_{\xi_1 + \xi_2}(t) = E\,[e^{it(\xi_1 + \xi_2)}] = E[e^{it\xi_1}\,e^{it\xi_2}].$$

From proposition 7, Section 6.3.1 we have

$$E[e^{it\xi_1}\,e^{it\xi_2}] = E[e^{it\xi_1}]\;E[e^{it\xi_2}],$$

thus

$$\varphi_{\xi_1+\xi_2}(t) = \varphi_{\xi_1}(t)\,\varphi_{\xi_2}(t).$$

Proposition 37. *If the first n absolute moments* $E[|\xi|^k]$ $(k = 1, 2, \ldots, n)$ *of the r.v.* ξ *exist, then the characteristic function* $\varphi_\xi(t)$ *is differentiable n times and we have*

$$\varphi_\xi^{(k)}(0) = i^k\,E[\xi^k], \qquad (k = 1, 2, \ldots, n). \tag{6.82''}$$

Let $F(x)$ be the distribution function of the r.v. ξ. If

$$E[|\xi|] = \int_{-\infty}^{+\infty} |x|\,\mathrm{d}F(x)$$

exists, then the integral

$$\int_{-\infty}^{+\infty} xe^{itx}\,\mathrm{d}F(x)$$

converges uniformly in t, thus

$$\varphi_\xi'(t) = i\int_{-\infty}^{+\infty} x\,e^{itx}\,\mathrm{d}F(x),$$

hence

$$\varphi'_\xi(0) = iE[\xi].$$

Repeating the operation, we obtain

$$\varphi_\xi^{(k)}(t) = i^k \int_{-\infty}^{+\infty} x^k e^{itx} \, dF(x).$$

By considering $t = 0$ we obtain relationship (6.82).

Proposition 38. *If for every natural p the absolute moment* $E[|\xi|^p]$ *exists, then the characteristic function* $\varphi_\xi(t)$ *of the r.v.* ξ *is expandable in a power series and we have*

$$\varphi_\xi(t) = \sum_{k=0}^{\infty}{}' \frac{(it)^k}{k!} E[\xi^k]. \tag{6.83}$$

We have

$$e^{itx} = 1 + \frac{itx}{1!} + \frac{(itx)^2}{2!} + \ldots + \frac{(itx)^{n-1}}{(n-1)!} + \frac{(itx)^n}{n!}\theta, \quad (0 < \theta < 1).$$

With this value we obtain

$$\varphi_\xi(t) = \int_{-\infty}^{+\infty} \left[1 + \frac{itx}{1!} + \ldots + \frac{(itx)^{n-1}}{(n-1)!} + \frac{(itx)^n}{n!}\theta\right] dF(x) =$$

$$= E[\xi^0] + \frac{it}{1!}E[\xi^1] + \ldots + \frac{(it)^{n-1}}{(n-1)!}E[\xi^{n-1}] + R_n,$$

where

$$R_n = \frac{(it)^n}{n!} \int_{-\infty}^{+\infty} \theta x^n \, dF(x).$$

Thus

$$|R_n| = \frac{|t|^n}{n!} \left| \int_{-\infty}^{+\infty} \theta \, x^n \, dF(x) \right| < \frac{|t|^n}{n!} \int_{-\infty}^{+\infty} |x^n| \, dF(x) = \frac{|t|^n}{n!} E[|\xi|^n].$$

Taking into account that

$$\lim_{n \to \infty} \frac{|t|^n}{n!} = 0$$

and that the absolute moment $E[|\xi|^p]$ exists for every natural p, we deduce that

$$\lim_{n \to \infty} R_n = 0,$$

thus

$$\varphi_\xi(t) = \sum_{k=0}^{\infty} \frac{(it)^k}{k!} E[\xi^k].$$

Proposition 39. *(The inversion formula.) Let ξ be a r.v. whose distribution function is F and the characteristic function is φ. If $x_1 < x_2$ are two continuity points of the function F we have*

$$F(x_2) - F(x_1) = \lim_{c \to \infty} \frac{1}{2\pi} \int_{-c}^{c} \frac{e^{-itx_1} - e^{-itx_2}}{it} \varphi(t)\, dt. \tag{6.84}$$

By taking into account the expression of the function $\varphi(t)$ we deduce that

$$\frac{1}{2\pi} \int_{-c}^{c} \frac{e^{-itx_1} - e^{-itx_2}}{it} \varphi(t)\, dt = \frac{1}{2\pi} \int_{-c}^{c} \frac{e^{-itx_1} - e^{-itx_2}}{it} \left(\int_{-\infty}^{+\infty} e^{ity}\, dF(y) \right) dt =$$

$$= \frac{1}{2\pi} \int_{-\infty}^{+\infty} \left(\int_{-c}^{c} \frac{e^{it(y - x_1)} - e^{it(y - x_2)}}{it}\, dt \right) dF(y).$$

But

$$\int_{-c}^{c} \frac{e^{it(y - x_1)}}{it}\, dt = \int_{0}^{c} \frac{e^{it(y - x_1)}}{it}\, dt + \int_{-c}^{0} \frac{e^{it(y - x_1)}}{it}\, dt =$$

$$= \int_{0}^{c} \frac{e^{it(y - x_1)}}{it}\, dt - \int_{0}^{c} \frac{e^{-it(y - x_1)}}{it}\, dt = 2 \int_{0}^{c} \frac{\sin (y - x_1) t}{t}\, dt.$$

Thus

$$\frac{1}{2\pi}\int_{-c}^{c}\frac{e^{-itx_1}-e^{-itx_2}}{it}\varphi(t)\,dt=$$

$$=\frac{1}{\pi}\int_{-\infty}^{+\infty}\left(\int_0^c\frac{\sin(y-x_1)t}{t}\,dt-\int_0^c\frac{\sin(y-x_2)t}{t}\,dt\right)dF(y).$$

By the change of variable $(y-x_1)\,t=u$ the first integral from the right-hand side becomes

$$\int_0^c\frac{\sin(y-x_1)t}{t}\,dt=\int_0^{c(y-x_1)}\frac{\sin u}{u}\,du.$$

By analogy

$$\int_0^c\frac{\sin(y-x_2)t}{t}\,dt=\int_0^{c(y-x_2)}\frac{\sin u}{u}\,du.$$

It follows that

$$\frac{1}{2\pi}\int_{-c}^{c}\frac{e^{-itx_1}-e^{-itx_2}}{it}\varphi(t)\,dt=\int_{-\infty}^{+\infty}\left(\frac{1}{\pi}\int_{c(y-x_2)}^{c(y-x_1)}\frac{\sin u}{u}\,du\right)dF(y),$$

thus

$$\lim_{c\to 0}\frac{1}{2\pi}\int_{-c}^{c}\frac{e^{-itx_1}-e^{-itx_2}}{it}\varphi(t)\,dt=$$

$$=\int_{-\infty}^{+\infty}\left(\lim_{c\to\infty}\frac{1}{\pi}\int_{c(y-x_2)}^{c(y-x_1)}\frac{\sin u}{u}\,du\right)dF(y).$$

On the other hand, by taking into account Euler's integral

$$\lim_{\substack{a\to-\infty\\ b\to\infty}}\int_a^b\frac{\sin u}{u}\,du=\pi,$$

we deduce that

$$I(y) = \lim_{c \to \infty} \frac{1}{\pi} \int_{c(y - x_2)}^{c(y - x_1)} \frac{\sin u}{u} \, du = \begin{cases} 1, & \text{if } x_1 < y < x_2, \\ \frac{1}{2}, & \text{if } y = x_1 \text{ or } y = x_2, \\ 0, & \text{if } y < x_1 \text{ or } y > x_2. \end{cases}$$

It follows that

$$\frac{1}{2\pi} \int_{-c}^{c} \frac{e^{-itx_1} - e^{-itx_2}}{it} \varphi(t) \, dt = \int_{-\infty}^{x_1} I(y) \, dF(y) + \int_{x_1}^{x_2} I(y) \, dF(y) +$$

$$+ \int_{x_2}^{\infty} I(y) \, dF(y) = \frac{1}{2} [F(x_1 + 0) - F(x_1 - 0)] + F(x_2 - 0) - F(x_1 + 0) +$$

$$+ \frac{1}{2} [F(x_2 + 0) - F(x_2 - 0)] = \frac{F(x_2 - 0) + F(x_2 + 0)}{2} - \frac{F(x_1 - 0) + F(x_1 + 0)}{2}.$$

But x_1 and x_2 are continuity points for the function $F(x)$, hence

$$\frac{F(x_2 - 0) + F(x_2 + 0)}{2} = F(x_2), \qquad \frac{F(x_1 - 0) + F(x_1 + 0)}{2} = F(x_1),$$

which proves formula (6.84).

Proposition 40. *(The unicity theorem.) The distribution function of a r.v. ξ is uniquely determined by the characteristic function of the r.v. ξ.*

Indeed, considering equality (6.84) where $x_2 = x$, $x_1 \to -\infty$ and by taking into account that $F(-\infty) = 0$, we obtain

$$F(x) = \lim_{y \to -\infty} \lim_{c \to \infty} \frac{1}{2\pi} \int_{-c}^{c} \frac{e^{-ity} - e^{-itx}}{it} \varphi(t) \, dt. \tag{6.84'}$$

We shall now determine the characteristic functions of some classical distributions.

Proposition 41. *The characteristic function of a binomial r.v. of order n and parameter p is*

$$\varphi_\xi(t) = [1 + p(e^{it} - 1)]^n$$

and the characteristic function of a Poisson distribution of parameter λ *is*

$$\varphi_\xi(t) = e^{\lambda(e^{it}-1)}.$$

By taking into account definition 14, Section 6.4.4, from equality (6.82′) we have in the case of a binomial r.v.

$$\varphi_\xi(t) = \sum_{k=0}^{n} \binom{n}{k} p^k q^{n-k} e^{itk} = \sum_{k=0}^{n} \binom{n}{k} (pe^{it})^k q^{n-k} = (pe^{it} + q)^n =$$

$$= [1 + p(e^{it} - 1)]^n.$$

If in the above inequality we put $np = \lambda$ and take the limit as $n \to \infty$, we obtain the characteristic function of a Poisson r.v.

$$\varphi_\xi(t) = \lim_{n \to \infty} \left[1 + \frac{\lambda}{n}(e^{it} - 1)\right]^n = e^{\lambda(e^{it}-1)}.$$

Proposition 42. *The characteristic function of a r.v. which is uniform in* $[-a, a]$ *is*

$$\varphi_\xi(t) = \frac{\sin at}{at}$$

and the characteristic function of a normal r.v. of parameters m, σ^2 *is*

$$\varphi_\xi(t) = e^{itm - \frac{1}{2}\sigma^2 t^2}.$$

The characteristic function of a Pearson r.v. χ^2 *with n degrees of freedom is*

$$\varphi_\xi(t) = (1 - 2it)^{-\frac{n}{2}}.$$

By taking into account definition 18, from formula (6.82″) we obtain

$$\varphi_\xi(t) = \int_{-a}^{+a} \frac{1}{2a} e^{itx}\,dx = \frac{1}{2ait}(e^{iax} - e^{-iax}) = \frac{\sin at}{at}.$$

In the case of a normal r.v. by taking into account definition 19, we deduce that

$$\varphi_\xi(t) = \frac{1}{\sigma\sqrt{2\pi}} \int_{-\infty}^{+\infty} e^{itx} e^{-\frac{(x-m)^2}{2\sigma^2}}\,dx.$$

Making a change of variable $x = \sigma\sqrt{2}\,u + m + i\sigma^2 t$, we obtain

$$\varphi_\xi(t) = \frac{1}{\sqrt{\pi}} \int_{-\infty}^{+\infty} e^{-u^2 + imt - \frac{t^2\sigma^2}{2}}\, du = \frac{e^{imt - \frac{t^2\sigma^2}{2}}}{\sqrt{\pi}} \int_{-\infty}^{+\infty} e^{-u^2}\, du =$$

$$= e^{imt - \frac{t^2\sigma^2}{2}}.$$

In the case of the Pearson r.v. χ^2, by taking into account definition 20, it follows that

$$\varphi_\xi(t) = \frac{1}{2^{\frac{n}{2}}\Gamma\left(\frac{n}{2}\right)} \int_0^\infty x^{\frac{n}{2}-1} e^{-\left(\frac{1}{2}-it\right)x}\, dx.$$

Making the change of variable $\left(\frac{1}{2} - it\right)x = u$, we obtain

$$\varphi_\xi(t) = \frac{1}{\Gamma\left(\frac{n}{2}\right)(1-2it)^{\frac{n}{2}}} \int_0^\infty u^{\frac{n}{2}-1} e^{-u}\, du = \frac{1}{\Gamma\left(\frac{n}{2}\right)(1-2it)^{\frac{n}{2}}}\Gamma\left(\frac{n}{2}\right) =$$

$$= (1 - 2it)^{-\frac{n}{2}}.$$

6.4.7 Sequences of random variables

Definition 22. We say that the sequence of r.v. $\xi_1, \xi_2, \ldots$ *converges in probability to the r.v.* ξ *(we denote this by* $\xi_n \xrightarrow{p} \xi$*) if for every number* $\varepsilon > 0, \eta > 0$ *there is a number* $N = N(\varepsilon, \eta) > 0$ *so that if* $n \geqslant N$ *we have*

$$P(\{\omega;\ |\xi_n - \xi| \geqslant \varepsilon\}) \leqslant \eta. \tag{6.85}$$

Let us now prove some properties about this type of convergence.

Proposition 43. *If the sequence of r.v.* $\{\xi_n\}$ *converges in probability to the r.v.* ξ *and to the r.v.* η *then*

$$P(\{\omega;\ \xi(\omega) \neq \eta(\omega)\}) = 0.$$

Indeed

$$|\xi-\eta| \leqslant |\xi-\xi_n| + |\xi_n-\eta|$$

thus

$$\{\omega;\ |\xi(\omega)-\eta(\omega) \geqslant \varepsilon\} \subset$$

$$\subset \left\{\omega;\ |\xi(\omega)-\xi_n(\omega)| \geqslant \frac{\varepsilon}{2}\right\} \cup \left\{\omega;\ |\eta(\omega)-\xi_n(\omega)| \geqslant \frac{\varepsilon}{2}\right\},$$

whence we deduce that

$$P(\{\omega;\ |\xi(\omega)-\eta(\omega)| \geqslant \varepsilon\}) \leqslant P\left(\left\{\omega;\ |\xi(\omega)-\xi_n(\omega)| \geqslant \frac{\varepsilon}{2}\right\}\right) +$$

$$+ P\left(\left\{\omega;\ |\eta(\omega)-\xi_n(\omega)| \geqslant \frac{\varepsilon}{2}\right\}\right).$$

But $\xi_n \xrightarrow{p} \xi$ and $\xi_n \xrightarrow{p} \eta$, hence it follows that

$$P(\{\omega;\ |\xi(\omega)-\eta(\omega)| \geqslant \varepsilon\}) = 0. \tag{6.86}$$

By taking into account that

$$P(\{\omega;\ |\xi(\omega)-\eta(\omega)| \neq 0\}) = P\left(\bigcup_{n=1}^{\infty}\left\{\omega;\ |\xi(\omega)-\eta(\omega)| \geqslant \frac{1}{n}\right\}\right) \leqslant$$

$$\leqslant \sum_{p=1}^{\infty} P\left(\left\{\omega;\ |\xi(\omega)-\eta(\omega)| \geqslant \frac{1}{n}\right\}\right),$$

from the relationship (6.86) it follows that

$$P(\{\omega;\ |\xi(\omega)-\eta(\omega)| > 0\}) = 0.$$

Proposition 44. *Let $\{\xi_n\}$ and $\{\eta_n\}$ be two sequences of r.v. and ξ, η two r.v. If $\xi_n \xrightarrow{p} \xi$ and $\eta_n \xrightarrow{p} \eta$ and α and β are two real constants then $\alpha\,\xi_n + \beta\eta_n \xrightarrow{p} \alpha\,\xi + \beta\eta$.*

If $\alpha\beta = 0$ the proposition is obvious. We shall suppose that $\alpha\beta \neq 0$. In this case we have

$$\{\omega;\ |\alpha\xi_n(\omega) + \beta\eta_n(\omega) - (\alpha\xi(\omega) + \beta\eta(\omega))| \geqslant \varepsilon\} \subset$$

$$\subset \left\{\omega;\ |\alpha|\,|\xi_n(\omega)-\xi(\omega)| \geqslant \frac{\varepsilon}{2}\right\} \cup \left\{\omega;\ |\beta|\,|\eta_n(\omega)-\eta(\omega)| \geqslant \frac{\varepsilon}{2}\right\} =$$

$$= \left\{\omega;\ |\xi_n(\omega)-\xi(\omega)| \geqslant \frac{\varepsilon}{2\,|\alpha|}\right\} \cup \left\{\omega;\ |\eta_n(\omega)-\eta(\omega)| \geqslant \frac{\varepsilon}{2\,|\beta|}\right\},$$

whence we deduce that

$$P(\{\omega;\ |\ \alpha\xi_n(\omega) + \beta\eta_n(\omega) - (\alpha\xi(\omega) + \beta\eta(\omega))\ |\ \geqslant \varepsilon\}) \leqslant$$

$$\leqslant P\left(\left\{\omega;\ |\ \xi_n(\omega) - \xi(\omega)\ |\ \geqslant \frac{\varepsilon}{2\,|\,\alpha\,|}\right\}\right) + P\left(\left\{\omega;\ |\ \eta_n(\omega) - \eta(\omega)\ |\ \geqslant \frac{\varepsilon}{2\,|\,\beta\,|}\right\}\right)$$

and taking into account that $\xi_n \xrightarrow{p} \xi$ and $\eta_n \xrightarrow{p} \eta$ it results that

$$\alpha\xi_n + \beta\eta_n \xrightarrow{p} \alpha\xi + \beta\eta.$$

Definition 23. *We say that the sequence of r.v. $\{\xi_n\}$ converges in distribution to the r.v. ξ (we denote this by $\xi_n \xrightarrow{d} \xi$) if at every continuity point x_0 of the distribution function of the r.v. ξ we have*

$$F_n(x_0) \to F(x_0), \tag{6.87}$$

where $F_n(x)$ is the distribution function of the r.v. ξ_n.

Proposition 45. *Let $\{\xi_n\}$ be a sequence of r.v. and ξ a r.v. If $\xi_n \xrightarrow{p} \xi$ then $\xi_n \xrightarrow{d} \xi$.*

Indeed, if x_0 is a continuity point of the function $F(x)$ and $\eta > 0$ then we can determine a number $\varepsilon > 0$ so that

$$F(x_0 + \varepsilon) - F(x_0 - \varepsilon) \leqslant \eta. \tag{6.88}$$

We have

$$F(x_0 - \varepsilon) = P(\{\omega;\ \xi(\omega) < x_0 - \varepsilon\}) = P(\{\omega;\ \xi(\omega) < x_0 - \varepsilon\} \cap$$

$$\cap\ \{\omega;\ \xi_n(\omega) < x_0) + P(\{\omega;\ \xi(\omega) < x_0 - \varepsilon\} \cap$$

$$\cap\ \{\omega;\ \xi_n(\omega) \geqslant x_0\}) \leqslant P(\{\omega;\ \xi_n(\omega) < x_0\}) + P(\{\omega;\ \xi(\omega) < x_0 - \varepsilon\} \cap$$

$$\cap\ \{\omega;\ \xi_n(\omega) \geqslant x_0\}) = F_n(x_0) + P(\{\omega; \xi(\omega) < x_0 - \varepsilon\} \cap \{\omega; \xi_n(\omega) \geqslant x_0\}) \leqslant$$

$$\leqslant F_n(x_0) + P(\{\omega;\ |\ \xi_n(\omega) - \xi(\omega)\ |\ \geqslant \varepsilon\}).$$

But, taking into account that $\xi_n \xrightarrow{p} \xi$ we have

$$P(\{\omega;\ |\ \xi_n(\omega) - \xi(\omega)\ |\ \geqslant \varepsilon\}) \leqslant \eta$$

thus

$$F(x_0 - \varepsilon) \leqslant \underline{\lim}\ F_n(x_0).$$

By analogy we have

$$F(x_0 + \varepsilon) \geqslant \overline{\lim}\ F_n(x_0).$$

But ε is chosen arbitrarily, thus from this and from inequality (6.88) it follows that:

$$\lim_{n \to \infty} F_n(x_0) = F(x_0).$$

This proposition has no converse, as we can notice from the following example.

Let there be a probability space $\{\Omega, \mathcal{K}, P\}$ where $\Omega = [0, 1]$ and let there be

$$\xi_n(\omega) = \begin{cases} 0, \text{ if } 0 \leqslant \omega \leqslant \dfrac{1}{2}, \\ 1, \text{ if} \dfrac{1}{2} < \omega \leqslant 1, \end{cases} \quad (n = 1, 2, \ldots)$$

and

$$\xi(\omega) = \begin{cases} 1, \text{ if } 0 \leqslant \omega \leqslant \dfrac{1}{2}, \\ 0, \text{ if} \dfrac{1}{2} < \omega \leqslant 1. \end{cases}$$

Then, for every natural number n we have

$$\{\omega; \mid \xi_n(\omega) - \xi(\omega) \mid = 1\} = [0, 1] = \Omega,$$

thus

$$P(\{\omega; \mid \xi_n(\omega) - \xi(\omega) \mid = 1\}) = 1, \ (n = 1, 2, \ldots).$$

It follows that the sequence $\{\xi_n\}$ does not converge in probability to ξ. On the other hand $\xi_n \xrightarrow{d} \xi$ since we have

$$F(x) = F_n(x) = \begin{cases} 0, \text{ if } x \leqslant 0, \\ \dfrac{1}{2}, \text{ if } 0 < x \leqslant 1, \\ 1, \text{ if } x > 1 \end{cases}$$

and

$$\lim_{n \to \infty} F_n(x) = F(x).$$

Definition 24. We say that the sequence of r.v. $\{\xi_n\}$ *converges in the mean of order* r *to the r.v.* ξ *if the absolute moments* $E[|\xi_n|^r]$, $(n = 1, 2, \ldots)$ *and* $E[|\xi|^r]$ *exist and if*

$$\lim_{n\to\infty} E[|\xi_n - \xi|^r] = 0.$$

Proposition 46. *If the sequence of r.v. converges in the mean of order r to the r.v.* ξ *then* $\xi_n \xrightarrow{p} \xi$.

Let us suppose that $\{\xi_n\}$ does not converge in probability to ξ. Then there are two numbers $\varepsilon > 0$, $\eta > 0$ and a sequence of natural numbers $n_i \to \infty$ so that

$$P(\{\omega;\ |\xi_{n_i}(\omega) - \xi(\omega)| \geqslant \varepsilon\}) \geqslant \eta$$

for every i.

Writing $A = \{\omega;\ |\xi_{n_i}(\omega) - \xi(\omega)| \geqslant \varepsilon\}$, we have $\Omega = A \cup \complement A$ and thus

$$\left[\int_\Omega |\xi_{n_i}(\omega) - \xi(\omega)|^r \,\mathrm{d}P(\omega)\right]^{\frac{1}{r}} \geqslant \left[\int_A |\xi_n(\omega) - \xi(\omega)|^r \,\mathrm{d}P(\omega)\right]^{\frac{1}{r}} \geqslant$$

$$\geqslant \left[\varepsilon^r P(\{\omega;\ |\xi_{n_i}(\omega) - \xi(\omega)| \geqslant \varepsilon\})\right]^{\frac{1}{r}} \geqslant \left(\varepsilon^r \eta\right)^{\frac{1}{r}} = \varepsilon\, \eta^{\frac{1}{r}},$$

that is

$$(E[|\xi_{n_i} - \xi|^r])^{\frac{1}{r}} \geqslant \varepsilon\, \eta^{\frac{1}{r}}. \tag{6.89}$$

On the other hand, by taking into account definition 24, we have

$$\lim_{n\to\infty} (E[|\xi_n - \xi|^r])^{\frac{1}{r}} = 0$$

thus

$$\lim_{i\to\infty} (E[|\xi_{n_i} - \xi|^r])^{\frac{1}{r}} = 0,$$

whence it follows that relationship (6.89) is impossible. Thus the assumption that $\{\xi_n\}$ does not converge in probability to ξ is absurd.

Proposition 47. *(Bernoulli's law of large numbers.) In a series of independent trials the relative frequency of the occurrence of the event A converges in probability to the probability P(A) as the number of trials tends to infinity.*

Let $\xi_n = \frac{k}{n}$ be the relative frequency of the occurence of the event A in a series of trials. That means that the r.v. has a binomial distribution and

$$E[n\ \xi_n] = np,\ D(n\xi_n) = \sqrt{npq}.$$

Let us apply Tchebycheff's inequality (6.79) to the r.v. $\zeta_n = n\ (\xi_n - p)$. By considering that $a = \varepsilon n$ we have

$$P(\{\omega;\ |\zeta_n| \geqslant \varepsilon\, n\}) \geqslant \frac{E[(n\xi_n - np)^2]}{\varepsilon^2\ n^2}.$$

But taking into account that $E[n\xi_n] = np$, we have

$$E[(n\xi_n - np)^2] = D^2(n\ \xi_n) = npq,$$

thus

$$P\ (\{\omega;\ |n\ \xi_n - np| \geqslant \varepsilon\ n\}) \leqslant \frac{pq}{n\varepsilon^2}$$

or

$$P(\{\omega;\ \ |\xi_n - p| \geqslant \varepsilon\}) \leqslant \frac{pq}{n\varepsilon^2},$$

whence it follows that

$$\xi_n \xrightarrow{p} p.$$

Proposition 48. *(Markov's theorem.) Let ξ_n be a sequence of r.v. which are mutually independent and so that $M_k = E[\xi_k]$ and $D_k = D(\xi_k)$, $(k = 1, 2, \ldots)$ are finite. We also suppose that:*
(a) *the limit*

$$\lim_{n\to\infty} \frac{\sum_{k=1}^{n} M_k}{n} = M$$

exists and is finite;
(b)

$$\lim_{n\to\infty} \frac{\sqrt{\sum_{k=1}^{n} D_k^2}}{n} = 0.$$

In these conditions

$$\frac{\sum_{k=1}^{n} \xi_k}{n} \xrightarrow{p} M.$$

In order to prove this theorem, let us denote by

$$\zeta_n = \frac{\sum_{k=1}^{n} \xi_k}{n}, \ \zeta_n^* = \frac{1}{n}\sum_{k=1}^{n} (\xi_k - M_k), \ S_n = \sqrt{\sum_{k=1}^{n} D_k^2}.$$

With these notations we have

$$E[\zeta_n^*] = 0, \ D(\zeta_n^*) = \frac{S_n}{n}.$$

Let us now apply Tchebycheff's inequality to the r.v. ζ_n^*. We have

$$P(\{\omega; |\zeta_n^*(\omega)| \geqslant a\}) \leqslant \frac{E[(\zeta_n^*)^2}{a^2}. \tag{6.90}$$

But, taking into account relationship (6.46), it follows that

$$E[(\zeta_n^*)^2] = D^2(\zeta_n^2) + (E[\zeta_n^*])^2 = \frac{S_n^2}{n^2},$$

thus, from condition (b), we have

$$\lim_{n\to\infty} E[(\zeta_n^*)^2] = 0.$$

From inequality (6.90) we obtain

$$\zeta_n^* \xrightarrow{p} 0.$$

But

$$\zeta_n^* = \zeta_n - \frac{\sum_{k=1}^{n} M_k}{n}$$

and from condition (a), for n large enough, we have

$$\left| \frac{\sum_{k=1}^{n} M_k}{n} - M \right| < \frac{\varepsilon}{2}$$

hence it follows that

$$| \zeta_n - M | > \varepsilon$$

only if $| \zeta_n^* | > \frac{\varepsilon}{2}$ · It follows that

$$\zeta_n \xrightarrow{p} M.$$

Proposition 49. *(Central limit theorem.) Let $\xi_1, \xi_2, \ldots, \xi_n, \ldots$ be a sequence of mutually independent r.v., which have the same distribution. We suppose that $M = E[\xi_n]$ and $D = D(\xi_n) > 0$ exists. If we write*

$$\zeta_n = \sum_{k=1}^{n} \xi_k, \quad \zeta_n^* = \frac{\zeta_n - E[\zeta_n]}{D(\zeta_n)}$$

then for every x the distribution function of the r.v. ζ_n^* tends to the normal distribution function

$$\Phi(x) = \frac{1}{\sqrt{2\pi}} \int_{-\infty}^{x} e^{-\frac{u^2}{2}} du.$$

Let us denote by $\Phi(t)$ the characteristic function of the r.v. $\eta_k = \xi_k - M$. By taking into account proposition 35, we deduce that

$$\varphi_{\xi_k}(t) = e^{iMt} \varphi(t)$$

and from the above relationship and proposition 36, it follows that

$$\varphi_{\zeta_n}(t) = e^{inMt} [(\varphi(t)]^n. \tag{6.91}$$

On the other hand, we have $E[\zeta_n] = nM$ and $D(\zeta_n) = D\sqrt{n}$, thus

$$\zeta_n^* = \frac{1}{D\sqrt{n}}\zeta_n - \frac{M\sqrt{n}}{D},$$

whence it follows that

$$\varphi_{\zeta_n^*}(t) = e^{-\frac{M\sqrt{n}}{D}it}\varphi_{\zeta_n}\left(\frac{1}{D\sqrt{n}}t\right)$$

or, taking into account equality (6.91)

$$\varphi_{\zeta_n^*}(t) = e^{-\frac{M\sqrt{n}}{D}it} e^{inM\frac{t}{D\sqrt{n}}}\left[\varphi\left(\frac{t}{D\sqrt{n}}\right)\right]^n,$$

that is

$$\varphi_{\zeta_n^*}(t) = \left[\varphi\left(\frac{t}{D\sqrt{n}}\right)\right]^n.$$

By taking into account formula (6.83) we obtain

$$\varphi\left(\frac{t}{D\sqrt{n}}\right) = 1 - \frac{t^2}{2n} + o\left(\frac{1}{n}\right)$$

thus

$$\lim_{n\to\infty}\varphi_{\zeta_n^*}(t) = e^{-\frac{t^2}{2}}$$

whence it follows that the distribution function of the r.v. ζ_n^* tends to $\Phi(x)$.

The following proposition is a limit theorem more general than the preceding one.

Proposition 50. (*Leapunov's theorem.*) *Let* $\{\xi_n\}$ *be a sequence o mutually independent r.v. so that the moments*

$$E[\xi_k] = E_k,\ D^2(\xi_k) = D_k^2 > 0,\ E[\,|\,\xi_k - E_k\,|^3] = H_k^3,\ (k = 1, 2, \ldots)$$

exist. Let us denote by

$$S_n = \sqrt{\sum_{j=1}^{n} D_j^2},\quad K_n = \sqrt{\sum_{j=1}^{n} H_j^3},\quad \zeta_n = \sum_{j=1}^{n}\xi_j,\quad \zeta_n^* = \frac{\zeta_n}{S_n}$$

and by $F_{(n)}(x)$ *the distribution function of the r.v.* ζ_n^*.

If Leapunov's condition is verified

$$\lim_{n\to\infty} \frac{K_n}{S_n} = 0$$

then

$$\lim_{n\to\infty} F_{(n)}(x) = \Phi(x), \quad (x \in R).$$

In order to prove this theorem it is sufficient to prove that if we denote the characteristic function of the r.v. ζ_n^* by $\varphi_{(n)}(t)$ then

$$\lim_{n\to\infty} \varphi_{(n)}(t) = e^{-\frac{t^2}{2}}, \quad (t \in R).$$

We denote the characteristic function of the r.v. ξ_j by $\varphi_j(t)$ and we have

$$\varphi_{(n)}(t) = \sum_{j=1}^{n} \varphi_j\left(\frac{t}{S_n}\right).$$

But for every u we have

$$e^{iu} = 1 + i\frac{u}{1} - \frac{u^2}{2} + \theta(u)\frac{u^3}{3},$$

where $|\theta(u)| \leqslant 1$, thus denoting the distribution function of the r.v. ξ_j by $F_j(x)$ we obtain

$$\varphi_j(t) = 1 - \frac{D_k^2}{2} t^2 + \frac{t^3}{3}\int_{-\infty}^{+\infty} \theta(ut)\, u^3\, \mathrm{d}F_j(u)$$

and thus

$$\varphi_j\left(\frac{t}{S_n}\right) = 1 + \omega_j^n(t),$$

where we have written

$$\omega_j^n(t) = -\frac{D_j^2}{S_n^2}\frac{t^2}{2} + \frac{1}{S_n^3}\frac{t^3}{3}\int_{-\infty}^{+\infty} \theta(ut)\, u^3\, \mathrm{d}F_j(u).$$

But $D_k \leqslant H_k$, $(k = 1, 2, \ldots)$ and thus

$$D_k \leqslant K_n, \ (k = 1, 2, \ldots).$$

On the other hand

$$\left|\int_{-\infty}^{+\infty} \theta(ut)\, u^3\, \mathrm{d}F_j(u)\right| \leqslant \int_{-\infty}^{+\infty} |u|^3\, \mathrm{d}F_j(u) = H_j^3 \leqslant K_n^3.$$

Thus

$$|\omega_j^n(t)| \leqslant \frac{K_n^2}{S_n^2} \cdot \frac{t^2}{2} + \frac{K_n^3}{S_n^3} \cdot \frac{|t|^3}{3}, \ (j \leqslant n),$$

whence, taking into account Leapunov's condition, it follows that the sequence $\{\omega_j^n(t)\}$ converges uniformly to zero in every interval $[-T, T]$.

Let us fix an interval $[-T, T]$ so that, if $n \geqslant N$, we have

$$|\omega_j^n(t)| \leqslant \frac{1}{2}$$

for every $t \in [-T, T]$, $(j = 1, \ldots, n)$. But, if $|z| \leqslant \frac{1}{2}$,

$$\ln(1 + z) = z + \theta_1(z)z^2,$$

with $|\theta_1(z)| \leqslant 1$. We deduce that

$$\ln \varphi_j \frac{t}{S_n} = \ln\,[1 + \omega_j^n(t)] = -\frac{D_j^2}{S_n^2}\frac{t^2}{2} + \frac{t^3}{3S_n^3}\int_{-\infty}^{+\infty} \theta(ut)\, u^3\, \mathrm{d}F_j(u) +$$

$$+ \theta_1(\omega_j^n)\left[\frac{D_j^4}{S_n^4}\frac{t^4}{4} + \frac{t^6}{36}\frac{1}{S_n^6}\left(\int_{-\infty}^{+\infty} \theta(ut)\, u^3\, \mathrm{d}F_j(u)\right)^2 -\right.$$

$$\left. - 2\,\frac{S_j^2}{S_n^2} \cdot \frac{t^2}{2} \cdot \frac{t^3}{6} \cdot \frac{1}{S_n^3}\int_{-\infty}^{+\infty} \theta(ut)\, u^3\, \mathrm{d}F_j(u)\right].$$

Thus

$$\left|\ln \varphi_j\left(\frac{t}{S_n}\right) - \left(-\frac{D_j^2}{S_n^2}\right)\frac{t^2}{2}\right| \leqslant L\left(\frac{H_j^3}{S_n^3} + \frac{H_j^6}{S_n^6} + \frac{D_j^4}{S_n^4} + \frac{D_j^2 H_j^3}{S_n^5}\right) \leqslant L_1 \frac{H_j^3}{S_n^3},$$

where L and L_1 are conveniently chosen. It follows that

$$\lim_{n\to\infty}\left|\sum_{j=1}^{n} \ln \varphi_j\left(\frac{t}{S_n}\right) + \frac{t^2}{2}\right| = 0,$$

where the limit is uniform in every interval $[-T, T]$. Thus

$$\lim_{n\to\infty} \ln \varphi_{(n)}(t) = -\frac{t^2}{2}$$

that is

$$\lim_{n\to\infty} \varphi_{(n)}(t) = e^{-\frac{t^2}{2}}.$$

6.5 ELEMENTS OF THE THEORY OF ERRORS

6.5.1 Distribution of the incidental errors

Let a be a quantity for which we determine by n measurements the values $a_1, a_2, \ldots, a_n$.

Definition 1. The quantities $\alpha_k = a - a_k$, $(k = 1, \ldots, n)$ *are called incidental errors in the n measurements.*

With this definition, let us prove the following proposition.

Proposition 1. *(Laplace and Gauss' theorem.) The distribution of the incidental errors is normal.*

Let us consider the r.v ξ whose values are α_k. We assume that this r.v. is of continuous type and let $f(x)$ be its density of distribution. Then we have

$$P(\xi < x) = \int_{-\infty}^{x} f(u)\, du, \tag{6.92}$$

$$P(x \leqslant \xi < x + k) = \int_{x}^{x+k} f(u)\, du, \tag{6.93}$$

$$\int_{-\infty}^{+\infty} f(u)\, du = 1. \tag{6.94}$$

Taking into account the mean value theorem for Riemann's integral, from formula (6.93), we deduce that

$$P(x \leqslant \xi < x + k) = kf(x + \theta k), \quad (0 < \theta < 1).$$

As the function $f(x)$ is continuous, we can approximate $f(x + \theta k)$ by $f(x)$ for k sufficiently small so that

$$P(x \leqslant \xi < x + k) = kf(x).$$

Thus

$$P(\alpha_k \leqslant \xi < \alpha_k + h_k) = h_k f(\alpha_k), \quad (k = 1, \ldots, n).$$

By taking into account that the n measurements are independent, the probability for the resulting errors to be contained in the intervals $[\alpha_k, \alpha_k + h_k]$ with the hypothesis that h_k, $(k = 1, \ldots, n)$ are small enough, is

$$p = f(\alpha_1)\, f(\alpha_2) \ldots f(\alpha_n)\, h_1 h_2 \ldots h_n$$

or

$$p = f(a - a_1)\, f(a - a_2) \ldots f(a - a_n)\, h_1 h_2 \ldots h_n .$$

If we set $\bar{a} = E[\xi]$ the probability p is at the same time the probability for a to be in a neighbourhood of $\bar{a}$. Since $\bar{a}$ is the most probable value of a it follows that p is maximum for $a = \bar{a}$.

Writing

$$\varphi(a) = f(a - a_1) f(a - a_2) \ldots f(a - a_n),$$

for p to be maximum it is necessary that $\varphi'(\bar{a}) = 0$, with the hypothesis that $f(x)$ is differentiable. We have

$$\frac{\varphi'(\bar{a})}{\varphi(\bar{a})} = \frac{f'(\bar{a} - a_1)}{f(\bar{a} - a_1)} + \frac{f'(\bar{a} - a_2)}{f(\bar{a} - a_2)} + \ldots + \frac{f'(\bar{a} - a_n)}{f(\bar{a} - a_n)} = 0. \tag{6.95}$$

By taking into account the expression of $E[\xi]$ it follows that

$$(\bar{a} - a_1) + (\bar{a} - a_2) + \ldots + (\bar{a} - a_n) = 0. \tag{6.96}$$

From formulae (6.95) and (6.96) it follows that

$$\sum_{k=1}^{n} \frac{f'(a - a_k)}{f(\bar{a} - a_k)} - c \sum_{k=1}^{n} (\bar{a} - a_k) = 0,$$

where c is an arbitrary constant, or by putting $x_k = \bar{a} - a_k$ we obtain

$$\sum_{k=1}^{n} \left[\frac{f'(x_k)}{f(x_k)} - cx_k \right] = 0. \tag{6.97}$$

But $a_1, a_2, \ldots, a_n$ are independent, hence $x_1, x_2, \ldots, x_n$ have the same property. We deduce that relationship (6.97) is satisfied if, and only if, every term of the sum is zero, that is

$$\frac{f'(x_k)}{f(x_k)} - cx_k = 0, \quad (k = 1, \ldots, n).$$

It follows that $f(x)$ is the solution of the differential equation

$$\frac{f'(x)}{f(x)} = cx$$

that is

$$f(x) = Ke^{\frac{1}{2}cx^2}.$$

By taking into account relationship (6.94) it follows that $c < 0$, thus writing $c = -\frac{1}{\sigma^2}$ we have

$$f(x) = Ke^{-\frac{x^2}{2\sigma^2}}.$$

With the above value, from relationship (6.94) we obtain

$$K \int_{-\infty}^{+\infty} e^{-\frac{u^2}{2\sigma^2}} du = 1,$$

whence

$$K = \frac{1}{\sigma\sqrt{2\pi}}.$$

Thus

$$f(x) = \frac{1}{\sigma\sqrt{2\pi}} e^{-\frac{x^2}{2\sigma^2}}. \tag{6.98}$$

It follows that ξ is a normal r.v. of parameters $0, \sigma^2$.

Definition 2. The function

$$\Theta(x) = \frac{2}{\sqrt{\pi}} \int_0^x e^{-u^2}\,du$$

is called the errors function.

Definition 3. If the distribution of the incidental errors of a measurement has the parameters 0, σ^2 *then the constant*

$$h = \frac{1}{\sigma\sqrt{2}}$$

is called the accuracy of the measurement.

By means of these definitions let us prove the following proposition.

Proposition 2. *The probability for the absolute value of the error of a measurement to be included in the interval* $[a, b)$, *with* $a > 0$, *is*

$$P(a \leqslant |\xi| < b) = \Theta(hb) - \Theta(ha), \tag{6.99}$$

where h is the accuracy of the measurement.

Taking into account formula (6.98) it follows that

$$P(a \leqslant \xi < b) = \frac{1}{\sigma\sqrt{2\pi}} \left[\int_0^b e^{-\frac{u^2}{2\sigma^2}}\,du - \int_0^a e^{-\frac{u^2}{2\sigma^2}}\,du \right].$$

By the change of variable $u = \sigma\sqrt{2}\,t$, we obtain

$$P(a \leqslant \xi < b) = \frac{1}{\sqrt{\pi}} \left(\int_0^{\frac{b}{\sigma\sqrt{2}}} e^{-t^2}\,dt - \int_0^{\frac{a}{\sigma\sqrt{2}}} e^{-t^2}\,dt \right)$$

that is

$$P(a \leqslant \xi < b) = \frac{1}{2}[\Theta(hb) - \Theta(ha)].$$

For reason of symmetry it follows that

$$P(a \leqslant |\xi| < b) = 2P(a \leqslant \xi < b)$$

whence we deduce relationship (6.99).

6.6 SAMPLE THEORY

6.6.1 The principle of the method

Let us consider a statistical collectivity which we shall call a population. Such a population is characterized in general by its distribution.

The basic problem of mathematical statistics is the following. How can we determine the distribution or at least its principal characteristics based on experiments we may make with the population, if we don't know a priori this distribution and if we can not determine this distribution? This determination is a very important method of mathematical statistics and is called the method of selection or the method of samples.

The principle of the method of samples is the following: given a population formed by N individuals, after n independent trials we obtain the results $x_1, x_2, \ldots, x_n$. We want to determine the characteristics of the population by means of the variables, x_i $(i = 1, 2, \ldots, n)$.

Definition 1. *We call theoretical distribution the distribution of the population, and the variables $x_1, x_2, \ldots, x_n$ are called the sample variables. A function $f(x_1, \ldots, x_n)$ is called the sample function. The distribution function of the sample $x_1, \ldots, x_n$ is called the empirical sample function.*

We shall now prove two propositions which will be necessary for the proof of a very important theorem of mathematical statistics.

Proposition 1. *If $\{E_n\}$ is a sequence of events so that the occurrence of every event E_{n+1}* implies the occurrence of E_n and *if*

$$E = \bigcap_{n=1}^{\infty} E_n,$$

then

$$P(E) = \lim_{n\to\infty} P(E_n).$$

By taking into account that $E_{n+1} \subseteq E_n$ it follows that

$$E_1 = (E_1 \cap \complement E_2) \cup (E_2 \cap \complement E_3) \cup \ldots \cup (E_{n-1} \cap \complement E_n) \cup E_n,$$

$$E_1 = \left[\bigcup_{n=1}^{\infty} (E_n \cap \complement E_{n+1})\right] \cup E.$$

As the events of the right-hand side are mutually exclusive we deduce that

$$P(E_1) = \sum_{j=1}^{n-1} P(E_j \cap \complement E_{j+1}) + P(E_n),$$

$$P(E_1) = \sum_{n=1}^{\infty} P(E_n \cap \complement E_{n+1}) + P(E). \tag{6.100}$$

We get

$$P(E) = P(E_n) - \sum_{k=n}^{\infty} P(E_k \cap \complement E_{k+1}). \tag{6.101}$$

But from formula (6.100) it follows that the series $\sum_{n=1}^{\infty} P(E_n \cap \complement E_{n+1})$ is convergent and hence its remainder tends to zero, that is

$$\lim_{n\to\infty} \sum_{k=n}^{\infty} P(E_k \cap \complement E_{k+1}) = 0.$$

From formula (6.101) we get

$$P(E) = \lim_{n\to\infty} P(E_n).$$

Proposition 2. *In the conditions of proposition 1, if*

$$P(E_j) = 1, \ (j = 1, 2, \ldots) \text{ then } P(E) = 1.$$

Let us first prove that $P(E) = 1$, for the case of a finite number of events $E_1, E_2, \ldots, E_n$.

If we have two events E_1, E_2 with $E_2 \subseteq E_1$ and with $P(E_1) = P(E_2) = 1$ then we have

$$E_1 \cup E_2 = E_1$$

thus

$$P(E_1 \cup E_2) = 1.$$

But, taking into account formula (6.17),

$$P(E_1 \cap E_2) = P(E_1) + P(E_2) - P(E_1 \cup E_2)$$

thus

$$P(E) = P(E_1 \cap E_2) = 1.$$

By a complete induction it follows that the property is true for a finite number of events: $E_1, E_2, \ldots, E_n$.

In order to prove this for the case of an infinite sequence of events let us notice that from the given hypothesis

$$E = \bigcap_{n=1}^{\infty} E_n = E_1 \cap (E_1 \cap E_2) \cup (E_1 \cap E_2 \cap E_3) \cap \ldots$$

and

$$E_1 \cap E_2 \cap \ldots \cap E_{n+1} \subseteq E_1 \cap E_2 \cap \ldots \cap E_n$$

thus we can apply proposition 1 from which we deduce that

$$P(E) = \lim_{n \to \infty} P(E_1 \cap E_2 \cap \ldots \cap E_n)$$

and by taking into account that $P(E_1 \cap E_2 \cap \ldots \cap E_n) = 1$, it follows that

$$P(E) = 1.$$

We shall now be able to prove the following proposition.

Proposition 3. *(Glivenko's theorem.) If $F(x)$ is a distribution function of a theoretical distribution and $F_n(x)$ is the empirical distribution function achieved in a sample of size n, then*

$$P(\sup_{-\infty < x < \infty} | F_n(x) - F(x) | \to 0) = 1. \tag{6.102}$$

If $r > 0$ is an integer, we denote by $x_{r,k}$ the least value of x for which

$$F(x - 0) = F(x) \leqslant \frac{k}{r} < F(x + 0), \ (k = 1, 2, \ldots, r). \tag{6.103}$$

Let ξ be a r.v. whose distribution function is $F(x)$. From equation (6.103) it follows that

$$P(\xi < x_{r,k}) = F(x_{r,k})$$

and from Bernoulli's law of large numbers we can write

$$P[\lim_{n \to \infty} F_n(x_{r,k}) = F(x_{r,k})] = 1, \ (k = 1, 2, \ldots, r). \tag{6.104}$$

If we denote by E'_k the following event

$$\lim_{n \to \infty} F_n(x_{r,k}) = F(x_{r,k})$$

then relationship (6.104) can be written as

$$P(E'_k) = 1, \qquad (k = 1, \ldots r). \tag{6.104'}$$

We also consider the event

$$E^r = E_1^r \cap E_2^r \cap \ldots \cap E_r^r,$$

that is the event

$$\max_{l \leqslant k \leqslant r} \left| \lim_{n \to \infty} F_n(x_{r,k}) - F(x_{r,k}) \right| = 1.$$

By taking into account relationships (6.104′) from proposition 2 we obtain

$$P(E^r) = 1. \tag{6.105}$$

Let us now denote by $E = \bigcap_{n=1}^{\infty} E^n$ and by S the event

$$\sup_{x \in R} | F_n(x) - F(x) | \to 0.$$

From relationship (6.105), proposition 2 yields

$$P(E) = 1. \tag{6.105'}$$

If $x_{r,k} \leqslant x \leqslant x_{r,k+1}$, then

$$F_n(x_{r,k} + 0) \leqslant F_n(x) \leqslant F_n(x_{r,k+1}),$$
$$F(x_{r,k} + 0) \leqslant F(x) \leqslant F(x_{r,k+1}),$$
$$0 \leqslant F(x_{r,k+1}) - F(x_{r,k} + 0) \leqslant \frac{1}{r},$$

whence

$$F(x_{r,k} + 0) - F(x_{r,k+1}) \leqslant F_n(x) - F(x) \leqslant F(x_{r,k+1}) - F(x_{r,k} + 0)$$

or

$$| F_n(x) - F(x) | \leqslant \max_{1 \leqslant k \leqslant r} | F_n(x_{r,k}) - F(x_{r,k}) | + \frac{1}{r} \cdot$$

It follows that

$$\sup_{x \in R} | F_n(x) - F(x) | \leqslant \max_{1 \leqslant k \leqslant r} | F_n(x_{r,k}) - F(x_{r,k}) | + \frac{1}{r} \cdot \tag{6.106}$$

But r is chosen arbitrarily, thus from equality (6.106) we deduce that $E \subseteq S$, that is

$$P(E) \leqslant P(S)$$

and by taking into account formula (6.105') it follows that

$$P(S) = 1,$$

which is Glivenko's relationship (6.102).

Definition 2. If $f(x_1, \ldots, x_n)$ is a sample function and if

$$\underset{n \to \infty}{f}(x_1, \ldots, x_n) \xrightarrow{p} A,$$

then $f(x_1, \ldots x_n)$ is an estimation of the unknown value A.

If

$$E[f(x_1, \ldots, x_n)] = A, \text{ and } \underset{n \to \infty}{D^2}(f(x_1, \ldots, x_n)) \to 0$$

then $f(x_1, \ldots, x_n)$ is an absolutly correct estimation of A.

If

$$E[f(x_1, \ldots, x_n)] = A + B(n), \quad \underset{n \to \infty}{D^2}(f(x_1, \ldots, x_n)) \to 0$$

and $\lim_{n \to \infty} B(n) = 0$, $f(x_1, \ldots, x_n)$ is a correct estimation of A.

6.6.2 Sample moments

Definition 3. Given the sample $x_1, \ldots, x_n$ the expression

$$\overline{M}_k = \frac{1}{n} \sum_{j=1}^{n} x_j^k, \quad (k = \text{natural number})$$

is a r.v. called the sample moment of order k. In the case when $k = 1$, the sample moment $\overline{M}_1$ is denoted by $\bar{x}$ and is called the sample mean.

The expression

$$\overline{m}_k = \frac{1}{n} \sum_{j=1}^{n} (x_j - \bar{x})^k$$

is a r.v. called the centred sample moment of order k.

We shall now prove some properties concerning these concepts.

Proposition 4. *If the first two moments of the theoretical distribution exist, then*

$$E[\bar{x}] = \alpha_1, \quad D^2(\bar{x}) = \frac{\sigma^2}{n}.$$

Indeed

$$E[\bar{x}] = E\left(\frac{\sum_{j=1}^{n} x_j}{n}\right) = \frac{1}{n}\sum_{j=1}^{n} E[x_j] = \frac{1}{n} n\alpha_1 = \alpha_1.$$

Also

$$E[\bar{x}^2] = E\left[\left(\frac{\sum_{j=1}^{n} x_j}{n}\right)^2\right] = \frac{1}{n^2}\sum_{j=1}^{n} E[x_i^2] + \frac{2}{n^2}\sum_{j<l}^{1\ldots n} E[x_j x_l] = \frac{\alpha_2}{n} + \frac{n-1}{n}\alpha_1^2,$$

thus

$$D^2(\bar{x}) = [E[\bar{x}^2] - (E[\bar{x}])^2 = \frac{\alpha_2}{n} + \frac{n-1}{n}\alpha_1^2 - \alpha_1^2 = \frac{\alpha^2 - \alpha_1^2}{n} = \frac{\sigma^2}{n}.$$

It follows that if the first two moments of the theoretical distribution exist then the sample mean is an absolutely correct estimation of the theoretical mean.

Proposition 5. *If the centred moments of second* (σ^2) *and third order* (ρ^3) *exist and if* $\sigma \neq 0$ *and* ρ *is finite and if the moments*

$$m_j = E\left[\frac{x_j}{n}\right], \quad \sigma_j^2 = E\left[\left(\frac{x_j}{n} - m_j\right)^2\right], \quad \rho_j^3 = E\left[\left(\frac{x_j}{n} - m_j\right)^3\right],$$

($j = 1, \ldots, n$) *exist then the distribution of the sample mean tends to the normal distribution.*

If we write $\xi_j = \frac{x_j}{n}$, the sample mean can be written

$$\bar{x} = \sum_{j=1}^{n} \xi_j.$$

By taking into account the calculus from proposition 4 we obtain

$$m_j = \frac{\alpha_1}{n}, \quad \sigma_j^2 = \frac{\sigma^2}{n^2}, \quad \rho_j^3 = \frac{\rho^3}{n^3} \quad (j = 1, \ldots, n)$$

thus

$$\lim_{n \to \infty} \frac{\sqrt[3]{\sum_{j=1}^{n} \rho_j^3}}{\sqrt{\sum_{i=1}^{n} \sigma_j^2}} = \lim_{n \to \infty} \frac{\left(\frac{\rho^3}{n^2}\right)^{1/3}}{\left(\frac{\sigma^2}{n}\right)^{1/2}} = \lim_{n \to \infty} \frac{\rho}{\sigma} \cdot \frac{1}{n^{1/6}} = 0.$$

Considering Leapunov's theorem it follows that the distribution of the sample mean tends to the normal distribution.

Proposition 6. *The mean value of the centred sample moment of order k is*

$$E[\overline{m}_k] = m_k + O\left(\frac{1}{n}\right),$$

where $O\left(\frac{1}{n}\right)$ *tends to zero as* $\frac{1}{n}$.

Indeed, taking into account definition 3, from the relationship

$$(x_j - \overline{x})^k = x_j^k - \binom{k}{1} \overline{x} x_j^{k-1} + \binom{k}{2} \overline{x}^2 x_j^{k-2} - \ldots$$

we deduce that

$$\overline{m}_k = \overline{M}_k - \binom{k}{} \overline{x}\, \overline{M}_{k-1} + \binom{k}{2} \overline{x}^2 \overline{M}_{k-2} - \ldots \tag{6.107}$$

Without lack of generality we can assume that $\alpha_1 = 0$, whence it follows that $\alpha_k = m_k$, $(k = 1, 2, \ldots)$.

Based on Schwarz's inequality we have

$$E[\overline{x}^s \overline{M}_{k-s}] \leqslant \sqrt{E[\overline{x}^{2s}]\, E[\overline{M}_{k-s}^2]}. \tag{6.108}$$

On the other hand

$$E[\overline{M}_k] = E\left[\frac{\sum_{j=1}^{n} x_j^k}{n}\right] = \frac{\sum_{j=1}^{n} E[x_j^k]}{n} = \alpha_k,$$

hence

$$E[\overline{M}_{k-s}^2] = \alpha_{k-s}^2 .$$

Also

$$E[\overline{x}^{2s}] = O\left(\frac{1}{n}\right)$$

thus from inequality (6.108) it follows that

$$E[\overline{x}^s \, \overline{M}_{k-s}] \leqslant O\left(\frac{1}{n^s}\right).$$

From formula (6.107) we obtain

$$E[\overline{m}_k] = m_k + O\left(\frac{1}{n}\right).$$

By analogy we can prove the following proposition.

Proposition 7. *The variance of the r.v.* $\overline{m}_k$ *is*

$$D^2(\overline{m}_k) = \frac{m_{2k} - m_k^2 - 2k m_{k-1} m_{k+1} + k^2 m_{2k} m_{k-1}^2}{n} + O\left(\frac{1}{n^2}\right)$$

and the covariance of the moments $\overline{M}_h$ *and* $\overline{M}_k$ *is*

$$E(\overline{M}_n - m_h)(\overline{M}_k - m_k) =$$

$$= \frac{m_{h+k} - m_h m_k - h m_{h-1} m_{k+1} - k m_{h+1} m_{k-1} + hk m_{h-1} m_{k-1}}{n} + O\left(\frac{1}{n^2}\right).$$

Let us now prove the following propositions.

Proposition 8. *(Hinchin's theorem.) Let* $\{x_n\}$, $(n = 1, 2, \ldots)$ *be a sequence of independent r.v. with the same distribution function* $F(x)$ *and a finite mean* M_1. *Then the sample mean*

$$\overline{x} = \frac{x_1 + x_2 + \ldots + x_n}{n}$$

tends in probability to M_1.

Let $\varphi(t)$ be the characteristic function of a r.v. x_k, $(k = 1, 2, \ldots)$. Then from proposition 36, Section 6.4.6, the characteristic function of the r.v. is $\left[\varphi\left(\frac{t}{n}\right)\right]^n$.

On the other hand, taking into account formula (6.83), we can write

$$\varphi(t) = 1 + iM_1 t + O(t)$$

thus for t fixed we have

$$\left[\varphi\left(\frac{t}{n}\right)\right]^n = \left[1 + iM_1\frac{t}{n} + O\left(\frac{1}{n}\right)\right]^n.$$

We deduce that

$$\lim_{n\to\infty}\left[\varphi\left(\frac{t}{n}\right)\right]^n = e^{itM_1},$$

whence it follows that

$$\bar{x} \xrightarrow{p} M_1 .$$

We shall now determine the sample mean for some classical statistical distributions.

Proposition 9. *If the theoretical distribution is a normal distribution of parameters m, σ^2 then the distribution of the sample mean is a normal distribution of parameters m, $\frac{\sigma^2}{m}$.*

We have

$$\varphi_{\bar{x}}(t) = E\left[e^{it\bar{x}}\right] = E\left[e^{i\frac{t}{n}\sum_{j=1}^{n} x_j}\right] = \prod_{j=1}^{n} E\left[e^{i\frac{t}{n}x_j}\right]. \tag{6.109}$$

But, taking into account that x_j are normal r.v. from proposition 42, Section 6.4.6, we deduce that

$$E\left[e^{i\frac{t}{n}x_i}\right] = e^{m\frac{t}{n}i - \frac{\sigma^2}{2}\frac{t^2}{n^2}}.$$

From formula (6.109) we obtain

$$\varphi_{\bar{x}}(t) = e^{m\tau i\,\frac{\sigma^2 t^2}{2n}}$$

whence from the same proposition 42, Section 6.4.6, it follows that $\varphi_{\bar{x}}(t)$ is the characteristic function of a normal distribution of parameters $m, \frac{\sigma^2}{n}\cdot$

Proposition 10. *If the theoretical distribution is a Poisson distribution of parameter λ, then the characteristic function of the sample mean is*

$$\varphi_{\bar{x}}(t) = e^{n\lambda(e^{i\frac{t}{n}} - 1)}.$$

We have

$$\varphi_{\bar{x}}(t) = E\left[e^{it\bar{x}}\right] = \prod_{j=1}^{n} E\left[e^{i\frac{t}{n}x_j}\right].$$

By taking into account that x_j are Poisson r.v. from proposition 42, Section 6.4.6, we obtain

$$\varphi_{\bar{x}}(t) = e^{n\lambda(e^{i\frac{t}{n}} - 1)}.$$

6.7 ESTIMATION THEORY

6.7.1 The estimation theory for families of distributions with a parameter

In some applications of mathematical statistics the distribution of the phenomenon is given by a known function where various parameters of unknown values occur. For example if the distribution of the phenomenon is normal, we have to determine the values of the parameters m and σ^2.

Definition 1. We say that the distribution expressed by a given function depending on some parameters is a specified distribution. When we know the numerical values of these parameters we have a completely specified distribution.

The determination of the numerical values of the parameters of a specified distribution, or their estimation is made by means of n independent trials from which are obtained the values $x_1, \ldots, x_n$ for the statistical variable which we study.

In the sequel we shall study the specified distributions which depend on a single parameter.

We denote by $f(x, \alpha)$ the density of probability of the specified distribution and suppose that $f(x, \alpha)$ is continuous and admits derivatives of a large enough order with respect to both variables for every x and α.

The functions $\bar{\alpha}(x_1, \ldots, x_n)$ by means of which we estimate the value of the parameter $\bar{\alpha}$ are called estimation functions.

Definition 2. If $\bar{\alpha}(x_1, \ldots, x_n)$ is an estimation function and if

$$\bar{\alpha}(x_1, \ldots, x_n) \underset{n\to\infty}{\overset{p}{\to}} \alpha$$

then we say that $\bar{\alpha}$ is a correct estimation function of α.

If

$$E[\bar{\alpha}(x_1, \ldots, x_n)] = \alpha, \qquad D^2(\bar{\alpha}(x_1, \ldots, x_n)) \underset{n\to\infty}{\to} 0,$$

then we say that $\bar{\alpha}$ is an absolutely correct estimation function of α.

Proposition 1. *Let $f(x, \alpha)$ be the densities of probability of a family of statistical distributions and*

$$\bar{\alpha}(x_1, \ldots, x_n)$$

an absolutely correct estimation function of α. Then

$$D^2(\bar{\alpha}(x_1, \ldots, x_n)) \geqslant \frac{1}{n \displaystyle\int_{-\infty}^{+\infty} \left[\frac{\partial \ln f(x, \alpha)}{\partial \alpha}\right]^2 f(x, \alpha)\, dx} \tag{6.110}$$

the equality takes place only if

$$f(x, \alpha) = e^{A'(\alpha)[L(x)-\alpha]+A(\alpha)+N(x)} \tag{6.111}$$

where $A'(\alpha) = \dfrac{dA(\alpha)}{d\alpha}$.

The function $\bar{\alpha}(x_1, \ldots, x_n)$ for which the minimum is obtained is expressed as

$$\bar{\alpha}(x_1, \ldots x_n) = \frac{L(x_1) + L(x_2) + \ldots + L(x_n)}{n}. \tag{6.112}$$

As the function $f(x, \alpha)$ is a density of probability we have

$$\int_{-\infty}^{+\infty} f(x, \alpha)\, dx = 1.$$

Differentiating with respect to α we obtain

$$\int_{-\infty}^{+\infty} \frac{\partial f(x, \alpha)}{\partial \alpha}\, dx = 0,$$

or

$$\int_{-\infty}^{+\infty} \frac{\partial \ln f(x, \alpha)}{\partial \alpha} f(x, \alpha)\, dx = 0,$$

that is

$$E\left[\frac{\partial \ln f(x, \alpha)}{\partial \alpha}\right] = 0. \tag{6.113}$$

By taking into account definition 2, as $\bar{\alpha}$ is an absolutely correct estimation function for α we have

$$E[\bar{\alpha}] = \alpha.$$

But, denoting by R^n the n-dimensional space of the points $P(x_1, \ldots, x_n)$ we have

$$E[\bar{\alpha}] = \int \cdots \int_{R^n} \bar{\alpha}(x_1, \ldots, x_n) \prod_{i=1}^{n} f(x_i, \alpha)\, dx_1 \ldots dx_n.$$

Thus

$$\int \cdots \int_{R^n} \bar{\alpha}(x_1, \ldots, x_n) \prod_{i=1}^{n} f(x_i, \alpha)\, dx_1 \ldots dx_n = \alpha.$$

Differentiating this relationship with respect to α we get

$$\int \cdots \int_{R^n} \bar{\alpha}(x_1, \ldots, x_n) \sum_{i=1}^{n} \frac{\partial \ln f(x_i, \alpha)}{\partial \alpha} \prod_{i=1}^{n} f(x_i, \alpha)\, dx_1 \ldots dx_n = 1.$$

By taking into account relationship (6.113), we deduce that

$$\int_{R^n}\!\!\cdots\!\int (\bar{\alpha}-\alpha)\sum_{i=1}^{n}\frac{\partial \ln f(x_i,\alpha)}{\partial \alpha}\prod_{i=1}^{n} f(x_i,\alpha)\,dx_1\ldots dx_n = 1,$$

that is

$$E\left[(\bar{\alpha}-\alpha)\sum_{i=1}^{n}\frac{\partial \ln f(x_i,\alpha)}{\partial \alpha}\right] = 1. \tag{6.114}$$

From Schwarz' inequality, we obtain

$$\left(E\left[(\bar{\alpha}-\alpha)\sum_{i=1}^{n}\frac{\partial \ln f(x_i,\alpha)}{\partial \alpha}\right]\right)^2 \leqslant E[(\bar{\alpha}-\alpha)^2]\,E\left[\left(\sum_{i=1}^{n}\frac{\partial \ln f(x_i,\alpha)}{\partial \alpha}\right)^2\right], \tag{6.115}$$

or from formula (6.114)

$$E[(\bar{\alpha}-\alpha)^2]\,E\left[\left(\sum_{i=1}^{n}\frac{\partial \ln f(x_i,\alpha)}{\partial \alpha}\right)^2\right] \geqslant 1,$$

which, taking into account formula (6.113), can be written as

$$D^2(\bar{\alpha}) \geqslant \frac{1}{D^2\left(\sum_{i=1}^{n}\frac{\partial \ln f(x_i,\alpha)}{\partial \alpha}\right)}. \tag{6.116}$$

But, since x_i are independent, we have

$$D^2\left(\sum_{i=1}^{n}\frac{\partial \ln f(x_i,\alpha)}{\partial \alpha}\right) = nD^2\left(\frac{\partial \ln f(x,\alpha)}{\partial \alpha}\right)$$

which, substituted in inequality (6.116), yields inequality (6.110).

On the other hand, Schwarz' inequality becomes an equality only if

$$K(\bar{\alpha}-\alpha) = \sum_{i=1}^{n}\frac{\partial \ln f(x_i,\alpha)}{\partial \alpha}, \tag{6.117}$$

where K is a function of α for every x_i. Thus we can consider $x_1 = x$, $x_2 = \ldots = x_n = 0$. With these values we obtain

$$\frac{\partial f(x,\alpha)}{\partial \alpha} = F(\alpha)\,Q(x) + G(\alpha).$$

Substituting this value in formula (6.117) we deduce that

$$\bar{\alpha}(x_1, \ldots, x_n) = \frac{F(\alpha)}{K(\alpha)} \sum_{i=1}^{n} Q(x_i) + \frac{nG(\alpha)}{K(\alpha)} + \alpha.$$

It follows that

$$F(\alpha) = k_1 K(\alpha), \frac{nG(\alpha)}{K(\alpha)} + \alpha = k_2, \quad (k_1, k_2 = \text{constant}),$$

hence

$$\alpha(x_1, \ldots, x_n) = k_1 \sum_{i=1}^{n} Q(x_i) + k_2.$$

We write

$$L(x) = nk_1 Q(x) + k_2$$

and we deduce that

$$\bar{\alpha}(x_1, \ldots, x_n) = \frac{1}{n} \sum_{i=1}^{n} L(x_i)$$

and

$$\frac{\partial f(x, \alpha)}{\partial \alpha} = \frac{K(\alpha)}{n} [L(x) - \alpha].$$

From this last relationship it follows that

$$\ln f(x, \alpha) = A'(\alpha) [L(x) - \alpha] + A(\alpha) + N(x)$$

which is formula (6.111).

Definition 3. An absolutely correct estimation function $\bar{\alpha}(x_1, \ldots, x_n)$ is called an efficient estimation function if

$$D^2(\bar{\alpha}(x_1, \ldots, x_n)) = \frac{1}{n \int_{-\infty}^{+\infty} \left[\frac{\partial \ln f(x, \alpha)}{\partial \alpha} \right]^2 f(x, \alpha) \, dx} \cdot \tag{6.118}$$

Let us now consider the density of probability

$$P(x_1, \ldots, x_n, \alpha) = \prod_{i=1}^{n} f(x_i, \alpha)$$

corresponding to the point of co-ordinates $x_1, \ldots, x_n$ from the space R^n.

Definition 4. The most likelihood value of α *is the value*

$$\alpha^* = \alpha^*(x_1, \ldots, x_n),$$

which depends on $x_1, \ldots, x_n$ *and for which the probability* $P(x_1, \ldots, x_n, \alpha)$ *is a maximum. The function* α^* *is called the maximum likelihood estimation function.*

Proposition 2. *If an efficient estimation function* $\bar{\alpha}$ *exists then it is given by the maximum likelihood estimation function* α^*.

The maximum of $P(x_1, \ldots, x_n, \alpha)$ is the same as the maximum of $\ln P(x_1, \ldots, x_n, \alpha)$, thus the function α^* has to verify the equation of maximum likelihood

$$\sum_{i=1}^{n} \frac{\partial \ln f(x_i, \alpha)}{\partial \alpha} = 0. \tag{6.119}$$

If an efficient estimation function $\bar{\alpha}$ exists then from definition 3 and proposition 1, we must have

$$\ln f(x, \alpha) = A'(\alpha)\,[L(x) - \alpha] + A(\alpha) + N(x),$$

$$\bar{\alpha} = \sum_{i=1}^{n} \frac{L(x_i)}{n}.$$

With this value of $\ln f(x, \alpha)$ from formula (6.118) we obtain

$$A''(\alpha^*) \sum_{i=1}^{n} [L(x_i) - \alpha^*] = 0$$

that is

$$A''(\alpha^*) \left[\frac{\sum_{i=1}^{n} L(x_i)}{n} - \alpha^* \right] = 0,$$

whence it follows that the only value of α^* which depends on $x_1, \ldots, x_n$ is

$$\alpha^* = \frac{\sum_{i=1}^{n} L(x_i)}{n}$$

that is $\alpha^* = \bar{\alpha}$.

6.8 EXERCISES

1. An urn contains M black balls and $N - M$ white balls. Drawings are effected without replacing the ball in the urn.

(i) what is the probability for a black ball to be obtained for the first time in the drawing of rank k;

(ii) what is the probability that in n drawings k black balls and $n - k$ white balls should be obtained.

2. r cards are distributed at random to n persons. What is the probability that m persons should not take any card $(0 < m < n)$?

3. An automatic machine produces two pieces a minute, and gives 40 bad pieces in the production of a shift. The process follows Poisson's law. Determine the probability that from five pieces at least two should be bad.

4. The following function is given:

$$f(x) = ax^2e^{-kx} \quad (k > 0,\ 0 \leqslant x < \infty)$$

(i) Determine a so that $f(x)$ be a density of distribution.

(ii) Calculate the corresponding distribution function.

(iii) Determine the probability that the random variable ξ should be contained in the interval $\left(0, \frac{1}{k}\right)$.

7

Elements of Information Theory

7.1 ENTROPY

7.1.1 Definition and main characteristics

Let Ω_n be a finite probability space whose elementary events are $\omega_1, \omega_2, \ldots, \omega_n$ with the corresponding probabilities $p_1, \ldots, p_n$ where

$$p_k \geqslant 0, \; (k = 1, \ldots, n), \; \sum_{k=1}^{n} p_k = 1. \tag{7.1}$$

Definition 1. (*C. E. Shannon.*) *The following expression is called the entropy of the finite probability space* Ω_n

$$H_n(\Omega_n) = H_n(p_1, \ldots, p_n) = -\sum_{k=1}^{n} p_k \log_2 p_k . \tag{7.2}$$

Let us now prove some properties of the entropy.

Proposition 1. *We have*

$$H_n(p_1, \ldots, p_n) \geqslant 0.$$

Indeed, since $p_k \leqslant 1$, we have $\log_2 p_k \leqslant 0$ and hence $H_n(\Omega_n) \geqslant 0$.

Proposition 2. *If for an index i we have $p_i = 1$, then*

$$H_n(p_1, \ldots, p_n) = 0.$$

Since $p_i = 1$ from formula (7.1) we have $p_k = 0$, $(k \neq 1)$. On the other hand we have

$$\lim_{p_k \to 0} p_k \log_2 p_k = 0,$$

hence from formula (7.2) we have

$$H_n(p_1, \ldots, p_n) = 0.$$

We shall now prove a relationship which will be useful for the proof of other properties of the entropy.

Proposition 3. *(J. L. Jensen's inequality.) If $y = f(x)$ is a convex function, defined on the interval $[a, b]$, we have the inequality*

$$\sum_{k=1}^{n} \lambda_k f(x_k) < f\left(\sum_{k=1}^{n} \lambda_k x_k\right) \tag{7.3}$$

where $x_1, \ldots, x_n$ are n arbitrary values from $[a, b]$ and $\lambda_1, \ldots, \lambda_n$ are non-negative numbers whose sum is 1.

The function $f(x)$ is convex on $[a, b]$ if for arbitrary points P, Q on its graph the arc PQ of the graph is situated above the chord PQ. Hence if we write $y = f(x)$, the points $M_k(x_k, y_k)$, $(k = 1, \ldots, n)$ are the vertices of a convex polygon inscribed in the graph of the function $y = f(x)$.

Let Q_2 be the point of the side $M_1 M_2$ so that

$$\frac{\overline{M_1 Q_2}}{\overline{Q_2 M_2}} = \frac{\lambda_2}{\lambda_1 + \lambda_2} : \frac{\lambda_1}{\lambda_1 + \lambda_2},$$

Q_3 the point on $M_3 Q_2$ so that

$$\frac{\overline{M_3 Q_3}}{\overline{Q_3 Q_2}} = \frac{\lambda_3}{\lambda_1 + \lambda_2 + \lambda_3} : \frac{\lambda_1 + \lambda_2}{\lambda_1 + \lambda_2 + \lambda_3}$$

....

Q_n the point on $M_n Q_{n-1}$ so that

$$\frac{\overline{M_n Q_n}}{\overline{Q_n Q_{n-1}}} = \frac{\lambda_n}{\lambda_1 + \lambda_2 + \ldots + \lambda_n} : \frac{\lambda_1 + \ldots + \lambda_{n-1}}{\lambda_1 + \ldots + \lambda_n}.$$

Hence the points $Q_2, Q_3, \ldots, Q_n$ have the co-ordinates

$$Q_2\left(\frac{\lambda_1 x_1 + \lambda_2 x_2}{\lambda_1 + \lambda_2}, \frac{\lambda_1 f(x_1) + \lambda_2 f(x_2)}{\lambda_1 + \lambda_2}\right),$$

$$Q_3\left(\frac{\lambda_1 x_1 + \lambda_2 x_2 + \lambda_3 x_3}{\lambda_1 + \lambda_2 + \lambda_3}, \frac{\lambda_1 f(x_1) + \lambda_2 f(x_2) + \lambda_3 f(x_3)}{\lambda_1 + \lambda_2 + \lambda_3}\right),$$

....

$$Q_n(\lambda_1 x_1 + \lambda_2 x_2 + \ldots + \lambda_n x_n, \lambda_1 f(x_1) + \lambda_2 f(x_2) + \ldots + \lambda_n f(x_n)).$$

Let us now denote by P and R the points where the line parallel to the axis Oy through the point Q_n intersects the graph of the function $y = f(x)$ and the axis Ox respectively. We have

$$\begin{aligned} RQ_n &= \lambda_1 f(x_1) + \lambda_2 f(x_2) + \ldots + \lambda_n f(x_n), \\ OR &= \lambda_1 x_1 + \lambda_2 x_2 + \ldots + \lambda_n x_n, \\ RP &= f(\lambda_1 x_1 + \lambda_2 x_2 + \ldots + \lambda_n x_n). \end{aligned}$$

Since Q_n is the interior of the convex polygon $M_1 M_2 \ldots M_n$, it follows that $\overline{RQ_n} < \overline{RP}$, that is

$$\sum_{k=1}^{n} \lambda_k f(x_k) < f\left(\sum_{k=1}^{n} \lambda_k x_k\right).$$

Proposition 4. *We have*

$$H_n(p_1, \ldots, p_n) \leqslant H_n\left(\frac{1}{n}, \ldots, \frac{1}{n}\right).$$

In order to prove this inequality we shall apply relationship (7.3) for $x_k = p_k$, $\lambda_k = \frac{1}{n}$, $(k = 1, \ldots, n)$, $f(x) = -x \log_2 x$. Thus we obtain

$$-\sum_{k=1}^{n} \frac{1}{n} p_k \log_2 p_k \leqslant -\left(\sum_{k=1}^{n} \frac{1}{n} p_k\right) \log_2 \left(\sum_{k=1}^{n} \frac{1}{n} p_k\right).$$

Since $\sum_{k=1}^{n} p_k = 1$, we have

$$\frac{1}{n} H_n(p_1, \ldots, p_n) \leqslant -\frac{1}{n} \log_2 \frac{1}{n} = \frac{1}{n} \log_2 n$$

or

$$H_n(p_1, \ldots, p_n) \leqslant \log_2 n.$$

But

$$H_n\left(\frac{1}{n}, \ldots, \frac{1}{n}\right) = -\sum_{k=1}^{n} \frac{1}{n} \log_2 \frac{1}{n} = \log_2 n,$$

hence

$$H_n(p_1, \ldots, p_n) \leqslant H_n\left(\frac{1}{n}, \ldots, \frac{1}{n}\right).$$

Proposition 5. *We have*

$$H_{n+1}(p_1, \ldots, p_n, 0) = H_n(p_1, \ldots, p_n).$$

Indeed,

$$H_{n+1}(p_1, \ldots, p_n, 0) = -\sum_{k=1}^{n} p_k \log_2 p_k - \lim_{p_{n+1} \to 0} p_{n+1} \log_2 p_{n+1} =$$

$$= -\sum_{k=1}^{n} p_k \log_2 p_k = H_n(p_1, \ldots, p_n).$$

Let us now consider two finite mutually independent probability spaces $\mathfrak{A}_n = \{A_1, \ldots, A_n\}$, $\mathfrak{B}_m = \{B_1, \ldots, B_m\}$. Writing $p_k = P(A_k)$, $(k = 1, \ldots, n)$ and $q_l = P(B_l)$, $(l = 1, \ldots, m)$, we have

$$\sum_{k=1}^{n} p_k = 1, \quad \sum_{l=1}^{m} q_l = 1. \tag{7.4}$$

Since the two spaces $\mathfrak{A}_n$ and $\mathfrak{B}_m$ are mutually independent we have

$$\pi_{kl} = P(A_k \cap B_l) = P(A_k) P(B_l) = p_k q_l. \tag{7.5}$$

The events $A_k \cap B_l$, $(k = 1, \ldots, n;\ l = 1, \ldots, n)$, form a new finite probability space denoted by $\mathfrak{A}_n \times \mathfrak{B}_m$.

We shall prove with respect to this field the following proposition.

Proposition 6. *If $\mathfrak{A}_n$ and $\mathfrak{B}_m$ are two independent finite probability spaces we have*

$$H_{nm}(\mathfrak{A}_n \times B_m) = H_n(\mathfrak{A}_n) + H_m(\mathfrak{B}_m).$$

Indeed taking formulae (7.4) and (7.5) into account in formula (7.2) we have

$$H_{nm}(\mathcal{A}_n \times \mathcal{B}_m) = -\sum_{k=1}^{n}\sum_{l=1}^{m} \pi_{kl}\log_2 \pi_{kl} = -\sum_{k=1}^{n}\sum_{l=1}^{m} p_k q_l(\log_2 p_k + \log_2 q_l) =$$

$$= -\sum_{l=1}^{m} q_l \sum_{k=1}^{n} p_k \log_2 p_k - \sum_{k=1}^{n} p_l \sum_{l=1}^{m} q_l \log_2 q_l =$$

$$= -\sum_{k=1}^{n} p_k \log_2 p_k - \sum_{l=1}^{m} q_l \log_2 q_l = H_n(\mathcal{A}_n) + H_m(\mathcal{B}_m).$$

Let us now consider the finite mutually dependent probability spaces $\mathcal{A}_n$ and $\mathcal{B}_m$ and denote by q_{kl} the probability that in the space $\mathcal{B}_n$ the elementary event B_l should occur provided that the elementary event A_k occurs in the space $\mathcal{A}_n$. Writing $\pi_{kl} = P(A_k \cap B_l)$ we have

$$\pi_{kl} = P(A_k \cap B_l) = P(A_k)\, P_{A_k}(B_l) = p_k q_{kl}. \tag{7.6}$$

Definition 2. We call the following expression the entropy of the space $\mathcal{B}_m$ *conditioned by* $A_k \in \mathcal{A}_n$

$$H_m(\mathcal{B}_m/A_k) = -\sum_{l=1}^{m} q_{kl}\log_2 q_{kl} \tag{7.7}$$

and the following expression the entropy of the space $\mathcal{B}_m$ *conditioned by the space* $\mathcal{A}_n$

$$H_m(\mathcal{B}_m/\mathcal{A}_n) = \sum_{k=1}^{n} p_k H_m(\mathcal{B}_m/A_k). \tag{7.8}$$

With this notation let us prove the following proposition.

Proposition 7. *If* $\mathcal{A}_n$ *and* $\mathcal{B}_m$ *are two arbitrary finite spaces of probability, we have*

$$H_{nm}(\mathcal{A}_n \times \mathcal{B}_m) = H_n(\mathcal{A}_n) + H_m(\mathcal{B}_m/\mathcal{A}_n). \tag{7.9}$$

Indeed, taking into account formulae (7.6), (7.7) and (7.8), we deduce that

$$H_{nm}(\mathcal{A}_n \times \mathcal{B}_m) = -\sum_{k=1}^{n}\sum_{l=1}^{m} \pi_{kl}\log_2 \pi_{kl} = -\sum_{k=1}^{n}\sum_{l=1}^{m} p_k q_{kl}(\log_2 p_k + \log_2 q_k) =$$

$$= -\sum_{k=1}^{n} p_k \log_2 p_k \sum_{l=1}^{m} q_{kl} - \sum_{k=1}^{n} p_k \sum_{l=1}^{m} q_{kl}\log_2 q_{kl} =$$

$$= -\sum_{k=1}^{n} p_k \log_2 p_k + \sum_{k=1}^{n} p_k H_m(\mathcal{B}_m/A_k) = H_n(\mathcal{A}_n) + H_m(\mathcal{B}_m/\mathcal{A}_n).$$

Proposition 8. *If $\mathcal{A}_n$ and $\mathcal{B}_m$ are two arbitrary finite probability spaces, we have*

$$H_m(\mathcal{B}_m/\mathcal{A}_n) \leqslant H_m(\mathcal{B}_m). \tag{7.10}$$

Writing Jensen's inequality (7.3) for $f(x) = -x \log_2 x$, $x_k = q_{kl}$, $\lambda_k = p_k$, $(k = 1, \ldots, n)$, we obtain

$$\sum_{k=1}^{n} p_k q_{kl} \log_2 q_{kl} \geqslant \left(\sum_{k=1}^{n} p_k q_{kl}\right) \log_2 \left(\sum_{k=1}^{n} p_k q_{kl}\right) = q_l \log_2 q_l .$$

Hence

$$\sum_{l=1}^{m} \left(\sum_{k=1}^{n} p_{kl}\, q_{kl} \log_2 q_{kl}\right) \geqslant \sum_{l=1}^{m} q_l \log_2 q_l . \tag{7.11}$$

But

$$\sum_{l=1}^{m} \sum_{k=1}^{n} p_k\, q_{kl} \log_2 q_{kl} = \sum_{k=1}^{n} p_k \sum_{l=1}^{m} q_{kl} \log_2 q_{kl} =$$

$$= - \sum_{k=1}^{n} p_k H_m(\mathcal{B}_m/\mathcal{A}_k) = - H_m(\mathcal{B}_m/\mathcal{A}_n)$$

and

$$\sum_{l=1}^{n} q_l \log_2 q_l = - H_m(\mathcal{B}_m)$$

which substituted in formula (7.9) yields

$$- H_m(\mathcal{B}_m/\mathcal{A}_n) \geqslant - H_m(\mathcal{B}_n)$$

that is

$$H_m(\mathcal{B}_m/\mathcal{A}_n) \leqslant H_m(\mathcal{B}_m).$$

Proposition 9. *If $\mathcal{A}_n$ and $\mathcal{B}_m$ are two arbitrary finite probability spaces, then*

$$H_{nm}(\mathcal{A}_n \times \mathcal{B}_m) \leqslant H_n(\mathcal{A}_n) + H_m(\mathcal{B}_n).$$

This property follows immediately from formula (7.9) if we take into account inequality (7.10).

Proposition 10. *If $\mathcal{A}_n$ and $\mathcal{B}_m$ are two arbitrary finite probability spaces, then*

$$H_{nm}(\mathcal{A}_n/\mathcal{B}_m) - H_{nm}(\mathcal{B}_m/\mathcal{A}_n) = H_n(\mathcal{A}_n) - H_m(\mathcal{B}_m). \tag{7.12}$$

From proposition 7 we have

$$H_{nm}(\mathcal{A}_n \times \mathcal{B}_m) = H_m(\mathcal{B}_m) + H_n(\mathcal{A}_n/\mathcal{B}_m).$$

Whence taking into account formula (7.9) we deduce formula (7.12).

Proposition 11. (*Hinchin's theorem.*) *Let $H_1(1), H_2(p_1, p_2), \ldots, H_n(p_1, \ldots, p_n), \ldots$ be a sequence of functions associated to the finite probability spaces consisting of one, two, ..., n, ... elementary events. We suppose that these functions satisfy the following conditions:*

(i) *$H_n(p_1, \ldots, p_n)$ is, for every n, a non-negative function defined and continuous on the domain $p_k \geqslant 0$, $(k = 1, \ldots, n)$,* and

$$\sum_{k=1}^{n} p_k = 1.$$

(ii) *$H_n(p_1, \ldots, p_n)$ is a symmetric function of n arguments.*

(iii) *$H_{n+1}(p_1, \ldots, p_n, 0) = H_n(p_1, \ldots, p_n)$.*

(iv) *$H(p_1, \ldots, p_n) \leqslant H_n\left(\frac{1}{n}, \ldots, \frac{1}{n}\right)$.*

(v) *Writing $p_k = \sum_{l=1}^{m} \pi_{kl}$, $(k = 1, \ldots, n)$, where $\pi_{kl} \geqslant 0$,*

$$(k = 1, \ldots, n; \quad l = 1, \ldots, m) \text{ and } \sum_{k=1}^{n} \sum_{l=1}^{m} \pi_{kl} = 1,$$

we have

$$H_{nm}(\pi_{11}, \pi_{12}, \ldots, \pi_{nm}) = H_n(p_1, \ldots, p_n) + \sum_{k=1}^{n} p_k H_m\left(\frac{\pi_{k1}}{p_k}, \ldots, \frac{\pi_{km}}{p_k}\right).$$

Then

$$H_n(p_1, \ldots, p_n) = -\lambda \sum_{k=1}^{n} p_k \log_2 p_k, \tag{7.13}$$

where λ is a positive constant.

Let us write

$$H_n\left(\frac{1}{n}, \ldots, \frac{1}{n}\right) = L(n).$$

From conditions (iii) and (iv) we deduce that, for every natural number n, we have

$$L(n) = H_{n+1}\left(\frac{1}{n}, \ldots, \frac{1}{n}, 0\right) \leqslant H_{n+1}\left(\frac{1}{n+1}, \ldots, \frac{1}{n+1}\right) = L(n+1),$$

that is

$$L(n) \leqslant L(n+1).$$

Hence by a recurrent procedure, it follows that if $a < b$ are two natural numbers, we have

$$L(a) \leqslant L(b). \tag{7.14}$$

We shall prove now by a complete induction that, if r and s are two natural numbers, then

$$L(r^s) = sL(r). \tag{7.15}$$

Indeed, applying condition (v) for

$$n = r, \quad m = r, \quad p_k = \frac{1}{r}, \qquad (k = 1, \ldots, r),\ \pi_{kl} = \frac{1}{r^2}, \quad (k,\ l = 1, \ldots, r),$$

we obtain

$$L(r^2) = 2L(r)$$

which is relationship (7.15) for $s = 2$.

We suppose now that equality (7.15) is true for $s - 1$ and we shall show that it is true also for s. Indeed applying condition (v) for

$$n = r,\ m = r^{s-1},\ p_k = \frac{1}{r}, \qquad (k = 1, \ldots, r),\ \pi_{kl} = \frac{1}{r^s},$$
$$(k = 1, \ldots, r; \quad l = 1, \ldots, r^{s-1}),$$

we obtain

$$L(r^s) = L(r) + L(r^{s-1}).$$

But $L(r^{s-1}) = (s - 1)\ L(r)$, hence

$$L(r^s) = sL(r).$$

Considering another pair of natural numbers t and u we have

$$L(t^u) = uL(t). \tag{7.15'}$$

Let r, t, u be three natural numbers and let us denote by n the non-negative integer for which

$$r^n \leqslant t^u \leqslant r^{n+1}. \tag{7.16}$$

Taking the logarithm of this inequality we deduce that

$$n \log_2 r \leqslant u \log_2 t < (n+1) \log_2 r,$$

hence

$$\frac{n}{u} \leqslant \frac{\log_2 t}{\log_2 r} < \frac{n}{u} + \frac{1}{u}\cdot \tag{7.17}$$

Taking into account relationships (7.16), formula (7.14) yields

$$L(r^n) \leqslant L(t^u) \leqslant L(r^{n+1})$$

or, from formula (7.15),

$$nL(r) \leqslant uL(t) \leqslant (n+1)\ L(r),$$

that is

$$\frac{n}{u} \leqslant \frac{L(t)}{L(r)} \leqslant \frac{n}{u} + \frac{1}{u}\cdot \tag{7.18}$$

From formulae (7.17) and (7.18) we obtain

$$\left| \frac{L(t)}{L(r)} - \frac{\log_2 t}{\log_2 r} \right| \leqslant \frac{1}{u}\cdot$$

But the left-hand side of this inequality is independent of u, and the second right-hand side tends to 0 as u tends to infinity, hence

$$\frac{L(t)}{L(r)} - \frac{\log_2 t}{\log_2 r} = 0,$$

or

$$\frac{L(t)}{\log_2 t} = \frac{L(r)}{\log_2 r}\cdot$$

Since r and t are arbitrary natural numbers, if we denote by λ the common value of the ratios, we deduce that

$$L(n) = \lambda \log_2 n. \tag{7.19}$$

Since for $p_k = \frac{1}{n}$, $(k = 1, \ldots, n)$, we have

$$\sum_{k=1}^{n} p_k \log_2 p_k = \sum_{k=1}^{n} \frac{1}{n} \log_2 \frac{1}{n} = -\log_2 n,$$

it follows that formula (7.19) proves formula (7.13) in the particular case $p_k = \frac{1}{n}$, $(k = 1, \ldots, n)$.

Let us suppose now that

$$p_k = \frac{m_k}{m}, \quad (k = 1, \ldots, n)$$

where m_k and m are natural numbers. Hence, taking into account that $\sum_{k=1}^{n} p_k = 1$, we have

$$\sum_{k=1}^{n} m_k = m.$$

Let us now define, for $k = 1, \ldots, n$ and $l = 1, \ldots, m$, the probabilities

$$q_{kl} = \begin{cases} 0 \text{ for } l = 1, 2, \ldots, \sum_{s=1}^{k-1} m_s, \\ \frac{1}{m_k} \text{ for } l = \sum_{s=1}^{k-1} m_s + 1, \sum_{s=1}^{k-1} m_s + 2, \ldots, \sum_{s=1}^{k} m_s, \\ 0, \text{ for } l = \sum_{s=1}^{k} m_s + 1, \sum_{s=1}^{k} m_s + 2, \ldots, m. \end{cases}$$

Hence we deduce that

$$\pi_{kl} = p_k\, q_{kl} = 0 \text{ or } \frac{1}{m}, \quad (k = 1, \ldots, n;\ l = 1, \ldots, m).$$

In this case we obtain from conditions (ii) and (iii)

$$H_m\left(\frac{\pi_{kl}}{p_k}, \ldots, \frac{\pi_{km}}{p_k}\right) = H_m(q_{k_1}, \ldots, q_{k_m}) = H_{m_k}\left(\frac{1}{m_k}, \ldots, \frac{1}{m_k}\right) =$$
$$= L(m_k), \quad (k = 1, \ldots, n)$$

and

$$H_{nm}(\pi_{11}, \pi_{12}, \ldots, \pi_{nm}) = H_m\left(\frac{1}{m}, \ldots, \frac{1}{m}\right) = L(m).$$

Taking into account these relationships from condition (v) it follows that:

$$L(m) = H(p_1, \ldots, p_n) + \sum_{k=1}^{n} p_k\, L(m_k). \tag{7.20}$$

But m_k and m are natural numbers, hence from formula (7.19) we deduce that

$$L(m) = \lambda \log_2 m, \; L(m_k) = \lambda \log_2 m_k = \lambda \log_2 (mp_k) =$$
$$= \lambda \log_2 p_k + \lambda \log_2 m.$$

With these values from formula (7.20) we obtain

$$H_n(p_1, \ldots, p_n) = -\lambda \sum_{k=1}^{n} p_k \log_2 p_k .$$

Thus we proved formula (7.13) for the case when p_k, $(k = 1, \ldots, n)$ are arbitrary non-negative rational numbers whose sum is 1. Since every irrational number is the limit of a sequence of rational numbers it follows that, by a passage, to the limit, formula (7.13) is also true in the case when p_k, $(k = 1, \ldots, n)$ are arbitrary non-negative numbers whose sum is equal to 1.

Taking

$$H\left(\frac{1}{2}, \frac{1}{2}\right) = 1,$$

which amounts to choosing as the unit of measure for the degree of non-determination, the non-determination which is contained in a finite probability space consisting of two equally probable events, formula (7.13) becomes identical to formula (7.2).

7.2 THE TRANSMISSION OF INFORMATION

7.2.1 The transmission of information without coding

A system for the transmission of the information can be represented by the general scheme

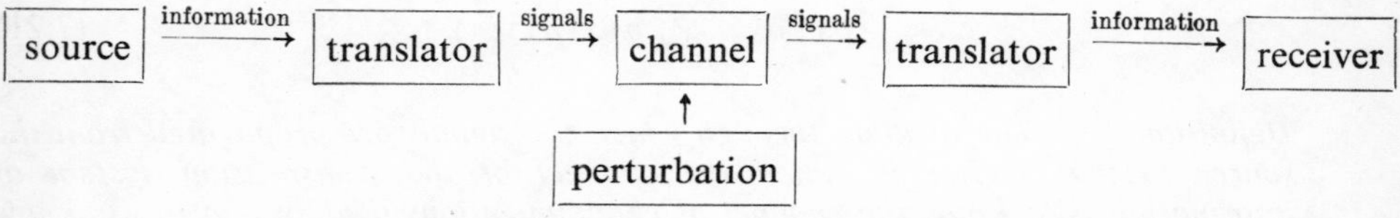

Figure 7.1 The general scheme of a transmission system of information.

The information sent by means of an input signal set is transformed by coding in other signals which are easier to be transmitted. The source is transformed through codification into other signals which can be transmitted more easily. These new signals are propagated through a channel of communication, then decoded and finally received. Various perturbations can occur in the communication channel.

We shall first study a simple case, when there is no coding and hence we have the following scheme

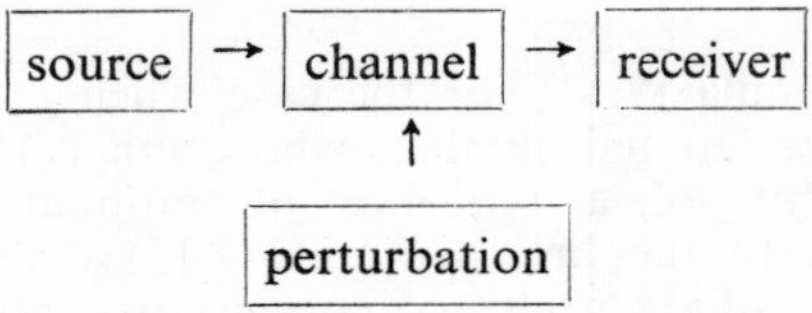

Figure 7.2 The scheme of a transmission system of information without coding.

Definition 1. *We call a transmission system of information the system consisting of two finite sets X, Y and a conditioned probability $p(y/x)$, defined on Y for every $x \in X$, and we denote it by $[X, p(y/x), Y]$.*

The set X is called the input signal, the set Y is called the output signal set and $p(y/x)$ is the probability that $y \in Y$ should be received if $x \in X$ is sent.

Definition 2. *Given the probability $p(x)$ for every $x \in X$, called the input probability, so that $\sum_{x \in X} p(x) = 1$, the probability space $\{X, x, p(x)\}$ is called the source of the transmission system of information $[X, p(y/x), Y]$.*

Definition 3. *Given the transmission system of information $[X, p(y/x), Y]$ and the probability $p(x)$ defined on X, we call the receiver of the system the probability space $\{Y, y, p(y)\}$ where the probability $p(y)$ of receiving the signals y is given by relationship*

$$p(y) = \sum_{x \in X} p(x)\, p(y/x). \tag{7.21}$$

Definition 4. *The medium through which the signals are propagated from the source to the receiver is called the channel of the transmission system of information. We know the channel of communications of a system if we know the probabilities $p(y/x)$ for all the signals $x \in X$, $y \in Y$. If $p(y/x)$ takes only* *the values* 0 *or* 1 *for every* $x \in X$, $y \in Y$, *the channel has no perturbation, otherwise the channel has perturbations.*

Definition 5. *Let us consider the transmission system of information* $[X, p(y/x), Y]$ *with the source* $\{X, x, p(x)\}$ *and the receiver* $\{Y, y, p(y)\}$. *The expressions* $H(X) = -\sum_{x \in X} p(x) \log_2 p(x)$, $H(Y) = -\sum_{y \in Y} p(y) \log_2 p(y)$ *are called the entropies corresponding to these spaces.*

Denoting by $p(x/y)$ the probability that the signal x be sent if the signal y has been received, the expression

$$H(X/y) = -\sum_{x \in X} p(x/y) \log_2 p(x/y) \tag{7.22}$$

represents the quantity of information which must be sent in order that the signal y should be received.

The average quantity of information necessary for the whole set of signals Y to be received is

$$H(X/Y) = \sum_{y \in Y} p(y) \, H(X/y).$$

Substituting here expression (7.22) we obtain

$$H(X/Y) = -\sum_{y \in Y} p(y) \sum_{x \in X} p(x/y) \log_2 p(x/y) =$$

$$= -\sum_{x \in X} \sum_{y \in Y} p(y) \, p(x/y) \log_2 p(x/y). \tag{7.23}$$

Denoting by $p(x, y)$ the probability that at the source the signal x be sent and at the receiver the signal y be received, we have

$$p(x, y) = p(y) \, p(x/y) = p(x) \, p(y/x). \tag{7.24}$$

Thus relation (7.24) becomes

$$H(X/Y) = -\sum_{x \in X} \sum_{y \in Y} p(x, y) \log_2 p(x/y). \tag{7.25}$$

The non-determination $H(X/Y)$ appears owing to the perturbations on the perturbations on the communication channel, thus it represents the average quantity of information which is lost on the channel, due to the perturbation, during the transmission. Hence it follows that if a quantity of information $H(X)$ is sent from the source, only the information

$$H(X) - H(X/Y)$$

reaches the receiver, and is also called the speed of the transmission of information. Taking into account formula (7.10), it follows that

$$0 \leqslant H(X) - H(X/Y) \leqslant H(X). \tag{7.26}$$

Given a transmission system $[X, p(y/x), Y]$, the probability $p(x)$, given for each $x \in X$, determines the entropies $H(X)$ and $H(X/Y)$. By modifying the probability $p(x)$ we modify the values of $H(X)$ and $H(X/Y)$ as well:

Definition 6. *We call the following value*

$$C = \max_{p(x)} [H(X) - H(X/Y)],$$

the capacity of a channel, when $p(x)$, $(x \in X)$ takes all the possible values.

With these definitions and notations figure 7.2 becomes

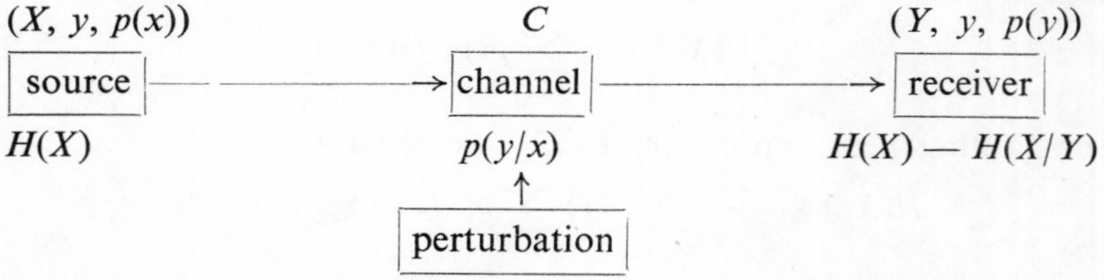

Figure 7.3 Mathematical model of the scheme of a transmission system of information without coding.

In the sequel, we shall represent a transmission system of information without coding in the form

$$\{X, x, p(x)\} \xrightarrow{p(y/x)} \{Y, y, p(y)\}.$$

7.2.2 Coding

In the communication systems we generally use the coding of the signals of the input set, that is their transformation into other signals easier to be transmitted.

For example the Morse code is used for transmission by means of the telegraph.

Let X be a set of signals x and A a set of signals a. Coding the given information by means of the signals $a \in A$ means realizing a correspondence between the sets X and A by indicating a sequence of signals from A which correspond to each signal $x \in X$. The signals $x \in X$ are called initial signals and the signals $a \in A$ are called simple singals. For instance in the Morse code the simple signals are line, point and empty space.

Let us consider a transmission system of the information $[X, p(y/x), Y]$. When we send successively a sequence of n signals x through the channel, we shall receive a sequence of n signals y. We shall denote by U the set of all the sequences of length n consisting of the signals $x \in X$ and by V the set of all the sequences of length n consisting of the signals $y \in Y$. We denote by u the elements of the set U and by v the elements of the set V that is

$$u = (x_1, x_2, \ldots, x_n), \qquad (x_k \in X, k = 1, \ldots, n),$$

$$v = (y_1, y_2, \ldots, y_n), \qquad (y_k \in Y, k = 1, \ldots, n).$$

Let $p(v/u)$ be the probability defined on the set V conditioned by the elements $u \in U$ given by the relationship

$$p(v/u) = p(y_1/x_1)\, p(y_2/x_2) \ldots p(y_n/x_n) \tag{7.27}$$

that is the probability of receiving the signal $v = (y_1, \ldots, y_n)$ if the signal $u = (x_1, \ldots, x_n)$ has been sent. This probability has been determined completely by the probability $p(y/x)$.

Definition 7. Given the transmission system of information $[X, p(y/x), Y]$, *the system* $[U, p(v/u), V]$ *is called the extension of length n.*

The passage from the system $[X, p(y/x), Y]$ to the system $[U, p(v/u), V]$ is called coding and the rule according to which sequences of signals $u \in U$ are associated to the signals $x \in X$ is called the code.

Let us prove the following proposition.

Proposition 1. *Let* $[X, p(y/x), Y]$ *be a transmission system of information, whose channel has the capacity C. Its extension of length n has the capacity nC.*

We shall first prove this property in the particular case when the probability of a sequence of length n of signals $x \in X$ is equal to the product of the probabilities of the signals belonging to the sequence. If we denote by $p_k(x)$, $(k = 1, \ldots, n)$, n probabilities admissible on X, that is

$$p_k(x) \geqslant 0, \ \sum_{x \in X} p_k(x) = 1, \ (k = 1, \ldots, n),$$

the assumption we made is written as

$$p(u) = p_1(x_1)\, p_2(x_2) \dots p_n(x_n). \tag{7.28}$$

Hence

$$H(U) = -\sum_{u \in U} p(u) \log_2 p(u) =$$

$$= -\sum_{x_1 \dots x_n \in X} p_1(x_1) \dots p_n(x_n) \log_2 [p_1(x_1) \dots p_n(x_n)] =$$

$$= -\sum_{x_1 \dots x_n \in X} p_1(x_1) \dots p_n(x_n) [\log_2 p_1(x_1) + \dots + \log_2 p_n(x_n)] =$$

$$= -\sum_{x_1 \in X} p_1(x_1) \log_2 p_1(x_1) \sum_{x_2 \in X} p_2(x_2) \dots \sum_{x_n \in X} p_n(x_n) -$$

$$- \sum_{x_1 \in X} p_1(x_1) \sum_{x_2 \in X} p_2(x_2) \log_2 p_2(x_2) \dots \sum_{x_n \in X} p_n(x_n) - \dots$$

$$\dots - \sum_{x_1 \in X} p_1(x_1) \dots \sum_{x_{n-1} \in X} p_{n-1}(x_{n-1}) \sum_{x_n \in X} p_n(x_n) \log_2 p_n(x_n).$$

But

$$\sum_{x_k \in X} p_k(x_k) = 1, \; (k = 1, \dots, n)$$

and

$$-\sum_{x_k \in X} p_k(x_k) \log_2 p_k(x_k) = H_k(X), \; (k = 1, \dots, n),$$

where $H_k(X)$ is the entropy of X corresponding to the probability p_k thus

$$H(U) = H_1(X) + H_2(X) + \dots + H_n(X) = \sum_{k=1}^{n} H_k(X). \tag{7.29}$$

Taking formula (7.25) into account we have

$$H(U/V) = -\sum_{u \in U} \sum_{v \in V} p(u, v) \log_2 p(u/v). \tag{7.30}$$

From formulae (7.24), (7.27) and (7.28) we deduce that

$$p(u, v) = p(u)\, p(v/u) = p_1(x_1) \dots p_n(x_n)\, p(y_1/x_1) \dots p(y_n/x_n) =$$

$$= p_1(x_1)\, p(y_1/x_1) \dots p_n(x_n)\, p(y_n/x_n) = p_1(x_1, y_1) \dots p_n(x_n, y_n).$$

Whence taking into account formulae (7.21) and (7.24) we obtain

$$p(v) = \sum_{u \in U} p(u)\, p(v/u) = \sum_{u \in U} p(u, v) = \sum_{x_1, \ldots, x_n \in X} p_1(x_1, y_1) \ldots p_n(x_n, y_n) =$$

$$= \sum_{x_1 \in X} p_1(x_1, y_1) \ldots \sum_{x_n \in X} p_n(x_n, y_n) = p_1(y_1) \ldots p_n(y_n).$$

With this value formula (7.24) yields

$$p(u/v) = \frac{p(u, v)}{p(v)} = \frac{p_1(x_1, y_1) \ldots p_n(x_n, y_n)}{p_1(y_1) \ldots p_n(y_n)} = p_1(x_1/y_1) \ldots p_n(x_n/y_n).$$

Substituting this value in formula (7.30) we have

$$H(U/V) =$$

$$= - \sum_{\substack{x_1, \ldots, x_n \in X \\ y_1, \ldots, y_n \in Y}} p_1(x_1, y_1) \ldots p_n(x_n, y_n) \log_2 [p_1(x_1/y_1) \ldots p_n(x_n/y_n)] =$$

$$= \sum_{\substack{x_1, \ldots, x_n \in X \\ y_1, \ldots, y_n \in Y}} p_1(x_1, y_1) \ldots p_n(x_n, y_n) [\log_2 p_1(x_1/y_1) + \ldots + \log_2 p_n(x_n/y_n)] =$$

$$= \sum_{\substack{x_1 \in X \\ y_1 \in Y}} p_1(x_1, y_1) \log_2 p_1(x_1/y_1) - \ldots - \sum_{\substack{x_n \in X \\ y_n \in Y}} p_n(x_n, y_n) \log_2 p_n(x_n/y_n) =$$

$$= H_1(X/Y) + \ldots + H_n(X/Y) = \sum_{k=1}^{n} H_k(X/Y), \qquad (7.31)$$

where we have written

$$H_k(X/Y) = - \sum_{\substack{x_k \in X \\ y_k \in Y}} p_k(x_k, y_k) \log_2 p_k(x_k/y_k), \quad (k = 1, \ldots, n).$$

According to definition 6, the capacity of the channel of the system $[U, p(v/u), V]$ is

$$\max_{p(u)} [H(U) - H(U/V)].$$

But

$$\max_{p_k(x)} [H_k(X) - H_k(X/Y)] = C,$$

hence, taking formulae (7.29) and (7.31) into account, we obtain

$$\max_{p(u)} [H(U) - H(u/V)] = \max_{p(u)} \left\{ \sum_{k=1}^{n} [H_k(X) - H_k(X/Y)] \right\} =$$
$$= \sum_{k=1}^{n} \left\{ \max_{p_k(x)} [H_k(X) - H_k(X/Y)] \right\} = nC.$$

We shall suppose now that the probability $p(u)$ is not equal to the product of the probabilities of the signals belonging to the sequence u. In this case let us denote by

$$p(u, v) = p((x_1, \ldots, x_n), (y_1, \ldots, y_n))$$

the probability that the sequence of signals $u = (x_1, \ldots x_n)$ should be transmitted and the sequence of signals $v = (y_1, \ldots, y_n)$, should be received. The probability

$$p(X, x_2, \ldots, x_n), (Y, y_2, \ldots, y_n))$$

is the probability that every signal from X should be transmitted and that every signal from Y should be received in the first moment, and that beginning with the second moment the signals $x_2, \ldots x_n$, should be transmitted and $y_2, \ldots, y_n$ should be received. Then we have

$$\sum_{x_1 \in X} \sum_{y_1 \in Y} p((x_1, \ldots, x_n), (y_1, \ldots, y_n)) = p((X, x_2, \ldots x_n), (Y, y_2, \ldots, y_n)).$$

Since this last probability no longer depends on the first moment it follows that

$$p((X, x_2, \ldots, x_n), (Y, y_2, \ldots, y_n)) = p((x_2, \ldots, x_n), (y_2, \ldots, y_n)).$$

By analogy

$$\sum_{x_k \in X} \sum_{y_k \in Y} p((x_1, \ldots, x_{k-1}, x_k, x_{k+1}, \ldots, x_n), (y_1, \ldots y_{k-1}, y_k, y_{k+1}, \ldots, y_n)) =$$
$$= p((x_1, \ldots, x_{k-1}, X, x_{k+1}, \ldots, x_n), (y_1, \ldots, y_{k-1}, Y, y_{k+1}, \ldots, y_n)) =$$
$$= p((x_1, \ldots, x_{k-1}, x_{k+1}, \ldots, x_n), (y_1, \ldots, y_{k-1}, y_{k+1}, \ldots, y_n)).$$

We also have

$$\sum_{\substack{x_1, \ldots, x_{k-1}, x_{k+1}, \ldots, x_n \in X \\ y_1, \ldots, y_{k-1}, y_{k+1}, \ldots, y_n \in Y}} p((x_1, \ldots, x_n), (y_1, \ldots, y_n)) =$$
$$= p((X, \ldots, X, x_k, X, \ldots, X) . (Y, \ldots, Y, y_k, Y, \ldots, Y)) =$$
$$= p_k(x_k, y_k). \quad (7.32)$$

Taking into account the relationships (7.12), (7.25), (7.27) and (7.32), we deduce that

$$H(U) - H(U/V) = H(V) - H(V/U) = H(V) + \sum_{\substack{u \in U \\ v \in V}} p(u, v) \log_2 (v/u) =$$

$$= -H(V) + \sum_{\substack{x_1, \ldots, x_n \in X \\ y_1, \ldots, y_n \in Y}} p((x_1, \ldots, x_n), (y_1, \ldots, y_n)) [\log_2 (p(y_1/x_1) \ldots p(y_n/x_n))] =$$

$$= H(V) + \sum_{\substack{x_1, \ldots, x_n \in X \\ y_1, \ldots, y_n \in Y}} p((x_1, \ldots, x_n), (y_1, \ldots, y_n)) [\log_2 p(y_1/x_1) + \ldots$$

$$\ldots + \log_2 p(y_n/x_n)] =$$

$$= H(V) + \sum_{\substack{x_1 \in X \\ y_1 \in Y}} \log_2 p(y_1/x_1) \sum_{\substack{x_2, \ldots, x_n \in X \\ y_2, \ldots, y_n \in Y}} p((x_1, \ldots, x_n), (y_1, \ldots, y_n)) + \ldots$$

$$\ldots + \sum_{\substack{x_n \in X \\ x_n \in Y}} \log_2 p(y_n/x_n) + \sum_{\substack{x_1, \ldots, x_{n-1} \in X \\ y_1, \ldots, y_{n-1} \in Y}} p((x_1, \ldots, x_n), (y_1, \ldots, y_n)) =$$

$$= H(V) + \sum_{\substack{x_1 \in X \\ y_1 \in Y}} p_1(x_1, y_1) \log_2 p(y_1/x_1) + \ldots$$

$$\ldots + \sum_{\substack{x_n \in X \\ y_n \in Y}} p_n(x_n, y_n) \log_2 p(y_n/x_n) = H(V) - \sum_{k=1}^{n} H_k(Y/X).$$

We shall now consider n probabilities $p_k(x)$, $(k = 1, \ldots, n)$, defined on X, that is

$$p_k(x) \geqslant 0, \; \sum_{k=1}^{n} p_k(x) = 1, \; (k = 1, \ldots, n)$$

and two probabilities $p(u)$ and $p^*(u)$ defined on U so that $p^*(u)$ is equal to the product of the probabilities of the signals belonging to sequences in U, that is

$$p^*(u) = p_1(x_1) \ldots p_n(x_n). \tag{7.33}$$

To the two probabilities p and p^* defined on U we associate $H(U) = H(U/V)$ and $H^*(U) - H^*(U/V)$ respectively. Taking formula (7.27) into account we have

$$p^*(u, v) = p^*(u) \; p(v/u)$$

or based on relationship (7.33) we obtain

$$p^*(u, v) = p_1(x_1) \dots p_n(x_n)\, p(y_1/x_1) \dots p(y_n/x_n) =$$
$$= p_1(x_1)\, p(y_1/x_1) \dots p_n(x_n)\, p(y_n/x_n) = p_1(x_1, y_1) \dots p_n(x_n, y_n).$$

Whence taking into account formula (7.32) we have

$$p_k^*(x_k, y_k) = p_k(x_k, y_k)$$

and

$$p_k^*(y_k) = p_k(y_k).$$

Thus we deduce that

$$H_k^*(Y/X) = H_k(Y/X)$$

that is

$$\sum_{k=1}^{n} H_k^*(Y/X) = \sum_{k=1}^{n} H_k(Y/X). \tag{7.34}$$

We also obtain, taking into account proposition 9, the following relationships

$$H(V) = H(Y, \dots, Y) \leqslant \sum_{k=1}^{n} H_k(Y) = \sum_{k=1}^{n} H_k^*(Y) = H^*(V). \tag{7.35}$$

From formulae (7.34) and (7.35) we have

$$H^*(V) - H^*(V/U) \geqslant H(V) - H(V/U),$$

hence

$$H^*(U) - H^*(U/V) \geqslant H(U) - H(U/V). \tag{7.36}$$

But

$$\max_{p^*(u)} [H^*(U) - H^*(U/V)] = nC,$$

thus, taking into account formula (7.36), we obtain

$$\max_{p(u)} [H(U) - H(U/V)] = nC.$$

In the sequel we shall prove two coding theorems for the case where perturbation is absent. First we shall prove some preliminary propositions.

Proposition 2. *For the coding to be possible, that is, for a family of N sequences of signals of length* $n_1, n_2, \dots, n_N$ *associated to the initial signals* $x_1, x_2, \dots, x_N$, *to exist it is necessary and sufficient that*

$$\sum_{k=1}^{N} Q^{-n_k} \leqslant 1 \tag{7.37}$$

where Q is the total number of simple signals.

Let us first prove the necessity of condition (7.37). In order to do it we shall denote by r_k the number of sequences of simple signals of length k from the coding we made. The number of the sequences of length l cannot exceed the total number of the simple signals, that is

$$r_1 \leqslant Q.$$

The sequences of length 2 are obtained from the rest of $Q - r$ simple signals which did not enter the r_1 sequences of length 1, hence

$$r_2 \leqslant (Q - r_1)\, Q = Q^2 - r_1 Q.$$

By analogy

$$r_3 \leqslant [(Q^2 - r_1 Q) - r_2]\, Q = Q^3 - r_1 Q^2 - r_2 Q,$$

$$\cdots\cdots\cdots\cdots\cdots\cdots$$

$$r_n \leqslant Q^n - r_1\, Q^{n-1} - \ldots - r_{n-1}\, Q.$$

Dividing the last inequality by Q^n, we obtain

$$\sum_{k=1}^{n} r_k Q^{-k} < 1. \qquad (7.38)$$

By hypothesis we assumed that there is a coding which associates to each initial signal $x_1, x_2, \ldots, x_N$ a sequence of simple signals of length $n_1, n_2, \ldots, n_N$. With this inequality (7.38) becomes identical to inequality (7.37).

Let us now prove the sufficiency of condition (7.37). We suppose that this relationship is satisfied and let r_j be the number of those numbers from $n_1, n_2, \ldots, n_N$, which are equal to j. Denoting by M the largest of the numbers $n_1, n_2, \ldots, n_N$, relationship (7.37) can be written

$$\sum_{j=1}^{M} r_j Q^{-j} < 1$$

hence we deduce that

$$r_1 \leqslant Q,$$

$$r_2 \leqslant Q^2 - r_1 Q = (Q - r_1)\, Q,$$

$$r_3 \leqslant Q^3 - r_1 Q^2 - r_2 Q = [(Q - r_1)\, Q - r_2]\, Q,$$

$$\cdots\cdots\cdots\cdots\cdots\cdots$$

It follows that we can choose r_1 sequences of one simple signal each, r_2 sequences of two simple signals each, $\ldots$, r_k sequences of k simple signals each, $\ldots$, so that none of these sequences coincide with the beginning of another longer sequence.

Proposition 3. *If $c_1, c_2, \ldots, c_n$ are arbitrary numbers and $q_1, q_2, \ldots, q_n$ are non-negative numbers for which $\sum_{k=1}^{n} q_k = 1$, then*

$$c_1^{q_1}\, c_2^{q_2} \ldots c_n^{q_n} \leqslant \sum_{k=1}^{n} q_k c_k .$$

We shall apply Jensen's inequality (7.3) for

$$\lambda_k = q_k , \quad f(c_k) = \log_2 c_k , \quad (k = 1, \ldots, n).$$

We obtain

$$\sum_{k=1}^{n} q_k \log_2 c_k \leqslant \log_2 \left(\sum_{k=1}^{n} q_k c_k \right) \cdot \tag{7.39}$$

But

$$\sum_{k=1}^{n} q_k \log_2 c_k = \sum_{k=1}^{n} \log_2 (c_k^{q_k}) = \log_2 (c_1^{q_1} c_2^{q_2} \ldots c_n^{q_n})$$

which substituted in inequality (7.39) yields

$$\log_2 (c_1^{q_1}\, c_2^{q_2} \ldots c_n^{q_n}) \leqslant \log_2 \left(\sum_{k=1}^{n} q_k c_k \right),$$

whence follows (7.39).

Proposition 4. *If the non-negative numbers p_k and q_k, $(k = 1, \ldots, n)$ satisfy the conditions*

$$\sum_{k=1}^{n} p_k = 1, \quad \sum_{k=1}^{n} q_k = 1, \tag{7.40}$$

then

$$-\sum_{k=1}^{n} q_k \log_2 q_k \leqslant -\sum_{k=1}^{n} q_k \log_2 p_k . \tag{7.41}$$

As the numbers p_k and q_k are non-negative and satisfy conditions (7.40), it follows that $0 \leqslant p_k \leqslant 1$, $0 \leqslant q_k \leqslant 1$, $(k = 1, \ldots, n)$. From proposition 3 we have, setting

$$c_k = \frac{p_k}{q_k}, \ k \leqslant n,$$

$$\left(\frac{p_1}{q_1}\right)^{q_1}\left(\frac{p_2}{q_2}\right)^{q_2}\cdots\left(\frac{p_n}{q_n}\right)^{q_n}\leqslant\sum_{k=1}^{n}\frac{q_k}{d_k}\cdot p_k=1$$

whence taking the logarithm we obtain

$$\sum_{k=1}^{n} q_k \log_2 \frac{p_k}{q_k} \leqslant \log_2 1 = 0$$

and hence relationship (7.41).

Proposition 5. *In order to effectuate a coding using Q simple signals, by means of which one could transmit a quantity $H(X)$ of information, it is necessary that the average length of the sequences of coding, associated to the initial signals from the set X bearing the information, should not be inferior to the number*

$$\frac{H(X)}{\log_2 Q}.$$

Indeed, given the set of initial signals X, consisting of the signals $x_1, x_2, \ldots, x_N$, let us suppose that we realized a coding by means of the simple signals $A = \{a_1, a_2, \ldots a_N\}$ by associating to each signal x_i a sequence of simple signals of length n_i, $(i = 1, \ldots, N)$. We have

$$\sum_{i=1}^{N}\frac{Q^{-n_i}}{\sum_{j=1}^{N} Q^{-n_j}} = \frac{\sum_{i=1}^{N} Q^{-n_i}}{\sum_{j=1}^{N} Q^{-n_j}} = 1.$$

Thus formula (7.41) yields

$$H(X) = -\sum_{i=1}^{N} p(x_i)\log_2 p(x_i) \leqslant -\sum_{i=1}^{N} p(x_i)\log_2 \frac{Q^{-n_i}}{\sum_{j=1}^{N} Q^{-n_j}} =$$

$$= \left[\log_2\left(\sum_{j=1}^{N} Q^{-n_j}\right)\right]\left(\sum_{i=1}^{N} p(x_i)\right) - \sum_{i=1}^{N} p(x_i)\log_2 Q^{-n_i} =$$

$$= \log_2\left(\sum_{j=1}^{N} Q^{-n_j}\right) + \left(\sum_{j=1}^{N} p(x_i)\, n_i\right)\log_2 Q.$$

From proposition 2, we have

$$\sum_{j=1}^{N}{}' Q^{-n_j} \leqslant 1,$$

hence by taking the logarithm we obtain

$$\log_2\left(\sum_{j=1}^{N}{}' Q^{-n_j}\right) \leqslant \log_2 1 = 0.$$

With this value the expression of the entropy $H(X)$ yields

$$H(X) \leqslant \left(\sum_{i=1}^{N} p(x_i)\, n_i\right) \log_2 Q.$$

Hence we have

$$\sum_{i=1}^{N} p(x_i)\, n_i \geqslant \frac{H(X)}{\log_2 Q},$$

which proves the property.

Proposition 6. (*Schannon and Fano's theorem.*) *Let X be a set of initial signals by means of which it is transmitted an information $H(X)$. Let us suppose that the coding of this information, by means of the set $A = \{a_1, a_2, \ldots, a_N\}$ of simple signals is such that we do not associate sequences of simple signals in each initial signal, but directly to the sequences of M initial signals. In this case, the average length of the sequences of M initial signals x divided by M, tends to $\dfrac{H(X)}{\log_2 Q}$, as M tends to infinity.*

Let us denote by $[Z, z, p(z)]$ a source of information consisting of K signals $z_1, z_2, \ldots, z_k$ by means of which the information $H(Z)$ is transmitted. We shall associate to the signal Z_i, $(i = 1, \ldots, K)$ an integer n such that

$$-\frac{\log_2 p(z_i)}{\log_2 Q} \leqslant n_i \leqslant -\frac{\log_2 p(z_i)}{\log_2 Q} + 1. \tag{7.42}$$

Thus we have

$$\sum_{i=1}^{K} Q^{-n_i} \leqslant \sum_{i=1}^{K} Q^{\frac{\log_2 p(z_i)}{\log_2 Q}},$$

or, taking into account that $\log_2 p(z_i) = \log_2 Q \log_a p(z_i)$,

$$\sum_{i=1}^{K} Q^{-n_i} \leqslant \sum_{i=1}^{K} Q^{\log_Q p(z_i)} = \sum_{i=1}^{K} p(z_i) = 1. \tag{7.43}$$

Taking into account proposition 2, inequality (7.43) permits to take the numbers n_i as lengths of the coding sequences for the signals $z_1, z_2, \ldots, z_K$. The average length of these sequences is

$$L = \sum_{i=1}^{K} p(z_i) n_i .$$

From inequalities (7.42) we deduce that

$$-\frac{1}{\log_2 Q} \sum_{i=1}^{K} p(z_i) \log_2 p(z_i) \leqslant \sum_{i=1}^{K} p(z_i) n_i <$$

$$< -\frac{1}{\log_2 Q} \sum_{i=1}^{K} p(z_i) \log_2 p(z) + \sum_{i=1}^{K} p(z_i),$$

that is

$$\frac{H(Z)}{\log_2 Q} \leqslant L \leqslant \frac{H(Z)}{\log_2 Q} + 1. \tag{7.44}$$

Let us now suppose that every signal from Z is a sequence of M signals from X, that is

$$z = (x_1, x_2, \ldots, x_M), \quad (x_i \in X; \ i = 1, 2, \ldots, M).$$

As the signals from X are assumed to be independent we have

$$p(z) = p(x_1)\, p(x_2) \ldots p(x_M).$$

Taking into account proposition 6, Section 7.1.1, we deduce

$$H(Z) = H(\underbrace{X, X, \ldots, X}_{M \text{ times}}) = MH(Z).$$

Denoting by L_M the average length corresponding to M from formula **(7.44)** we obtain

$$\frac{MH(X)}{\log_2 Q} \leqslant L_M \leqslant \frac{MH(X)}{\log_2 Q} + 1$$

or

$$\frac{H(X)}{\log_2 Q} \leqslant \frac{L_M}{M} < \frac{H(X)}{\log_2 Q} + \frac{1}{M},$$

hence

$$\lim_{M\to\infty} \frac{L_M}{M} = \frac{H(X)}{\log_2 Q}.$$

7.3 EXERCISES

1. Let $f(x)$ be a differentiable function on the interval $[0, A]$ with $f(0) = 0$ and $|f'(x)| \leqslant B$. Find an upper bound of the information necessary to determine the value of $f(x)$ for every $x \in [0, A]$ up to a given $\varepsilon > 0$.

2. We have n apparently identical coins. One is false and we know it is heavier than the other. We dispose of a balance with two scales and no marked weights. How many weighings do we need to determine the false coins.

3. Let $\mathfrak{P} = \{p_k\}$ be an arbitrary distribution with

$$\sum_{k=1}^{\infty} kp_k = \lambda > 1.$$

Then

$$\sum_{k=1}^{\infty} p_k \log_2 \frac{1}{p_k} = I_1(\mathfrak{P})$$

is maximum for

$$p_k = \frac{1}{\lambda}\left(1 - \frac{1}{\lambda}\right)^{k-1}.$$

4. Let ξ be a positive random variable with the absolutely continuous distribution and with $E[\xi] = \lambda > 0$. Show that the information of ξ is maximum when the distribution of ξ is exponential.

8

Linear Programming

8.1 DEFINITIONS, FUNDAMENTAL PROPERTIES

8.1.1 Definitions, fundamental properties

The general problem of linear programming is to determine the optimum (minimum or maximum) of a linear function of n variables x_j, $(j = 1, \ldots, n)$ which are linked by linear relationships (equations or inequations) called constraints.

The constraints are of two types:

(i) of the type $x_j \geqslant 0$ (or $x_j \leqslant 0$) imposed on the variables;

(ii) any other type.

We suppose that the inequations have always the same sense and otherwise we multiply those with a different sense by -1.

From the algebraic point of view the general problem of linear programming is

$$\sum_{j=1}^{n} a_{ij}x_j \geqslant d_i, \qquad (i = 1, \ldots, p), \tag{8.1}$$

$$\sum_{j=1}^{n} a_{ij}x_j = d_i, \qquad (i = p + 1, \ldots, m), \tag{8.2}$$

$$x_j \geqslant 0, \qquad (j = 1, \ldots, n), \tag{8.3}$$

$$\min \text{ (or max) } z = \sum_{j=1}^{n} c_j x_j, \tag{8.4}$$

where the coefficients a_{ij}, c_j and d_i are constant.

Every system of values of the variables x_j, $(j = 1, \ldots, n)$ which satisfies the system of constraints (8.1), (8.2) and (8.3) is called program or feasible solution. A finite program, that is a program whose components are all finite, is called an optimal program if it optimizes (minimizes or maximizes) function (8.4) (the objective function).

In the sequel we shall use some notations which will simplify the writing.

$M = \{1, \ldots, m\}$,	— the set of the indices of the constraints themselves;
$N = \{1, \ldots, n\}$,	— the set of the indices of the variables;
$\mathbf{A} = \lVert a_{ij} \rVert$, $i \in M$, $j \in N$,	— matrix of the $(m \times n)$th order;
$\mathbf{a}_j$	— the column of rank j of the matrix $\mathbf{A}$;
$\boldsymbol{\alpha}_i$	— the row of rank i of the matrix $\mathbf{A}$;
$\mathbf{x} = \lVert x_j \rVert$, $j \in N$,	— column vector with n components called the activities level vector;
$\mathbf{c} = \lVert c_j \rVert$, $j \in N$,	— row vector with n components called the costs vector;
$\mathbf{d} = \lVert d_i \rVert$, $i \in M$,	— column vector with m components.

By means of these notations the general problem of the linear programming can be written as

$$\begin{cases} \boldsymbol{\alpha}_i\, \mathbf{x} \geqslant d_i, & i \in M_1 \subset M, \\ \boldsymbol{\alpha}_i\, \mathbf{x} = d_i, & i \in M - M_1, \\ x_j \geqslant 0, & j \in N, \\ \min \text{ (or max) } z = \mathbf{cx}. \end{cases} \tag{8.5}$$

By several simple operations we can write the system (8.5) in other equivalent forms which we shall use in different cases.

Indeed, if we denote by $f(x)$ a linear form of the variables $x_1, \ldots, x_n$, we have

$$\min f(x) = -\max [-f(x)]$$

thus every maximization problem can be considered as a minimization problem and conversely.

Every equation of the form

$$\boldsymbol{\alpha}_i\, \mathbf{x} = d_i$$

can be replaced by two inequalities

$$\boldsymbol{\alpha}_i\, \mathbf{x} \geqslant d_i,$$
$$-\boldsymbol{\alpha}_i\, \mathbf{x} \geqslant -d_i.$$

Every inequality

$$\boldsymbol{\alpha}_i\, \mathbf{x} \geqslant d_i$$

can be replaced by the equation

$$\boldsymbol{\alpha}_i\, \mathbf{x} - x_i^* = d_i$$

where

$$x_i^* \geqslant 0.$$

The variables x_i^* are called slack variables.

In this way we can pass from the general form (8.5) to one of the following forms

$$\begin{cases} \mathbf{A}_c \mathbf{x}_c \geqslant \mathbf{d}_c, \\ \mathbf{x}_c \geqslant 0, \\ \min z = \mathbf{c}_c \mathbf{x}_c, \end{cases} \tag{8.6}$$

$$\begin{cases} \mathbf{A}_d \mathbf{x}_d = \mathbf{d}, \\ \mathbf{x}_d \geqslant 0, \\ \min\ z = \mathbf{c}_d \mathbf{x}_d\,, \end{cases} \tag{8.7}$$

$$\begin{cases} \boldsymbol{\alpha}_i \mathbf{x} \geqslant d_i\,, & i \in M_1 \\ \boldsymbol{\alpha}_i \mathbf{x} = d_i\,, & i \in M - M_1 \\ \mathbf{x} \geqslant 0, & \\ \min\ z = \mathbf{c}_t \mathbf{x}. & \end{cases} \tag{8.8}$$

The form (8.6) is called the canonical form, (8.7) the standard form, and (8.8) the mixed form.

One can immediately verify that

$$\mathbf{A}_c = \begin{Vmatrix} a_{11} & \dots & a_{1n} \\ \vdots & & \vdots \\ a_{m1} & \dots & a_{mn} \\ -a_{p+1,1} & \dots & -a_{p+1,n} \\ \vdots & & \vdots \\ -a_{m1} & \dots & -a_{mn} \end{Vmatrix} \quad \mathbf{x}_c = \mathbf{x}, \quad \mathbf{d}_c = \begin{Vmatrix} d_1 \\ \vdots \\ d_m \\ -d_{p+1} \\ \vdots \\ -d_m \end{Vmatrix}, \quad \mathbf{c}_c = \pm\, \mathbf{c};$$

$$\mathbf{A}_d = \begin{Vmatrix} a_{n1} & \dots & a_1 & -1 & \dots & 0 \\ \vdots & & \vdots & \vdots & & \vdots \\ a_{p1} & \dots & a_{pn} & 0 & \dots & -1 \\ a_{p+1,1} & \dots & a_{p+1,n} & 0 & \dots & 0 \\ \vdots & & \vdots & \vdots & & \vdots \\ a_{m1} & \dots & a_{mn} & 0 & \dots & 0 \end{Vmatrix}, \quad \mathbf{x}_d = \begin{Vmatrix} x_1 \\ \vdots \\ x_n \\ x_1^* \\ \vdots \\ x_p^* \end{Vmatrix},$$

$$\mathbf{c}_d = \| \pm c_1 \dots \pm c_n \ \underbrace{0 \ \dots \ 0}_{p} \|,$$

$$\mathbf{c}_t = \pm\, \mathbf{c}.$$

In the sequel, when we shall not specify otherwise, we shall refer to the mixed form

$$\begin{cases} \boldsymbol{\alpha}_i\mathbf{x} \geqslant d_i, & i \in M_1, \\ \boldsymbol{\alpha}_i\mathbf{x} = d_i, & i \in M - M_1, \\ \quad \mathbf{x} \geqslant 0, & \\ \min\ z = \mathbf{cx}. & \end{cases} \tag{8.8'}$$

Two vector spaces are attached to this form: the m-dimensional space R^m of the column vectors $\mathbf{a}_j$, called the constraints space, and the n-dimensional space R^n, called the activities space to which belongs, in particular, the activities levels vector $\mathbf{x}$.

In the case of the standard form, written for the sake of simplicity as follows:

$$\mathbf{Ax} = \mathbf{d}, \tag{8.9}$$

$$\mathbf{x} \geqslant 0, \tag{8.9'}$$

$$\min\ z = \mathbf{cx}, \tag{8.9''}$$

the situation in which the system (8.9) admits more than one solution (that is an infinity of solutions) is interesting to analyse, since otherwise it is not possible to optimize the linear form $z = \mathbf{cx}$. To this purpose it is necessary that:

$$m < n, \ \text{rank}\ (\mathbf{A}) = \text{rank}\ (\mathbf{A},\ \mathbf{d}) = m, \tag{8.9'''}$$

where $(\mathbf{A},\ \mathbf{d})$ is the matrix obtained from $\mathbf{A}$ by adding the column $\mathbf{d}$.

In these conditions, every set of m vectors $\mathbf{a}_j$ of the matrix $\mathbf{A}$ which are linearly independent form a basis $\mathbf{B}$ of the linear system (8.9). These vectors form a non singular sub-matrix of mth order of the matrix $\mathbf{A}$ and conversely. In the following we shall denote this sub-matrix by $\mathbf{B}$ too.

The variables associated to the columns of a basis $\mathbf{B}$ are called basic variables. They form a vector which we shall denote by $\mathbf{x}^B$. The other variables are called secondary variables, and they form a sub-vector which we shall denote by $\mathbf{x}^R$. The vectors corresponding to the secondary variables form a sub-matrix of $\mathbf{A}$ denoted by $\mathbf{R}$. We shall also denote by $\mathbf{c}^B$ the sub-vector of $\mathbf{c}$ corresponding to the basic variables and by $\mathbf{c}^R$ the sub-vector corresponding to the secondary variables.

We have

$$\mathbf{A} = (\mathbf{B},\ \mathbf{R}),\ \mathbf{x} = (\mathbf{x}^B,\ \mathbf{x}^R),\ \mathbf{c} = (\mathbf{c}^B,\ \mathbf{c}^R).$$

With these notations the system (8.9), (8.9′), and (8.9″) can be written as:

$$\begin{cases} (\mathbf{B},\ \mathbf{R})\ (\mathbf{x}^B,\ \mathbf{x}^R) = \mathbf{d}, & (8.10) \\ \mathbf{x}^B \geqslant 0,\ \mathbf{x}^R \geqslant 0, & (8.10') \\ \min\ z = (\mathbf{c}^B,\ \mathbf{c}^R)\ (\mathbf{x}^B,\ \mathbf{x}^R). & (8.10'') \end{cases}$$

After performing the calculus, relationships (8.10) and (8.10″) become

$$\mathbf{B}\mathbf{x}^B + \mathbf{R}\mathbf{x}^R = \mathbf{d}, \tag{8.11}$$

$$\min\ z = \mathbf{c}^B\mathbf{x}^B + \mathbf{c}^R\ \mathbf{x}^R. \tag{8.11'}$$

As the matrix **B** is a non-singular square matrix, it admits an inverse $\mathbf{B}^{-1}$. Multiplying each side of equation (8.11) by $\mathbf{B}^{-1}$ we obtain

$$\mathbf{x}^B = \mathbf{B}^{-1}\ \mathbf{d} - \mathbf{B}^{-1}\ \mathbf{R}\mathbf{x}^R. \tag{8.12}$$

Since $\mathbf{x}^R = 0$, we obtain

$$\begin{cases} \mathbf{x}^B = \mathbf{B}^{-1}\ \mathbf{d}, \\ \mathbf{x}^R = 0, \end{cases} \tag{8.13}$$

which is called a basic solution associated with the basis **B.** The basic solution is called degenerate if some components of the vector $\mathbf{x}^B$ are zero.

A basic solution which satisfies the condition $\mathbf{B}^{-1}\mathbf{d} \geqslant 0$ is called a basic program.

Taking into account the values of equation (8.13), the linear form (8.11) becomes

$$z = \mathbf{c}^B\mathbf{B}^{-1}\ \mathbf{d}. \tag{8.14}$$

A basic program, for which the linear form (8.14) is minimum, is called an optimal basic program.

Let us now prove the following proposition.

Proposition 1. (*The fundamental theorem of the linear programming.*) *Given a problem of linear programming in the standard form:*

(a) *If it admits at least one finite program, then it admits at least a basic program;*

(b) *If it admits at least one finite optimal progam, it admits at least an optimal basic program.*

First we shall prove point (a) of this theorem.

Thus we suppose that the system (8.9) admits some finite program, that is a program whose variables have all finite values. We also suppose that the variables have been renumbered so that the first $k \leqslant n$ variables have positive values and the last $n - k$ variables are zero. Let $\mathbf{A}_1$ be the matrix formed with the first k columns of the matrix $\mathbf{A}$ (corresponding to the variables which are not zero), that is

$$\mathbf{A}_1 = (\mathbf{a}_1, \mathbf{a}_2, \ldots, \mathbf{a}_k).$$

Based on the hypothesis from equation (8.9) it follows that

$$\sum_{j=1}^{k} x_j \mathbf{a}_j = \mathbf{d}. \tag{8.15}$$

We have to consider two cases.

(i) rank $\mathbf{A}_1 = k$, (thus $k \leqslant m$). In this case the vectors $\mathbf{a}_1, \ldots, \mathbf{a}_k$ are linearly independent.

By taking into account that rank $\mathbf{A} = m$, it follows that there is at least a non-singular square sub-matrix $\mathbf{A}^{(m)}$ of rank m in $\mathbf{A}$. Let

$$\mathbf{A}^{(m)} = (\mathbf{a}_{\alpha_1}, \mathbf{a}_{\alpha_2}, \ldots, \mathbf{a}_{\alpha_m}).$$

The columns of the matrix $\mathbf{A}$ are a basis of the vector space R^m. It follows that we can express every vector $\mathbf{a}_j$, $(j = 1, \ldots, k)$ as a linear function of $\mathbf{a}_{\alpha i}$. the coefficients being non null. Hence

$$\mathbf{a}_1 = \sum_{i=1}^{m} \lambda_i \mathbf{a}_{\alpha i}$$

and we may suppose for example that $\lambda_1 \neq 0$.

It follows that $\mathbf{a}_1, \mathbf{a}_{\alpha_2}, \ldots, \mathbf{a}_{\alpha_m}$ forms a basis for R^m whence

$$\mathbf{a}_2 = \mu_1 \mathbf{a}_1 + \sum_{i=2}^{m} \lambda_i' \mathbf{a}_{\alpha_i}$$

and at least one of the coefficients λ_i', $(i = 2, \ldots, m)$ is non null since $\mathbf{a}_1$ and $\mathbf{a}_2$ are linearly independent. Let $\lambda_2' \neq 0$. Then $\mathbf{a}_1, \mathbf{a}_2, \mathbf{a}_{\alpha_3}, \ldots, \mathbf{a}_{\alpha_m}$ is a basis for R^m. Repeating this operation it follows that $\mathbf{a}_1, \mathbf{a}_2, \ldots, \mathbf{a}_k, \mathbf{a}_{\alpha_{k+1}}, \ldots$ $\ldots, \mathbf{a}_{\alpha_m}$ is a basis for R^m.

The program $x_1, x_2, \ldots, x_k, x_{k+1} = 0, \ldots, x_n = 0$ is therefore a basic program. This program is degenerate if $k < m$ and non-degenerate if $k = m$.

(ii) rank $\mathbf{A}_1 < k$, (thus $k > m$). In this case the vectors $\mathbf{a}_1, \mathbf{a}_2, \ldots, \mathbf{a}_k$ are linearly dependent thus

$$\sum_{j=1}^{k} \lambda_j \mathbf{a}_j = 0 \tag{8.16}$$

and we may suppose that at least one coefficient λ_j is positive.

Let us write

$$\frac{x_r}{\lambda_r} = \min\left(\frac{x_j}{\lambda_j}\right),$$

where the index j takes all the values for which $\lambda_j > 0$.

Taking into account equation (8.15) and formula (8.16), it follows that

$$\sum_{j=1}^{k}\left(x_j - \frac{\lambda_j}{\lambda_r}\, x_r\right)\mathbf{a}_j = \sum_{j=1}^{k} x_j\mathbf{a}_j - \frac{x_r}{\lambda_r}\sum_{j=1}^{k}\lambda_j\mathbf{a}_j = \mathbf{d}.$$

Thus the new variables

$$x_j' = x_j - \frac{\lambda_j}{\lambda_r}\, x_r, \quad (j = 1, \ldots, k)$$

form a new program which has at most $k - 1$ variables, strictly positive, since $x_r' = 0$.

If the vectors associated with these positive variables are linearly dependent then we continue the operation. After p such operations, the $k - p$ vectors associated with the non-null variables are linearly independent. Since $\mathbf{d} \neq 0$ we have $p \leqslant k - 1$.

Thus we come back to case (i).

We shall prove now point (b) of the theorem. We suppose that the system (8.9), (8.9′), (8.9″) admits every optimal finite program. With the notation we used in (a) we have to consider two cases also:

(i) rank $\mathbf{A}_1 = k$. In this case we proved above that the considered program is a basic program degenerate or non-degenerate.

(ii) rank $\mathbf{A}_1 = l < k$. In this case there is at least one system of l column vectors $\mathbf{a}_j$ of the matrix $\mathbf{A}$ which are linearly independent. We suppose that we renumbered the variables so that these l vectors should be the first l columns of the matrix $\mathbf{A}_1$, consequently the first l columns of the matrix $\mathbf{A}$.

The linear system deduced from the system $\mathbf{Ax} = \mathbf{d}$ by taking the variables $x_{k+1}, x_{k+2}, \ldots, x_n$ equal to zero and considering $x_{l+1}, x_{l+2}, \ldots, x_k$ as parameters is a Cramer system which has the solution

$$x_j = \alpha_j + \sum_{i=1}^{k-l} \beta_{ji}\, x_{l+1} \quad (j = 1, \ldots, l). \tag{8.17}$$

The positive values of the variables $x_1, x_2, \ldots, x_k$ from the optimal program obviously verify the relationships (8.17). Substituting $x_1, \ldots, x_l$ for the values obtained from formulae (8.17) and $x_{k+1}, \ldots, x_n$ for zero in the linear form (8.9″), we obtain

$$z = \alpha_0 + \sum_{i=1}^{k-l} \beta_i\, x_{l+i}\,.$$

This form becomes an optimum form when we give to the variables $x_{l+1}, \ldots, x_k$ their values from the optimal program. If in equation (8.17) the variable x_{l+i} has a variation δx_{l+i} we obtain new values for x_j. Since the initial values of the variables x_{l+i}, $(i = 1, \ldots, k - l)$ and x_j, $(j = 1, \ldots \ldots, l)$ are strictly positive, we can always choose the variations δx_{l+i} small enough, so that the new variables should remain strictly positive and thus form a new program if we add the values

$$x_{k+1} = x_{k+2} = \ldots = x_n = 0.$$

On the other hand δz has the sign of β_i or $-\beta_i$ depending on whether δx_{l+i} is positive or negative. It follows that $\beta_i = 0$ for every i and hence $z = \alpha_0$ since z is optimum for $\delta x = 0$. Therefore the new program is also optimal. Consequently the variable x_{l+i} can decrease until it vanishes or one of the variables x_j vanishes, and we obtain a new optimal program which has at most $k - 1$ non zero variables. Repeating the previous operations, after p such operations the $k - p$ vectors associated to the $k - p$ positive variables are linearly independent thus the matrix formed by means of these vectors has the rank $k - p$. This occurs for $p \leqslant k - 1$ since $\mathbf{d} \neq 0$. Thus we come back to case (i).

8.2 THE SIMPLEX METHOD

8.2.1 Fundamental theorems

Let us consider the general problem of linear programming written in the standard form (8.9), (8.9′) and (8.9″) with conditions (8.9‴). In this case, as we noticed in the previous paragraph, the system (8.9) admits a basis **B.** We shall denote by $\mathbf{a}_{j_1}, \ldots, \mathbf{a}_{jm}$ the vectors which form the basis **B,** with $I = \{j_1, \ldots, j_m\}$ and $J = N - I$.

As the system (8.9) is written in the form (8.12) we shall write

$$\mathbf{B}^{-1}\,\mathbf{R} = \mathbf{Y} = \|\,\mathbf{y}_j\,\| = \|\,y_{sj}\,\|, (s \in I,\ j \in J), \tag{8.18}$$

$$\mathbf{B}^{-1}\,\mathbf{d} = \overline{\mathbf{x}}^B = \|\,\overline{x}_s\,\|, \quad (s \in I). \tag{8.19}$$

From equation (8.18) we deduce that

$$\mathbf{R} = \mathbf{BY},$$

hence

$$\mathbf{a}_j = \mathbf{B}\mathbf{y}_j = \sum_{s\in I} y_{sj}\,\mathbf{a}_s\,, \qquad (j \in J). \tag{8.20}$$

The system (8.12) can be written as

$$\mathbf{x}^B + \sum_{j\in J} x_j\,\mathbf{y}_j = \overline{\mathbf{x}}^B \tag{8.21}$$

or

$$x_s + \sum_{j\in J} y_{sj}\,x_j = \overline{x}_s\,, (s \in I). \tag{8.22}$$

We shall call the system (8.12), (8.21) or (8.22) an explicit system.

The linear form

$$z = \mathbf{c}^B\,\mathbf{x}^B + \mathbf{c}^R\mathbf{x}^R, \tag{8.23}$$

or

$$z = \sum_{s\in I} c_s\,x_s + \sum_{j\in J} c_j\,x_j$$

can be expressed only by means of the secondary variables. Indeed, taking into account equation (8.21) in equation (8.23), we obtain

$$\sum_{j\in J} (c_j - \mathbf{c}^B\,\mathbf{y}_j)\,x_j = z - \mathbf{c}^B\,\overline{\mathbf{x}}^B.$$

Writing

$$\mathbf{c}^B\mathbf{y}_j = z_j\,, \; (j \in J) \tag{8.24}$$

and

$$\mathbf{c}^B\,\overline{\mathbf{x}}^B = \overline{z},$$

it follows that

$$z - \overline{z} = \sum_{j\in J} (c_j - z_j)\,x_j\,. \tag{8.25}$$

Proposition 1. *Given a basic program associated with a basis* **B**, *if for an index $k \in J$ we have $z_k - c_k > 0$ and $y_k \geqslant 0$ then we can form a class of programs in which $m+1$ variables can be strictly positive, the variable x_k can take every non-negative value and thus z can be as small as we wish. It follows that we do not have a finite minimal program.*

Taking into account that $y_k \leqslant 0$, from equation (8.21) it follows that from a basic program $\overline{\mathbf{x}}^B$ we obtain another program if x_k has an arbitrary positive value $\overline{x}_k$ and if the other secondary variables are null. In this way we obtain the program

$$\overline{\mathbf{x}}'^B = \overline{\mathbf{x}}^B - x_k\,\mathbf{y}_k \geqslant \overline{\mathbf{x}}^B.$$

From equation (8.25) we obtain

$$\bar{z}' = \bar{z} - (z_k - c_k)\, x_k$$

hence if $x_k \to +\infty$,then $\bar{z}' \to -\infty$.

Proposition 2. *Given a basic program associated to a basis* **B** *if for an index* $k \in J$ *we have* $z_k - c_k > 0$ *and for at least an index* $s \in I$ *we have* $y_{sk} > 0$, *then denoting by* l *the index for which*

$$\frac{\bar{x}_l}{y_{lk}} = \min_{s:y_{sk} > 0} \left(\frac{\bar{x}_s}{y_{sk}} \right)$$

the basic solution associated with the basis **B′** *deduced from* **B** *by substituting the column* $\mathbf{a}_k$ *for* $\mathbf{a}_l$ *is a new basic program which yields for* z *the value* $\bar{z}' \leqslant \bar{z}$.

If we consider the solution of the system (8.22) obtained by assigning to x_k a value $\bar{x}_k > 0$, the other secondary variables are null, and we deduce

$$\bar{x}'_s = x_s - y_{sk}\,\bar{x}_k, \quad (s \in I). \tag{8.26}$$

From the hypothesis concerning the sign of y_{sk} it follows that there is at least an index s for which $\bar{x}'_s < \bar{x}_s$. In order for $\bar{\mathbf{x}}'^B$ to be a program, that is to be non-negative, it is necessary and sufficient that

$$\bar{x}_s - y_{sk}\,\bar{x}_k \geqslant 0, \quad (s \in I),$$

that is

$$\bar{x}_k \leqslant \min_{s:y_{sk}>0} \left(\frac{\bar{x}_s}{y_{sk}} \right).$$

For the new program to have, at most, m non-zero variables it is sufficient to take

$$\bar{x}_k = \frac{\bar{x}_l}{y_{lk}} = \min_{s:y_{sk}>0} \left(\frac{\bar{x}_s}{y_{sk}} \right)$$

Thus from equation (8.26) we have $\bar{x}'_l = 0$.

Also, taking into account equation (8.20) and the relationship $y_{lk} \neq 0$ it follows that **B′** is a basis.

By means of the previous operations we obtain a new program in which all the variables associated with the vectors which do not belong to **B′**, are null. This new program is a basic program associated with **B′**.

Taking into account equation (8.25) it follows that the new value of z is

$$\bar{z}' = z - (x_k - c_k)\,\frac{\bar{x}_l}{y_{lk}} \leqslant z,$$

where the inequality is strict if, and only if, $\bar{x}_l \neq 0$.

Proposition 3. *Given a basic program associated with a basis* **B**, *the necessary and sufficient condition for the program to be minimal is that for every index j associated with a secondary variable we should have* $z_j - c_j \leqslant 0$.

The necessity of the condition follows immediately from propositions 1 and 2.

Let us now prove the sufficiency of the condition.

Indeed, relationship (8.25) is true for all the solutions of the system (8.22). As the only variables from relationship (8.25) are x_j and z, the minimum of z with the conditions $x_j \geqslant 0$, $(j \in J)$ which are necessary for the solution to be a program, is attained for $x_j = 0$, $(j \in J)$ since, by hypothesis, all the coefficients $z_j - c_j$ are negative or null.

The basic program $\bar{\mathbf{x}}^B$ is minimal and the minimum value of z is

$$\bar{z} = \mathbf{c}^B \bar{\mathbf{x}}^B.$$

From the above theorem we obtain the following proposition immediately.

Proposition 4. *Given a minimal basic program and the quantities* $z_j - c_j \leqslant 0$ *associated with the secondary variables of the program, the necessary and sufficient condition for another program to be minimal is that,* $z_j - c_j = 0$, $(j \in J)$ *should follow from* $x_j > 0$.

Also, from the above proposition and from the fact that a solution associated with a determined basis is unique, we have the following proposition.

Proposition 5. *A necessary and sufficient condition for a minimal basic program to be the only minimal program of the problem is that for every* $j \in J$ *we have* $z_j - c_j < 0$.

8.2.2 The primal simplex algorithm

We suppose that we were able to determine a basic program $\bar{\mathbf{x}}^B$ of the problem of programming given in the standard form and that we calculated the matrix **Y** and the differences $z_i - c_i$. Then the propositions 1, 2 and 3 from Section 8.2.1 enable us to know if we have to stop the calculus, because we have already an optimal program, or because there is no finite optimal program, or if we have to determine a new basic program.

These approaches of calculus form the primal simplex algorithm. Shortly this algorithm consists of the following steps:

(a) Determine an initial basic program $\mathbf{x}^B$. Let I be the set of the indices of the columns of the matrix **A** which belong to the basis **B** and $J = N - I$;

(b) Calculate by means of formulae (8.18) and (8.24) the matrix $\mathbf{Y} = \| y_{sj} \|$, $(s \in I, j \in J)$ and the quantities $z_j - c_j$, $(j \in J)$;

(c) Examine the signs of the differences $(z_j - c_j)$, $(j \in J)$;

(i) if $z_j - c_j \leqslant 0$ [or $\geqslant 0$] for every $j \in J$, the considered program is minimal [or maximal];

(ii) otherwise, let $J_1 \subset J$ be the set of the indices j for which

$$z_j - c_j > 0 \text{ [or } < 0];$$

(d) The components of the vectors $\mathbf{y}_j$, $(j \in J_1)$ are examined

(i) if $\mathbf{y}_j \leqslant 0$ for at least an index $j \in J_1$ we have no finite minimal [maximal] program;

(ii) otherwise determine the index k for which $| z_k - c_k | = \max\limits_{j \in J_1} | z_j - c_j |$, (the entrance criterion in basis) and the index for which

$$\frac{x_l}{y_{lk}} = \min_{s: y_{sk} > 0} \left(\frac{x_s}{y_{sk}} \right), \text{ (the bottom-row criterion);}$$

(e) The new basis $\mathbf{B}'$ deduced from $\mathbf{B}$ by substituting the vector a_k for a_l is formed. The new program $\mathbf{x}'^B$ associated with the basis $\mathbf{B}'$ and also the new values $\mathbf{Y}'$ and $z_j' - c_j$ of $\mathbf{Y}$ and $z_j - c_j$ respectively are calculated. One repeats the application of the algorithm starting with step (c).

In order to pass from the matrix $\mathbf{B}^{-1}$ to the matrix $(\mathbf{B}')^{-1}$ it is convenient to use a simple linear transformation which consists of eliminating the variable x_k from $m - 1$ equations of the system (8.23). This elimination is called pivoting.

Let $\mathbf{a}_k$ be the vector which enters the basis and $\mathbf{a}_l$ the vector which leaves the basis. In this way the secondary variable x_k becomes a basic variable and the basic variable x_l becomes a secondary variable. The sets of indices I and J become $I' = I - \{l\} + \{k\}$ and $J' = J + \{l\} - \{k\}$ respectively.

The coefficient $y_{lk} > 0$ from the system (8.22) is used as pivot for the elimination of x_k from the other $m - 1$ equations of the system. The function of rank l can be written

$$x_k + \frac{x_l}{y_{lk}} + \sum_{j \in J - \{k\}} \frac{y_{lj}}{y_{lk}} x_j = \frac{\bar{x}_l}{y_{lk}} \cdot \tag{8.27}$$

Using the value of x_k from this equation in the other equations (8.22) we obtain

$$x_s - \frac{y_{sk}}{y_{lk}} x_l + \sum_{j \in J - \{k\}} \left(y_{sj} - \frac{y_{lj}}{y_{lk}} y_{sk} \right) x_j = \bar{x}_s - \frac{y_{sk}}{y_{lk}} \bar{x}_l, \ (s \in I - \{l\}). \tag{8.28}$$

From the left-hand side of equations (8.27) and (8.28) we obtain the formulae of transformation of the variables y_{sj}:

$$\begin{cases} y'_{sj} = y_{sj} - \dfrac{y_{lj}}{y_{lk}} y_{sk}, \ (s \in I - \{l\} = I' - \{k\}), \\ y'_{kj} = \dfrac{y_{lj}}{y_{lk}} \cdot \end{cases} \tag{8.29}$$

One can immediately verify that these formulae are true for every $j \in N$. For example for $j = l$ the formulae (8.29) are written

$$y'_{sl} = y_{sl} = \frac{y_{ll}}{y_{lk}} y_{sk} = -\frac{y_{sk}}{y_{lk}}, \ (s \neq k),$$

$$y'_{kl} = \frac{y_{ll}}{y_{lk}} = \frac{1}{y_{lk}}$$

and again we come across the coefficients of x_l from equations (8.27) and (8.28). From the right-hand side of the equations (8.27) and (8.28) we obtain the transformation formulae of the variables $\bar{x}_s$, $(s \in I')$:

$$\begin{cases} \bar{x}'_s = x_k - \dfrac{x_l}{y_{lk}} y_{sk}, \ (s \in I' - \{k\}), \\ \bar{x}'_k = \dfrac{\bar{x}_l}{y_{lk}} \cdot \end{cases} \tag{8.30}$$

These formulae are also applied for $s = l$ in which case we find x'_l. Substituting in equation (8.25) x_k for its value given by equation (8.27), we deduce that

$$\sum_{j \in J - \{k\}} (c_j - z_j) x_j + (c_k - z_k) \left[-\frac{x_l}{y_{lk}} - \sum_{j \in J - \{k\}} \frac{y_{lj}}{y_{lk}} x_j \right] =$$

$$= z - \bar{z} - (c_k - z_k) \frac{\bar{x}_l}{y_{lk}} \tag{8.31}$$

whence we have

$$\begin{cases} z'_j - c_j = (z_j - c_j) - \dfrac{y_{lj}}{y_{lk}} (z_k - c_k), \ (j \in J'), \\ \bar{z}' = \bar{z} - \dfrac{\bar{x}_l}{y_{lk}} (z_k - c_k). \end{cases} \tag{8.32}$$

If we agree to write

$$y_{s,0} = \bar{x}_s, \ y_{0,j} = z_j - c_j, \ \bar{z} = y_{0,0}$$

the equations (8.29), (8.30) and (8.32) can be written in the form

$$y'_{sj} = y_{sy} - \frac{y_{lj}}{y_{lk}}\, y_{sk}, \quad (s \in I - \{l\} + \{0\}, j \in N + \{0\}),$$

$$y'_{kj} = \frac{y_{lj}}{y_{lk}}, \qquad (j \in N + \{0\}). \tag{8.33}$$

Now we can summarise the rules for the primal simplex algorithm in the following table:

			c_1	c_2	$\cdots$	c_s	$\cdots$	c_j	$\cdots$	c_n
			x_1	x_2	$\cdots$	x_s	$\cdots$	x_j	$\cdots$	x_n
	z	$\bar{z}$	$z_1 - c_1$	$z_2 - c_2$	$\cdots$	0	$\cdots$	$z_j - c_j$	$\cdots$	$z_n - c_n$
c_{j_1}	x_{j_1}	$\bar{x}_{j_1}$	y_{j_11}	y_{j_12}	$\cdots$	0	$\cdots$	y_{j_1j}	$\cdots$	y_{j_1n}
c_{j_2}	x_{j_2}	$\bar{x}_{j_2}$	y_{j_21}	y_{j_22}	$\cdots$	0	$\cdots$	y_{j_2j}	$\cdots$	y_{j_2n}
$\vdots$	$\vdots$	$\vdots$	$\vdots$	$\vdots$	$\cdots$	$\cdots$	$\cdots$	$\vdots$	$\cdots$	$\vdots$
c_{js}	x_s	$\bar{x}_s$	y_{s1}	y_{s2}	$\cdots$	1	$\cdots$	y_{sj}	$\cdots$	y_{sn}
$\vdots$	$\vdots$	$\vdots$	$\vdots$	$\vdots$	$\cdots$	$\cdots$	$\cdots$	$\vdots$	$\cdots$	$\vdots$
c_{j_m}	x_{j_m}	$\bar{x}_{j_m}$	y_{j_m1}	y_{j_m2}	$\cdots$	0	$\cdots$	y_{j_mj}	$\cdots$	y_{j_mn}

8.2.3 Initial basic program

In order to use the simplex algorithm we have to know an initial basic program. Also before getting such a program we have to know if the system of constraints written in the standard form is compatible and what its rank is, that is we have to calculate the ranks of the matrices **A** and (**A**, **d**).

We shall give a method of calculus by means of which we can artificially form the initial basic program, a method of calculus which permits us to avoid the calculus of the rank of the two matrices.

This method consists of multiplying by —1 the equations whose right-hand side is negative so that we obtain $\mathbf{d} \geqslant 0$, and then of adding a necessary and sufficient number of unit column vectors to the matrix **A** in order to turn it into an extended matrix $\mathbf{A}^a$ having an unit sub-matrix $\dot{\mathbf{I}}^{(m)}$ of mth order. The number of these vectors is, at most, m.

In this way, by eventually rearranging the equations we can write the extended system of constraints in the form

$$\sum_{j=m-p+1}^{n} a_{ij}\, x_j = d_i, \qquad (i = 1, \ldots, p),$$

$$x_{i-p} + \sum_{j=m-p+1}^{n} a_{ij} x_j = d_i, \qquad (i = p+1, \ldots, m),$$

where $d_i \geqslant 0$, $(i = 1, \ldots, m)$.

In order to form the matrix $\dot{\mathbf{I}}^{(m)}$ we consider the unit vectors p associated with p variables $x_1^a, x_2^a, \ldots, x_p^a$ which are called artificial variables and we consider the equations

$$x_i^a + \sum_{j=m-p+1}^{n} a_{ij} x_j = d_i, \qquad (i = 1, \ldots, p).$$

In this way the initial problem is replaced by

$$\begin{cases} \mathbf{A}^a \left\| \begin{matrix} \mathbf{x}^a \\ \\ \mathbf{x} \end{matrix} \right\| = \mathbf{d}, & (8.34) \\ \mathbf{x}, \mathbf{x}^a \geqslant 0, & \\ \min\ z = \mathbf{cx}, & (8.34') \end{cases}$$

where

$$\mathbf{A}^a = \| \dot{\mathbf{I}}^{(m)}\ \mathbf{A}' \|,$$

$$\mathbf{A}' = \| \mathbf{a}_{m-p+1}\ \mathbf{a}_{m-p+2} \ldots \mathbf{a}_n \|,$$

$$\mathbf{x}^a = \| x_1^a\ x_2^a \ldots x_p^a \|,$$

$$\mathbf{d} \geqslant 0.$$

Taking into account that $\dot{\mathbf{I}}^{(m)}$ is a sub-matrix of $\mathbf{A}^a$ and that rank $\dot{\mathbf{I}}^m = m$ it follows that rank $(\mathbf{A}^a) = m$ and thus the primal simplex algorithm can be applied to problem (8.34), (8.34′) without any hypothesis on the rank of the matrix **A**.

It is obvious that

$$\mathbf{x}^a = \mathbf{d} \tag{8.35}$$

forms a basic program for equations (8.34) and (8.34′).

Thus we can apply the simplex method considering program (8.35) as the initial basic program. This program will be a program for the initial problem if in the final result all the artificial variables are zero. In order to obtain this there are several methods; we shall state two of them: the penalty method and the two-phase method.

8.2.3.1 The penalty method

The condition of optimization given by equation (8.34′) is replaced by the condition

$$\min\ z = M \sum_{i=1}^{n} x_i^a + \sum_{j=1}^{n} c_j\, x_j, \tag{8.36}$$

where M is a positive constant arbitrarily large.

The problem (8.34) (8.36) is called the augmented problem.

Every program of the initial problem is also a program of the augmented problem. Since M is sufficiently large, if there is a program for the initial problem, the optimal program of the augmented problem will not contain a non zero artificial variable.

Conversely, every program of the augmented problem which does not contain any strictly positive artificial variable, is a program of the initial problem.

It follows that the simplex algorithm can be applied to the augmented problem if the optimization conditions are satisfied. If there remain non-zero artificial veriables then the initial problem has no program. Otherwise the optimal program of the augmented problem is also an optimal program of the initial problem.

In order to simplify the calculus we might distinguish two steps.

(i) *The first step* — The simplex algorithm is applied to the augmented problem till we come across one of the following two situations:

(a) There is no artificial variable in the basis. Then we have a program for the initial problem and we pass to the second step.

(b) The basis contains artificial variables, all zero, and the coefficient of M from each difference $z_j - c_j$ is negative or zero. The program determined in this way is a degenerate program of the initial problem. If we have $z_j - c_j \leqslant 0$ for every j the program is optimal. Otherwise we pass to the second step.

(c) The basis contains at least a strictly positive artificial variable, and the coefficient of M from each difference $z_j - c_j$ is negative or zero. In this case the initial problem has no program.

(ii) *The second step* — In applying the simplex algorithm we have to go on until an optimal program (if it exists) is obtained.

8.2.3.2 The two-phase method

When this method can be applied, we use the form (8.34) of the system of constraints.
First phase — The objective function (8.34′) is substituted for:

$$\min \zeta = \sum_{i=1}^{p} x_i^a . \tag{8.37}$$

The problem defined by equations (8.34) and (8.37) is called the auxiliary problem.

Every basic program of the initial problem is a basic program for the auxiliary problem with $\mathbf{x}^a = 0$. Thus, the existence of such a problem implies min ζ=0. Conversely, every program of the auxiliary problem for which $\zeta = 0$, or, which is the same thing, in which all the artificial variables are zero, is a program of the initial problem. Thus the simplex algorithm can be applied untill one of the following situations is met:

(a) $\zeta = 0$ and the basis obtained does not contain an artificial variable. The program obtained is the basic program for the initial problem. We can pass to the situation (a) of the second phase.

(b) $\zeta = 0$ but the basis contains at least an artificial variable. The program obtained is a program for the initial problem. We can pass to situation (b) of the second phase.

(c) ζ is minimal and strictly positive. The initial problem has no program.
Second phase — We come back to the initial objective function (8.34).

The matrix $\mathbf{Y}$ obtained at the end of the first phase is used with no modifications for the first iteration of the second phase.

The new differences $z_j - c_j$ with respect to equation (8.34′) are calculated directly by means of formulae (8.24) taking into account that if the program contains artificial variables equal to zero, the costs associated with these variables are zero.

The program we obtain at the end of the first phase is used as a starting program for the second phase in which we can meet two situations:

(a) The simplex algorithm is applied to the column vectors of the initial matrix $\mathbf{A}$ untill we find the optimal program.

(b) The simplex algorithm is applied to a restricted problem which differs from the problem (8.34) and (8.34′) by the supplementary condition

$x_j = 0$ for all the indices j which correspond to the differences $z_j - c_j < 0$ from the end of the first phase. The algorithm is applied untill we get an optimal program or untill we reach the conclusion that we have no finite optimal program.

8.3 DUALITY

8.3.1 Fundamental theorems

Let us consider the general problem of linear programming written in the canonical form

$$\begin{cases} \mathbf{Ax} \geqslant \mathbf{d}, \\ \mathbf{x} \geqslant 0, \\ \min\ z = \mathbf{cx} \end{cases} \tag{8.38}$$

and let us denote by $\mathbf{u} = \| u_i \|$, $(i \in M)$ a row vector with m components.

Definition 1. The problem

$$\begin{cases} \mathbf{uA} \leqslant \mathbf{c}, \\ \mathbf{u} \geqslant 0, \\ \max\ w = \mathbf{ud} \end{cases} \tag{8.39}$$

is called the dual problem of problem (8.38).

In the sequel we shall call problem (8.38) the primal problem, the constraints from formulae (8.39) will be called dual constraints and the variables u_i will be called dual variables.

We notice that if we consider, problem (8.39) as the primal problem then its dual problem is (8.38).

We shall now prove several fundamental theorems about the dual systems (8.38) and (8.39). To this purpose we shall first prove some auxiliary propositions.

Proposition 1. *For every matrix* **A** *and every vector* **d**, *the necessary and sufficient condition for the system*

$$\mathbf{Ax} = \mathbf{d} \tag{8.40}$$

to have no solution, is that the system

$$\begin{aligned} \mathbf{uA} &= 0, \\ \mathbf{ud} &= \boldsymbol{\alpha}, \quad (\boldsymbol{\alpha} \neq 0) \end{aligned} \tag{8.41}$$

has a solution.

Indeed if system (8.38) has a solution $\bar{\mathbf{x}}$, then for every $\mathbf{u}$ we have

$$\mathbf{uA\bar{x}} = \mathbf{ud},$$

whence it follows that (8.41) has no solutions.

If equation (8.40) has no solution then the vector $\mathbf{d}$ is not a linear combination of the columns $\mathbf{a}_j$, $(j = 1, \ldots, n)$ of the matrix $\mathbf{A}$.

Let s be the rank of the matrix $\mathbf{A}$. We suppose the we rearranged the columns of the matrix $\mathbf{A}$ so that the vectors $\mathbf{a}_1, \ldots, \mathbf{a}_s$ be linearly independent. It follows that the system

$$\begin{cases} \mathbf{ua}_j = 0, \ (j = 1, \ldots, s) \\ \mathbf{ud} \ \ = \alpha \end{cases}$$

has at least one solution $\mathbf{u}$ for every α, because the matrix determined by the vectors $\mathbf{a}_j$ $(j = 1, \ldots, s)$ and $\mathbf{d}$ has the rank $s + 1$. On the other hand, since rank $\mathbf{A} = s$ we have

$$\mathbf{a}_k = \sum_{j=1}^{s} \mu_j \mathbf{a}_j, \quad (k = s + 1, \ldots, n)$$

that is

$$\bar{\mathbf{u}}\mathbf{a}_k = \sum_{j=1}^{s} \mu_j \bar{\mathbf{u}}\mathbf{a}_j = 0,$$

whence it follows that $\bar{\mathbf{u}}$ is a solution of system (8.41).

Proposition 2. *For every matrix* $\mathbf{A}$ *and for every vector* $\mathbf{d}$, *the necessary and sufficient condition for system* (8.40) *to have a non-negative solution is that the system*

$$\begin{cases} \mathbf{uA} \geqslant 0, \\ \mathbf{ud} < 0 \end{cases} \tag{8.42}$$

should have no solution.

If system (8.40) has a solution $\bar{\mathbf{x}} \geqslant 0$, then for every u we have

$$\mathbf{u\,A\,\bar{x}} = \mathbf{u\,d},$$

whence it follows that equation (8.41) has no solution.

Let us now suppose that system (8.40) has non-negative solutions.

If system (8.40) has no solution then, by applying proposition 1 for $\alpha < 0$, it follows that system (8.42) has a solution.

We shall now consider the case when system (8.40) has a negative solution. In this case, we shall prove by a complete induction on the number n of columns of the matrix $\mathbf{A}$ that system (8.42) has a solution.

For $n = 1$, the system with one unknown $\mathbf{x}\mathbf{a}_1 = \mathbf{d}$ has a solution $\bar{x} < 0$ from the hypothesis. It follows that $\mathbf{u} = -\mathbf{d}$ is a solution of system (8.42) because

$$\mathbf{u}\mathbf{a}_1 = -\frac{1}{\bar{x}}\mathbf{d}^2 > 0,$$

$$\mathbf{u}\mathbf{d} = -\mathbf{d}^2 < 0.$$

We suppose that the property is true for a matrix $\mathbf{A}$ with $n - 1$ columns and we shall prove that the property is true also for a matrix with n columns.

From the hypothesis the system

$$\sum_{j=1}^{n} x_j \mathbf{a}_j = \mathbf{d} \tag{8.43}$$

has no non-negative solution.

Hence the system

$$\sum_{j=1}^{n-1} x_j \mathbf{a}_j = \mathbf{d}$$

has no non-negative solution, because such a solution together with $x_n = 0$ would be a non-negative solution of the sistem (8.43).

Consequently if this property is supposed to be true for a matrix with $n-1$ columns, it follows that there is a vector $\bar{u}$ so that

$$\begin{cases} \bar{\mathbf{u}}\mathbf{a}_j \geqslant 0, & (j = 1, \ldots, n-1), \\ \bar{\mathbf{u}}\mathbf{d} < 0. \end{cases}$$

Hence we have to analyse two cases:

I $\bar{\mathbf{u}}\mathbf{a}_n \geqslant 0$. In this case $\bar{\mathbf{u}}$ is a solution of system (8.42) and the theorem (s)proved.

(II) $\bar{\mathbf{u}}\mathbf{a}_n < 0$. In this case let us denote by

$$\mathbf{a}'_j = \mathbf{a}_j + \lambda_j \mathbf{a}_n, \ (j = 1, \ldots, n-1), \tag{8.44}$$

where

$$\lambda_j = -\frac{\bar{\mathbf{u}}\,\mathbf{a}_s}{\bar{\mathbf{u}}\mathbf{a}_n} \geqslant 0$$

and

$$\mathbf{d}' = \mathbf{d} + \lambda_0\ \mathbf{a}_n \tag{8.44'}$$

where

$$\lambda_0 = -\frac{\bar{\mathbf{u}}\,\mathbf{d}}{\bar{\mathbf{u}}\,\mathbf{a}_n} < 0.$$

With these notations the system

$$\sum_{j=1}^{n-1} x'_j\ \mathbf{a}'_j = \mathbf{d}' \tag{8.45}$$

can be written as follows

$$\sum_{j=1}^{n-1} \mathbf{x}'_j\,\mathbf{a}_j + \left[\sum_{j=1}^{n-1} \lambda_j \mathbf{x}'_j - \lambda_0\right] \mathbf{a}_n = \mathbf{d}. \tag{8.46}$$

From equation (8.46) it follows that system (8.45) cannot have a non-negative solution in x'_j unless system (8.40) has a non-negative solution. Thus from the hypothesis, there is a vetor $\bar{\mathbf{u}}'$ so that

$$\begin{cases} \bar{\mathbf{u}}'\,\mathbf{a}_j \geqslant 0, & (j = 1, \ldots, n-1), \\ \bar{\mathbf{u}}'\,\mathbf{d}' < 0. \end{cases}$$

But, writing

$$\mathbf{u} = \bar{\mathbf{u}}' - \frac{\bar{\mathbf{u}}'\,\mathbf{a}_n}{\bar{\mathbf{u}}\,\mathbf{a}_n}\,\bar{\mathbf{u}}$$

and taking into account equation (8.44) and (8.44'), we have

$$\begin{cases} \mathbf{u}\mathbf{a}_j = \mathbf{u}'\mathbf{a}'_j \geqslant 0, & (j = 1, \ldots, n-1), \\ \mathbf{u}\mathbf{a}_n = 0, & \\ \mathbf{u}\mathbf{d} \ \ = \bar{\mathbf{u}}'\mathbf{d}' < 0, & \end{cases}$$

which proves the theorem.

Proposition 3. *For every matrix* $\mathbf{A}_1$ *with m rows and for every vector* $\mathbf{d}$, *the necessary and sufficient condition for the system*

$$\mathbf{A}_1\ \mathbf{x}_1 \leqslant \mathbf{d} \tag{8.47}$$

to have a non-negative solution is that the system

$$\begin{cases} \mathbf{uA}_1 \geqslant \mathbf{d} \\ \mathbf{ud} \geqslant 0, \end{cases} \tag{8.48}$$

should have no non-negative solution.

In order to prove this proposition it is sufficient to apply proposition 2 by writing

$$\mathbf{x} = \left\| \begin{matrix} \mathbf{x}_1 \\ \mathbf{y}_1 \end{matrix} \right\|,$$ (where $\mathbf{y}_1$ is a variable vector with m components)

and

$\mathbf{A} = \| \mathbf{A}_1\ \mathbf{I} \|$ (where $\mathbf{I}$ is a unit matrix of mth order).

It follows that we can come across one of the following two situations

(i) Either the system

$$\| \mathbf{A}_1\ \mathbf{I} \| \cdot \left\| \begin{matrix} \mathbf{x} \\ \mathbf{y} \end{matrix} \right\| = \mathbf{d}$$

has a solution $\bar{\mathbf{x}}_1 \geqslant 0, \mathbf{y}_1 \geqslant 0$, whence it follows that system (8.47) has a solution $\bar{\mathbf{x}}_1 \geqslant 0$.

(ii) Or the system

$$\begin{aligned} \mathbf{uA}_1 &\geqslant 0, \\ \mathbf{uI} &\geqslant 0, \\ \mathbf{ud} &< 0 \end{aligned}$$

has a solution, whence it follows that system (8.48) has a solution $\mathbf{u} \geqslant 0$.

Proposition 4. *The selfdual system*

$$\begin{aligned} K\mathbf{x} &\geqslant 0, \\ \mathbf{x} &\geqslant 0, \end{aligned} \tag{8.49}$$

where $\mathbf{K}$ *is an antisymmetric matrix* ($\mathbf{t_K} = -\mathbf{K}$ *where* $\mathbf{t_K}$ *is the transpose of the matrix* $\mathbf{K}$) *has at least one solution* $\bar{\mathbf{x}}$ *for which*

$$\mathbf{K}\bar{\mathbf{x}} + \bar{\mathbf{x}} \geqslant 0.$$

From proposition 3 it follows that there is always a solution $\bar{\mathbf{x}}_i$ so that if we denote by $\mathbf{x}_i$ the row of rank i of the matrix $\mathbf{K}$ we obtain

$$\mathbf{x}_i\bar{\mathbf{x}}_i + \bar{x}_{ii} > 0.$$

Thus it is sufficient to consider

$$\mathbf{d} = -\mathbf{e}_i = (0, \ldots, 0, -1, 0, \ldots, 0),$$

$$\mathbf{A} = -\mathbf{K} = \mathbf{t_K}.$$

If system (8.47) has a non-negative solution, then there is a vector $\bar{\mathbf{x}}_i \geqslant \mathbf{0}$ so that

$$-\mathbf{K}\bar{\mathbf{x}}_i \leqslant -\mathbf{e}_i$$

that is

$$\begin{cases} \mathbf{K}\bar{\mathbf{x}}_i \geqslant 0, \\ \mathbf{x}_i\bar{\mathbf{x}}_j \geqslant 1. \end{cases}$$

If system (8.48) has a non-negative solution, then there is a vector $\bar{\mathbf{x}}_i \geqslant 0$ so that

$$\begin{cases} \bar{\mathbf{x}}_i\mathbf{t}_A = \mathbf{K}\mathbf{x}_i \geqslant 0, \\ -\bar{x}_{ii} < 0. \end{cases}$$

This result is true for every row of the matrix $\mathbf{K}$ by considering successively as vector $\mathbf{d}$ the vectors $-\mathbf{e}_1, -\mathbf{e}_2, \ldots \cdot, -\mathbf{e}_n$. Hence the vector

$$\bar{\mathbf{x}} = \sum_{i=1}^{n} \bar{\mathbf{x}}_i$$

is a solution of system (8.49) so that

$$\mathbf{x}_i\bar{\mathbf{x}} + \bar{\mathbf{x}}_i \geqslant \mathbf{x}_i\bar{\mathbf{x}}_i + \bar{x}_{ii} > 0, \quad (i = 1, \ldots, n)$$

We shall now come back to the dual problems (8.38) and (8.39).

Proposition 5. *If $\bar{\mathbf{x}}$ and $\bar{\mathbf{u}}$ are dual programs we have*

$$\mathbf{cx} \geqslant \mathbf{ud}. \tag{8.50}$$

Indeed, taking into account equations (8.38) and (8.39) we can write

$$\mathbf{c}\bar{\mathbf{x}} \geqslant \bar{\mathbf{u}}\mathbf{A}\bar{\mathbf{x}} \geqslant \bar{\mathbf{u}}\mathbf{d}.$$

Proposition 6. *If* $\bar{\mathbf{x}}$ *and* $\bar{\mathbf{u}}$ *are dual programs for which*

$$\mathbf{c}\bar{\mathbf{x}} = \bar{\mathbf{u}}\mathbf{d}, \tag{8.51}$$

then the two programs are optimal.

Indeed, let us suppose that $\bar{\mathbf{x}}$ is not optimal. Hence there is another program $\bar{\mathbf{x}}'$ for which $\bar{\mathbf{x}}' < \bar{\mathbf{x}}$. We can write

$$\mathbf{c}\bar{\mathbf{x}}' < \mathbf{c}\bar{\mathbf{x}}$$

or taking into account equation (8.51),

$$\mathbf{c}\mathbf{x}' < \mathbf{u}\mathbf{d}$$

which contradicts equation (8.50).

By analogy we can obtain the same result for $\bar{\mathbf{u}}$.

Proposition 7. *The system of linear homogeneous inequations*

$$\mathbf{A}\mathbf{x} - t\mathbf{d} \geqslant 0, \ \mathbf{x} \geqslant 0, \tag{8.52}$$

$$-\mathbf{u}\mathbf{A} + t\mathbf{c} \geqslant 0, \ \mathbf{u} \geqslant 0, \tag{8.52'}$$

$$\mathbf{u}\mathbf{d} - \mathbf{c}\mathbf{x} \geqslant 0, \ t \geqslant 0, \tag{8.52''}$$

has at least one solution $(\mathbf{x}_0, \mathbf{u}_0, t_0)$ *so that*

$$\mathbf{A}\mathbf{x}_0 - t_0\mathbf{d} + \mathbf{u}_0 > 0, \tag{8.53}$$

$$-\mathbf{u}_0\mathbf{A} + t_0\mathbf{c} + \mathbf{x}_0 > 0, \tag{8.53'}$$

$$\mathbf{u}_0\mathbf{d} - \mathbf{c}\mathbf{x}_0 + t_0 > 0. \tag{8.53''}$$

This property follows immediately by applying proposition 4 to the system

$$\left\| \begin{matrix} 0 & \mathbf{A} & -\mathbf{d} \\ -t_{\mathbf{A}} & 0 & t_{\mathbf{c}} \\ t_{\mathbf{d}} & -\mathbf{c} & 0 \end{matrix} \right\| \cdot \left\| \begin{matrix} t_u \\ \mathbf{x} \\ t \end{matrix} \right\| \geqslant 0,$$

$$\left| \begin{matrix} t_{\mathbf{u}} \\ \mathbf{x} \\ t \end{matrix} \right| \geqslant 0,$$

in which the matrix of the coefficients is antisymmetric.

Proposition 8. *For every matrix* $\mathbf{A}$ *and for every vectors* $\mathbf{c}$ *and* $\mathbf{d}$ *the following statements are mutually exclusive*:

(a) *the dual problems* (8.38) *and* (8.39) *have a pair of optimal programs* $(\bar{\mathbf{x}}, \bar{\mathbf{u}})$ *so that*

$$\mathbf{c}\bar{\mathbf{x}} = \bar{\mathbf{u}}\mathbf{d};$$

(b) (i) *none of the two problems has an optimal program*;

(ii) *at least one of the two problems has no program;*

(iii) *if one of the problems has at least a program, the set of its programs is not bounded and there is a finite optimum for this problem.*

Proposition 7 establishes the existence of the solution $(\mathbf{x}_0, \mathbf{u}_0, t_0)$ with $t_0 \geqslant 0$. Hence we have to consider two cases which are mutually exclusive.

(a) $t_0 > 0$. In this case the values $\bar{\mathbf{x}} = \dfrac{\mathbf{x}_0}{t_0}$, $\bar{\mathbf{u}} = \dfrac{\mathbf{u}_0}{t_0}$, are a solution of the homogeneous system of equations (8.52), (8.52′) and (8.52″). From equations (8.52) and (8.52′) it follows that $(\bar{\mathbf{x}}, \bar{\mathbf{u}})$ are dual programs, from equation (8.52″) and from propositions 5 and 6 it follows that these programs are optimal and that

$$\mathbf{c}\bar{\mathbf{x}} = \bar{\mathbf{u}}\mathbf{d}.$$

(b) $t_0 = 0$. We suppose that there is a pair of programs $\bar{\mathbf{x}}$, $\bar{\mathbf{u}}$. Since $\mathbf{x} = \mathbf{x}_0$, $\mathbf{u} = \mathbf{u}_0$, $t = 0$ is a solution of the system (8.52), (8.52′) and by hypothesis $\bar{\mathbf{x}}$ and $\bar{\mathbf{u}}$ verify the relations:

$$\mathbf{A}\bar{\mathbf{x}} \geqslant \mathbf{d}, \quad \bar{\mathbf{x}} \geqslant 0,$$

$$\bar{\mathbf{u}}\mathbf{A} \leqslant \mathbf{c}, \quad \mathbf{u} \geqslant 0,$$

then

$$\mathbf{c}\mathbf{x}_0 \geqslant \bar{\mathbf{u}}\mathbf{A}\mathbf{x}_0 \geqslant 0 \geqslant \mathbf{u}_0\mathbf{A}\bar{\mathbf{x}} \geqslant \mathbf{u}_0\mathbf{d}, \tag{8.54}$$

which contradicts relationship (8.53″) where $t = 0$ and consequently proves part b, (ii) of the theorem.

We shall now suppose that there is a program $\bar{\mathbf{x}}$ of the primal problem and let $\lambda \geqslant 0$ be a scalar as large as we wish. We have

$$\bar{\mathbf{x}} + \lambda\mathbf{x}_0 > 0$$

and, using relationship (8.52) for $\mathbf{x} = \mathbf{x}_0$, $t = 0$, we obtain

$$\mathbf{A}(\bar{\mathbf{x}} + \lambda\mathbf{x}_0) \geqslant \mathbf{A}\mathbf{x} \geqslant \mathbf{d},$$

which shows that for every value of λ, $\bar{\mathbf{x}} + \lambda\mathbf{x}_0$ is a program. On the other hand from equation (8.52″) with $t = 0$ and from equation (8.54) it follows that

$$\mathbf{cx}_0 < \mathbf{u}_0\mathbf{d} \leqslant 0,$$

a relationship which proves that the expression

$$\mathbf{c}(\bar{\mathbf{x}} + \lambda\mathbf{x}_0) = \mathbf{c}\bar{\mathbf{x}} + \lambda\mathbf{cx}_0$$

can be as small as we wish if we consider λ large enough. In this way part b, (iii) of the theorem is obtained. Finally b, (i) follows directly from b, (ii) and b,(iii).

We shall now prove the fundamental theorem of duality.

Proposition 9. (*The theorem of existence.*) *Given a pair of dual problems, one and only one of the following statements is true:*

(i) *none of the two problems has a program;*

(ii) *one of the problem has no program and the other has at least one program but which is not optimal;*

(iii) *the two problems have optimal programs.*

This theorem follows immediately from proposition 8.

Proposition 10. (*The theorem of duality.*) *Given a pair of dual programs, a necessary and sufficient condition for a program* $\bar{\mathbf{x}}$ (*or* $\bar{\mathbf{u}}$) *of one of the problem to be optimal is that a program* $\bar{\mathbf{u}}$ (*or* $\bar{\mathbf{x}}$) *of the other problem exists so that*

$$\mathbf{c}\bar{\mathbf{x}} = \bar{\mathbf{u}}\mathbf{d}.$$

Then the program $\bar{\mathbf{u}}$ (or $\bar{\mathbf{x}}$) is an optimal program and the above relationship is verified for every pair of optimal programs.

The sufficiency of the condition follows immediately from proposition 6 and the necessity from proposition 8.

Proposition 11. (*The weak theorem of complementarity.*) *Given a pair of dual problems, a necessary and sufficient condition for the programs* $\bar{\mathbf{x}}$ *and* $\bar{\mathbf{u}}$ *to be optimal is that they verify the relationships*

$$\bar{\mathbf{u}}(\mathbf{A}\bar{\mathbf{x}} - \mathbf{d}) = 0, \tag{8.55}$$

$$(\mathbf{c} - \bar{\mathbf{u}}\mathbf{A})\bar{\mathbf{x}} = 0. \tag{8.55'}$$

Indeed, as $\bar{\mathbf{x}}$ and $\bar{\mathbf{u}}$ are programs we have

$$\alpha \equiv \bar{\mathbf{u}}(\mathbf{A}\bar{\mathbf{x}} - \mathbf{d}) \geqslant 0,$$

$$\beta \equiv (\mathbf{c} - \bar{\mathbf{u}}\mathbf{A})\,\bar{\mathbf{x}} \geqslant 0. \tag{8.56}$$

Hence

$$\alpha + \beta = \mathbf{c}\bar{\mathbf{x}} - \bar{\mathbf{u}}\mathbf{d}. \tag{8.57}$$

From proposition 10 it follows that a necessary and sufficient condition for $\bar{\mathbf{x}}$ and $\bar{\mathbf{u}}$ to be optimal is that the right-hand side of relationship (8.57) be null, that is, taking into account equations (8.56) and (8.56′), we obtain proposition 12.

Proposition 12. (*The strong theorem of the complementarity.*) *Given a pair of dual problems, both of them having programs, there is at least one pair of optimal programs* $\bar{\mathbf{x}}$ *and* $\bar{\mathbf{u}}$ *for which*

$$(\mathbf{A}\bar{\mathbf{x}} - \mathbf{d}) + \bar{\mathbf{u}} > 0, \tag{8.58}$$

$$(\mathbf{c} - \bar{\mathbf{u}}\mathbf{A}) + \bar{\mathbf{x}} > 0. \tag{8.58'}$$

We are in case (a) of proposition 8 since both problems have programs. Let ($\mathbf{x}_0$, $\mathbf{u}_0$, $t_0 > 0$) be the solution given by proposition 7. Then

$$\bar{\mathbf{x}} = \frac{\mathbf{x}_0}{t_0}, \quad \bar{\mathbf{u}} = \frac{\mathbf{u}_0}{t_0}$$

are dual programs (according to the proof of (a) in proposition 8). Whence taking into account equation (8.53) and (8.53′) we have relationship (8.58) and (8.58′).

8.3.2 The duality and the simplex method

We consider a problem of linear programming in the standard form (8.38). We suppose that $m < n$ and rank $\mathbf{A} = m$. For the dual problem, only the dual variables associated with the primal constraints which were initially equations, will not be constrained to be non-negative. In the optimal table of the simplex method we have

$$z_j - c_j \leqslant 0$$

or, from equations (8.18) and (8.24)

$$\mathbf{c}^{\mathbf{B}}\,\mathbf{B}^{-1}\,\mathbf{a}_j \leqslant c_j, \tag{8.59}$$

for every *i*. We add a slack variable $-x_i^*$ for all the constraints which are initially inequations with the sign $\geqslant$ (let M_1 be the set of indices of these constraints). Hence we suppose that the matrix **A** contains the vector $-\mathbf{e}_i = (0, \ldots, 0, -1, 0, \ldots, 0)$, having the component of the rank i equal to -1, for every $i \in M_1$. Thus

$$\mathbf{c}^B \mathbf{B}^{-1}(-\mathbf{e}_i) \leqslant 0, \ (i \in M_1). \tag{8.60}$$

From equations (8.59) and (8.60) it follows that the vector $\mathbf{c}^B \mathbf{B}^{-1}$ forms a dual program.

On the other hand denoting by $\mathbf{x}^{-B}$ the primal optimal program, we have

$$\mathbf{B}\mathbf{x}^{-B} = \mathbf{d}$$

whence

$$\mathbf{c}^B \mathbf{B}^{-1} \mathbf{d} = \mathbf{c}^B \mathbf{x}^{-B}.$$

From this and from proposition 6 it follows that $\mathbf{c}^B \mathbf{B}^{-1}$ is a dual optimal program.

Let us now suppose that the primal problem (8.39) has a known basis

$$\mathbf{B} = \| \mathbf{a}_{j_1} \ldots \mathbf{a}_s \ldots \mathbf{a}_{jm} \|, \ (\{j_1, \ldots, s, \ldots, j_m\} = I)$$

so that the vector

$$\bar{\mathbf{u}} = \mathbf{c}^B \mathbf{B}^{-1}$$

is a dual program, that is

$$\bar{\mathbf{u}}\mathbf{a}_j \leqslant c_j, (j \in J = N - I) \tag{8.61}$$

where

$$z_j - c_j \leqslant 0.$$

If $\bar{\mathbf{u}}$ is not optimal, the basic solution $\bar{\mathbf{x}}^B$ is not a program for the primal problem. In this case there is at least an index $s \in I$ for which $\bar{\mathbf{x}}_s < 0$.

Conversely we shall prove that, if $\bar{x}_l < 0$ $(l \in I)$, we can find another dual solution $\bar{\mathbf{u}}'$ better than $\bar{\mathbf{u}}$.

Let β_s be the row of index s of the matrix $\mathbf{B}^{-1}$ and

$$\bar{\mathbf{u}}' = \bar{\mathbf{u}} - \theta\beta_l.$$

We have

$$\bar{\mathbf{u}}'\,\mathbf{d} = \bar{\mathbf{u}}\mathbf{d} - \theta x_l \tag{8.62}$$

whence

$$\bar{\mathbf{u}}'\,\mathbf{d} \geqslant \bar{\mathbf{u}}\mathbf{d}$$

for every $\theta > 0$.

On the other hand we have

$$\bar{\mathbf{u}}'\,\mathbf{a}_j = \bar{\mathbf{u}}\mathbf{a}_j - \theta\beta_l\,\mathbf{a}_j, \quad (j \in N)$$

whence

$$\bar{\mathbf{u}}'\,\mathbf{a}_j = \bar{\mathbf{u}}\mathbf{a}_j - \theta y_{lj} = z_j - \theta y_{lj}, \qquad (j \in J), \tag{8.63}$$

$$\bar{\mathbf{u}}'\,\mathbf{a}_s = \bar{\mathbf{u}}\mathbf{a}_s = z_s = c_s, \qquad (s \in I - \{l\}), \tag{8.63'}$$

$$\bar{\mathbf{u}}'\,\mathbf{a}_l = \bar{\mathbf{u}}\mathbf{a}_l - \theta = z_l - \theta = c_l - \theta. \tag{8.63''}$$

We have to analyse two cases:

(i) $y_{lj} \geqslant 0$, $(j \in J)$.

From equations (8.63), (8.63′), (8.63″) and (8.61) it follows that $\bar{\mathbf{u}}'$ is a dual program for every pozitive value of θ, and taking into account equation (8.62) as $\theta \to \infty$ it follows that the dual form has no maximum. Hence from proposition 9 it follows that primal problem has no program.

(ii) $y_{lj} < 0$ for at least an index $j \in J$.

From equations (8.63), (8.63′), (8.63″) it follows that $\bar{\mathbf{u}}'$ is a dual program if, and only if,

$$z_j - \theta y_{lj} \leqslant c_j, \quad (j \in J),$$

that is, if

$$\theta \leqslant \min_{j:y_{lj}<0} \left[\frac{z_j - c_j}{y_{lj}}\right], \qquad (j \in J).$$

If we consider

$$\theta = \frac{z_k - c_k}{y_{lk}} = \min_{j:y_{lj}<0} \left[\frac{z_j - c_i}{y_{lj}}\right], \qquad (j \in J), \tag{8.64}$$

then from equation (8.63) it follows that

$$\bar{\mathbf{u}}'\mathbf{a}_k = c_k. \tag{8.65}$$

Let us consider the matrix $\mathbf{B}'$ deduced from $\mathbf{B}$ by the substitution of $\mathbf{a}_k$ by $\mathbf{a}_l$. Since $y_{lk} \neq 0$ this matrix forms a basis. Therefore the set of indices of the columns of this basis is

$$I' - I - \{l\} + \{k\}.$$

From equations (8.63′) and (8.65) we deduce that

$$\bar{\mathbf{u}}'\mathbf{B}' = c'^B$$

whence

$$\bar{\mathbf{u}}' = \mathbf{c}'^B\,(\mathbf{B}')^{-1}. \tag{8.66}$$

Let $\bar{\mathbf{x}}'^B = (\mathbf{B}')^{-1}\,\mathbf{d}$ be the new basic solution. We can have two cases: (1) if $\bar{\mathbf{x}}'^B \geqslant 0$, this solution is an optimal basic program of the primal problem;

(2) if $\bar{\mathbf{x}}'_s < 0$ for at least an index $s \in I'$ we may continue the previous process defining a new dual solution $\bar{\mathbf{u}}''$ better than $\bar{\mathbf{u}}'$.

In order to obtain a maximal increase of the dual form for each new iteration of the process, we have to determine the index so that the expression

$$-\,\bar{x}_l\left(\frac{z_k - c_k}{y_{lk}}\right)$$

in which the choice of the index k follows from equation (8.64), should be maximal. In practice, as in the case of the primal algorithm this choice is replaced by the simpler choice

$$\bar{x}_l = \sin_{s:\bar{x}_s<0} [\bar{x}_s], \quad (s \in I).$$

Now we can state the dual simplex algorithm:

(a) We determine a basis **B** of the primal problem so that

$$z_j - c_j \leqslant 0 \quad [\geqslant 0], \quad (j \in J),$$

where J is the set of indices of the columns of the matrix **A** associated with the secondary vectors, z_j are given by equation (8.24) and **Y** is given by equation (8.18), where $I = N - J$;

(b) We study the sign of $\bar{\mathbf{x}} = \mathbf{B}^{-1}\,\mathbf{d}$:

if $\bar{\mathbf{x}} \geqslant 0$, it is a minimal maximal program; in the contrary case let $I_1 \subset I$ be that set of indices s for which $\bar{\mathbf{x}} < 0$.

(c) We notice the sign of the quantities y_{sj}, $(s \in I_1,\ j \in J)$:

if $y_{sj} \geqslant 0$ for at least an index $s \in I_1$ and all the indices $j \in J$ then the primal problem has no program;

if $y_{sj} < 0$ for at least an index $j \in J$ and all the indices $s \in I_1$, we determine l by means of the relationship

$\bar{x}_l = \min_{s \in I_1} [\bar{x}_s]$, (the entrance criterion) and k from the relationship

$$\left|\frac{z_k - c_k}{y_{lk}}\right| = \min_{\substack{j:y_{lj}<0\\ j\in J}} \left|\frac{z_j - c_j}{y_{cj}}\right| \text{(the bottom-row criterion).}$$

(d) We consider the new basis $\mathbf{B}'$ obtained from **B** by replacing $\mathbf{a}_k$ with $\mathbf{a}_l$. We calculate the new values $\mathbf{x}'^B$, $\mathbf{Y}'$ and z'_j by means of formulae (8.33) and we repeat the algorithm starting with (b).

8.4 EXERCISES

1. Determine

$$\max z = 3x_1 + 2x_2 + x_3 + x_4 + 5x_5$$

with the constraints

$$\begin{cases} 3x_1 + x_3 - x_5 = 3, \\ x_1 + x_2 - 3x_4 = -12, \\ x_2 + x_3 + x_5 = 4. \end{cases}$$

2. Determine

$$\max z = 60x_1 + 60x_2 + 90x_3 + 90x_4$$

with the constraints

$$\begin{cases} x_1 + x_2 + x_3 + x_4 \leqslant 15, \\ 7\ x_1 + 5\ x_2 + 3\ x_3 + 2\ x_4 \leqslant 1, \\ 3\ x_1 + 5\ x_2 + 10\ x_3 + 15\ x_4 \leqslant 1\cdot \end{cases}$$

Give the dual program.

3. Calculate

$$\min z = 15\ x_1 + 33\ x_2$$

with the constraints

$$3\ x_1 + 2\ x_2 \geqslant 6,$$

$$6\ x_1 + x_2 \geqslant 6,$$

$$x_2 \geqslant 1,$$

$$x_j \geqslant 0,\ (j = 1,\ 2).$$

4. A factory produces, at a given machine, working 45 hours a week, three different products P_1, P_2 and P_3. The article P_1 produces a net income of \$ 4, the article P_2 a net income of \$ 12 and the article P_3 produces a net income of \$ 3, the hour output of the machine are for each product: 50, 25 and 75 respectively. At last the possible sales are thus limited: 1,000 P_1 objects, 500 P_2 objects and 1,500 P_3 objects. In what way must one distribute the production so that the income of the factory should be maximum?

References

1. A n g o t, A., **Compléments de mathématique à l'usage des ingénieurs de l'électrotechnique et des télécommunications,** Editions de la revue d'optique, 1961.
2. B e r s, L., J o h n. F., S c h e c h t e r, M., **Partial Differential Equations,** vol. 3, Wiley-Interscience, New York, 1964.
3. B l a n c-L a p i e r r e, A., F o r t e t, R., **Théorie des functions aléatoires,** Masson et Cie, Paris, 1953.
4. C i u c u, G., **Elemente de teoria probabilităţilor şi statistică matematică (Elements of probability theory and statistical mathematics),** Editura Didactică şi Pedagogică, Bucureşti, 1963.
5. C o n s t a n t i n e s c u, I., C o n d r e a, D., N i c o l a u, E. **Teoria informaţiei (Information Theory),** Editura tehnică, Bucureşti, 1968.
6. C o u r a n t, R. and H i l b e r t, D., **Methods of Mathematical Physics.** vol. 1 and 2, Wiley Interscience, New York, 1962—1965.
7. C r a m e r, H., **Mathematical Methods of Statics,** Princeton University Press, New York, 1946.
8. C r a m e r, H., **The Elements of Probability Theory and Some of its Applications,** Wiley, New York, 1955.
9. C r i s t e s c u, R., **Matematici superioare (Higher mathematics),** Editura Didactică şi Pedagogică, Bucureşti, 1963.
10. C r i s t e s c u, R., **Elemente de analiză funcţională şi introducere în teoria distribuţiilor (Elements of mathematical analysis and introduction to the theory of distribution),** Editura tehnică, Bucureşti, 1966.
11. C r i s t e s c u, R., M a r i n e s c u, G., **Unele aplicaţii ale teoriei distribuţiilor (Some applications of the theory of distribution),** Editura Academiei R.S.R., Bucureşti, 1966.
12. C u l l m a n n, G., D e n i s- P a p i n, M., K a u f m a n n, A., **Elements de calcul informationnel,** Dunod, Paris, 1960.
13. D a n t z i g, G. B., **Linear Programming and Extensions,** Princeton University Press, New York, 1963.
14. F e l l e r, W., **An Introduction to Probability Theory and its Applications,** Wiley, New York, 1957.
15. G a r s o u x, J., **Espaces vectoriels topologiques et distributions,** Dunod, Paris, 1963.
16. G a s s, S. I., **Linear Programming: Method and Applications,** McGraw-Hill Book Comp. Inc., New York, 1964.
17. G e r r e t s e n, J. C. H., **Tangente und Flächeninhalt,** Vandenbrock und Ruprecht in Götingen, 1964.
18. G u i a ş u, S., T h e o d o r e s c u R., **Teoria matematică a informaţiei (Mathematical Theory of Information),** Editura Academiei R.S.R., Bucureşti, 1966.
19. H a d l e y, G. F., **Linear Programming,** Addison-Wesley, Reading, (Mass), 1961.

20. Haimovici, A., **Ecuaţiile fizicei matematice şi elemente de calcul variaţional (The Equations of Mathematical Physics and Elements of the Calculus of Variation),** Editura Didactică şi Pedagogică, Bucureşti, 1966.
21. Hille, E., Phillips, S. R., **Functional Analysis and Semigroups,** Amer. Math. Soc. Col. Publ., 1957.
22. Iacob, C., **Curs de matematici superioare (Lectures on Higher Mathematics),** Editura tehnică, Bucureşti, 1957.
23. Yaglom, A. U., Yaglom, I. M., **Veroyatnosti i informatsiya,** Gosizdat, Moskva, 1960.
24. Iosifescu, M., Mihoc, Gh., Theodorescu, R., **Teoria probabilităţilor şi statistica matematică (Probability Theory and Mathematical Statistics),** Editura tehnică, Bucureşti, 1966.
25. John, F., **Partial Differential Equations,** New York University, 1952—1953.
26. Karlin, S., **Mathematical Methods and Theory in Games,** Programming and Economics, Adison-Wesley, Reading, (Mass.), 1959.
27. Kaufmann, A., **Méthodes et modèles de la recherche opérationnelle,** vol. 1—2, Dunod, Paris, 1962.
28. Loéve, M., **Probability Theory,** van Nostrand, New York, 1955.
29. Marinescu, G., **Espaces vectoriels pseudotopologiques et théorie des distributions,** VEB Deutscher Verlag des Wissenschaften, Berlin, 1963.
30. Mihoc, Gh., Ionescu, H., **Bazele matematice ale programării lineare (The Mathematical Bases of Linear Programming),** Editura Ştiinţifică, Bucureşti, 1965.
31. Mihoc, Gh., Nădejdie, I., **Programarea matematică (Mathematical programming),** vol. 1—2, Editura Ştiinţifică, Bucureşti, 1966—1967.
32. Mihoc, Gh., Urseanu, V., **Matematici aplicate în statistică (Mathematics Applied to Statistics),** Editura Academiei R.P.R., Bucureşti, 1962.
33. Miranda, G., **Equazioni alle derivate parziale di tipo ellittico,** Spinger Verlag, Berlin-Göttingen-Heidelberg, 1955.
34. Moisil, Gr. C., **Ecuaţii cu derivate parţiale. Problema geometrizării (Equations with Partial Derivatives. Geometrisation Problems),** partea I—II (lithographic print), Acad. R.P.R., Inst.de Matematică, Bucureşti, 1950.
35. Moisil, Gr. C., **Ecuaţiile fizicii matematice (Equations of Mathematical Physics),** (litographic print), Min. Inv. Univ. C. I. Parhon, Fac. Mat. Fiz., Bucureşti, 1955.
36. Nicolescu, J. L., Stoka, M. I., **Matematici pentru ingineri (Mathematics for Engineers),** vol. I, Editura tehnică, Bucureşti, 1962.
37. Nicolescu, M., **Analiza matematică (Mathematical Analysis),** vol. 1—3, Editura tehnică, Bucureşti, 1958—1960.
38. Nicolescu, M., Dinculeanu, N., Marcus, S., **Manual de analiză matematică (Treatise of Mathematical Analysis),** Editura Didactică şi Pedagogică, Bucureşti, 1962—1964.
39. Onicescu, O., **Curs de teoria probabilităţilor (Lectures on Probability Theory),** Editura tehnică, Bucureşti, 1956.
40. Onicescu, O., **Teoria probabilităţilor şi aplicaţii (Probability Theory and Applications),** Editura Didactică şi Pedagogică, Bucureşti, 1963.
41. Onicescu, O., Mihoc, Gh., **Lecţii de statistică matematică (Lectures on Mathematical Statistics),** Editura tehnică, Bucureşti, 1958.
42. Onicescu, O. ş.a., **Calculul probabilităţilor şi aplicaţii (Probability Calculus and Applications),** Editura Academiei R.P.R., Bucureşti, 1956.
43. Petrovsky, I. G., **Lectures on Partial Differential Equations,** Interscience Publishers, Inc., New York, 1954.

44. Reicher, C. ş. a., **Theoria probabilităţilor (Probability Theory)**, Editura Didactică şi Pedagogică, Bucureşti, 1967.
45. Renyi, A., **Calcul de probabilités**, Dunod, Paris, 1966.
46. Roşculeţ M., **Analiza matematică (Mathematical Analysis)**, vol. 1—2, Editura Didactică şi Pedagogică, Bucureşti, 1964—1965.
47. Rudeanu, S., **Axiomele laticelor şi ale algebrelor booleene (Axioms of Latices and Boolean Algebra)**, Editura Academiei R.P.R., Bucureşti, 1963.
48. Schmetterer, L., **Einführung in die mathematische Statistik**. Springer-Verlag, Wien, 1956.
49. Schwarz, L., **Méthodes mathématiques pour les sciences physiques**, Hermann, Paris, 1961.
50. Shwarz, L., **Cours d'annalyse**, vol. 1—12, Hermann, Paris, 1967.
51. Simonnard, M., **Programmation linéaire**, Dunod, Paris, 1962.
52. Sobolev, S. L., **Uravneniya matematicheskoi fiziki**, Gosizdat, Moskva, 1950.
53. Stoka, M., Theodorescu, R., **Probabilitate şi geometrie, (Probability and Geometry)**, Editura Ştiinţifică, Bucureşti, 1966.
54. Stoka, M., Theodorescu, R., **Einführung in die Monte-Carlo-Methode**, VEB Deutscher Verlag der Wissenschaften, Berlin, 1971.
55. Şabac, *I.* Gh., **Matematici speciale (Mathematics for Engineers)**, vol. 1—2, Editura Didac*tică şi* Pedagogică, Bucureşti, 1964—1965.
56. Teodorescu, N., **Metode vectoriale în fizica matematică, (Vector methods in Mathematical Physics)**, vol. 1—2, Editura tehnică, Bucureşti, 1954.
57. Teodorescu, N., Olaru, V., **Curs de ecuaţiile fizicii matematice (Lectures on the Equations of Mathematical Physics)**, Editura Didactică şi Pedagogică, Bucureşti, 1963.
58. Teodorescu, N., Olaru, V. **Ecuaţiile fizicii matematice, (Lectures on the equations of Mathematical Physics)**, vol. 2, Editura Didactică şi Pedagogică, Bucureşti, 1965.
59. Tihonov, A. N., Samarski, A. A., **Uravnenya matematicheskoi fiziki**, Gosizdat, Moskva, 1953.
60. Tricomi, G. F., **Equazioni a derivate parziali**, Editura Cremonese, Roma, 1957.
61. Van der Waerden, B. L., **Mathematische Statistik**, Springer-Verlag, Berlin-Göttingen-Heidelberg, 1957.
62. Vayda, S., **Mathematical Programming**, Addison-Wesely, Reading, (Mass) 1961.
63. Vladimirov, B. C., **Uravneniya matematicheskoi fiziki**, Nauka, Moskva, 1967.

INDEX